The Metallurgy of Lead
and the Desilverization of Base Bullion

by H. O. Hofman

with an introduction by Kerby Jackson

Introduction

It has been over a century since H.O. Hofman released his important publication "The Metallurgy of Lead and the Desilverization of Base Bullion". First released in 1896, this work has been unavailable to the mining community since those days, with the exception of expensive original collector's copies and poorly produced digital editions.

It has often been said that "*gold is where you find it*", but even beginning prospectors understand that their chances for finding something of value in the earth or in the streams of the Golden West are dramatically increased by going back to those places where gold and other minerals were once mined by our forerunners. Despite this, much of the contemporary information on local mining history that is currently available is mostly a result of mere local folklore and persistent rumors of major strikes, the details and facts of which, have long been distorted. Long gone are the old timers and with them, the days of first hand knowledge of the mines of the area and how they operated. Also long gone are most of their notes, their assay reports, their mine maps and personal scrapbooks, along with most of the surveys and reports that were performed for them by private and government geologists. Even published books such as this one are often retired to the local landfill or backyard burn pile by the descendents of those old timers and disappear at an alarming rate. Despite the fact that we live in the so-called "Information Age" where information is supposedly only the push of a button on a keyboard away, true insight into mining properties remains illusive and hard to come by, even to those of us who seek out this sort of information as if our lives depend upon it. Without this type of information readily available to the average independent miner, there is little hope that our metal mining industry will ever recover.

Though this volume may not at first seem to be of great importance to gold miners, I feel that those miners with an interest in smelting and refining their finds, especially those recovered from lodes, will find the processes outlined to be of great value.

This important volume and others like it, are being presented in their entirety again, in the hope that the average prospector will no longer stumble through the overgrown hills and the tailing strewn creeks without being well informed enough to have a chance to succeed at his ventures.

Please note that at times it is necessary to rearrange illustration plates in these texts. Any illustrations not found in their original sequence may be found following the index.

Kerby Jackson
Josephine County, Oregon
January 2015

TABLE OF CONTENTS.

PART I.—INTRODUCTORY.

TABLE OF CONTENTS.

TABLE OF CONTENTS.

LIST OF ILLUSTRATIONS.

LIST OF ILLUSTRATIONS.

CHAPTER I.

HISTORICAL[1] AND STATISTICAL[2] NOTICE.

§ 1. **Introductory Remarks.**—Lead was probably known at a very early date. The oldest people of whom we have any record —the Egyptians—used it in glossing their pottery, and the abundance of silver among the ancients suggests the presence of argentiferous lead ores in many places. We know that lead was mined in considerable quantities by the Greeks and Romans. The mines of Laurion, in Attica, opened up again by a French company in 1863, flourished especially in the fifth century B.C. The Romans extracted large amounts of ore from the mines in the southeastern part of Spain, which had been opened in the third century B.C. by Hannibal and which form the main lead district of Spain to-day. They also carried on mining operations in England and along the Rhine from Bâle to Cologne. About the year 1000 A.D. the celebrated German silver-lead mines of the Hartz Mountains, Saxony and Silesia, and those of Austria were discovered.

§ 2. **Lead in the United States.**—At the present day the lead mines of the United States occupy the first place. In this country lead mining dates as far back as the beginning of the seventeenth century, when lead was mined and smelted near Falling Creek,

[1] Pulsifer, W. H., "Notes for a History of Lead," New York, 1888.

[2] C. Kirchhoff, in "United States Geological Survey: Mineral Resources of the United States, 1882," and following, Washington, 1883.

Va. During colonial times lead mines were operated mainly in North Carolina, New York, and the New England States, but on a small scale and not very successfully. The mines principally mentioned are the Washington mine, Davidson County, N. C.; the Rossie mine, St. Lawrence County, N. Y.; and the mines near Middletown, Conn., and Southampton, Mass. None of them were worked continuously, and to-day the East produces hardly any lead ore.

The lead ores of Missouri were discovered by Le Sueur in 1700 or 1701, and were first worked in 1720; of these the Mine La Motte, of Madison County, which is worked to-day, was the first discovered. Rumors of lead in the upper Mississipi Valley were afloat as early as 1766, and in 1788 Dubuque obtained from the Indians a grant for a lead mine on the place where the city named after him now stands.

From these two districts the bulk of the lead of the United States came until the great mines of the West were opened up in 1867. The total product of the United States in 1825 was only 1,500 net tons. This increased steadily until 1848, when 28,000 tons were produced. After falling off considerably for a little over twenty years, the production of the Mississippi Valley increased again, and reached 31,351 tons in 1890. As the total product of the United States for the same year was 161,754 tons, this district produces 19.36 per cent. of the entire output. All the rest came from the Western States and Territories, which produce argentiferous lead ores, while those of the Mississippi Valley are practically free from silver.

Argentiferous lead was first produced, according to O. H. Hahn, in 1866 or 1867, near Helena, Mont., and at Oreana, Nev. In 1870 the mines of Eureka, Nev., discovered in 1864, were reopened and the treatment of ores was begun. Next came Utah, where smelters were erected in 1870, followed by Colorado, which came into prominence in 1878. Later Idaho, New Mexico, Arizona, and California were added to the list. Colorado is to-day the greatest producer. Then follow Idaho and Utah, furnishing together about half as much as Colorado; the three together produce about 60 per cent. of all the argentiferous lead of the United States.

§ 3. **The World's Production.**—The world's production in 1884 [1] was as follows:

---

[1] "Mineral Resources of the United States," 1885, p. 264.

COUNTRIES.	METRIC TONS.
Spain	116,293
Germany	102,584
England	40,716
France	7,500*
Italy	14,000*
Greece	8,000*
Belgium	8,000*
Austria	11,391 †
Hungary	1,800*
Russia	500*
Sweden	400*
United States	126,907
Mexico	1,000*
Turkey	600*
Australia	5,000*
South America	2,000*
TOTAL	446,691

* Estimated.
† Including litharge, estimated at 80 per cent. lead.

The price of common pig-lead in New York averaged for the years 1886 to 1890 inclusive per pound avoirdupois 4.32 cents. The two other principal lead markets are St. Louis and Chicago, where the lead is cheaper than in New York by the cost of transportation.

CHAPTER II.

PROPERTIES[1] OF LEAD AND OF SOME OF ITS COMPOUNDS.

§ 4. **Lead.**—Lead has a bluish-gray color; on a freshly-cut surface it shows a considerable lustre, but loses it quickly when exposed to atmospheric air. It does not crystallize readily. When it is cooled slowly, as in the Pattinson process (§ 91), bundles of small, imperfect octahedrons form. Also, when refined lead is poured at the correct temperature into a warm mould and is allowed to cool, fern-like crystalline aggregates appear at the surface. It is the heaviest of all base metals. Reich's figure, 11.37, as specific gravity for pure lead at 0° C. (water at 4° C. being unity), is the one generally accepted. The specific gravity of lead will vary slightly, according as it is cooled quickly or slowly, hammered or rolled. Commercial lead has a lower specific gravity than 11.37 on account of the impurities contained in it. Lead is very soft, especially when allowed to cool and solidify slowly. It is harder if cooled quickly and if it contains slight admixtures of other metals, such as copper, arsenic, antimony, zinc, etc. The grade of commercial lead is often approximately determined by the resistance it offers to scratches with the finger-nail and the facility with which it makes a gray streak on paper. Lead is very malleable ; it is rolled into sheets and hammered into foil. In the form of filings it becomes a solid mass if subjected to a pressure of thirteen tons to the square inch, and liquefies at two and one-half times this pressure (Roberts-Austin [2]).

Lead is not sufficiently ductile to be drawn into fine wire ; its tenacity, according to Karmarsch, is inferior to that of most ductile

[1] Condensed for the most part from Percy, "Metallurgy of Lead," London, 1870, pp. 8-98.

[2] *Engineering and Mining Journal*, November 3, 1888 ; *Scientific American Supplement*, numbers 675, 676, 677.

metals. It fuses at 325° C. (Le Chatelier [1]), boils at between 1450 and 1600° C. (Carnelly and Williams [2]), but cannot be distilled. The latent heat of lead is 5.369; the coefficient of cubical dilation for 1° C., 0.000089; the linear coefficient about one-third of the cubical. The specific heat between 10° and 100° C. is 0.0314; with silver as 100, the conductivity for heat at 12° C. is 8.5, and for electricity 10.7.

Lead undergoes no change in perfectly dry air nor in water that is free from air; its surface becomes, however, dull by oxidation when it is exposed to the atmospheric air on account of the moisture which it contains. Similarly, it is oxidized by water that is not free from air. If melted in contact with air it oxidizes and becomes covered with an iridescent pellicle, said to be the suboxide, Pb_2O; this gradually changes to the oxide, PbO, and if the heating be prolonged sufficiently, the red oxide, Pb_3O_4, is obtained. The other two oxides which lead forms are the sesquioxide, Pb_2O_3, and the peroxide, PbO_2.

The best solvent of lead is dilute nitric acid. Dilute hydrochloric and sulphuric acids have little or no action; boiling concentrated hydrochloric acid and sulphuric acid of 66° B. dissolve it slowly. Organic acids—acetic, tartaric, citric acids—attack it in contact with air.

§ 5. **Lead Oxide,** *Pb O.*—Of the different oxides enumerated, the oxide is the one that is metallurgically interesting and of importance. It is obtained on a large scale as massicot and litharge, which have different physical properties. Massicot, an amorphous yellow powder, is formed by heating lead on a flat hearth to a low red heat, removing the film of suboxide as fast as it forms, and oxidizing it to yellow oxide. If the temperature be raised to the melting-point—that is, to a bright red heat—and the fused oxide cooled, it solidifies as crystalline litharge. On a large scale litharge is obtained by cupelling argentiferous lead. It crystallizes in orthorhombic octahedrons, and is soft and greasy to the touch. While molten it is transparent and orange-colored; when cold it is opaque, and its color varies from yellow to red according to the rate at which it has cooled; quick cooling promotes the yellow, slow cooling the red color. Yellow litharge is produced on a large scale by allowing it to run from the furnace over an iron plate and

[1] *Engineering and Mining Journal,* October 11, 1890.
[2] *Journal of the Chemical Society,* xxxv., p. 563.

chilling it with water, if necessary; it is thus obtained in small lumps. The red, flaky variety is formed by allowing the running litharge to collect in front of the furnace in cakes of from one to one and one-half tons in weight, and to cool slowly. The inner part of a cake will swell up and form flakes of red litharge; the outer and lower parts, having cooled quickly, will remain solid and have a yellow color. This swelling is caused by the giving off of oxygen, which molten litharge absorbs. In solidifying quickly in small lumps the oxygen only makes the surface uneven; in cooling slowly in large lumps the outer solid crust obstructs the passage of the oxygen. This prevents the inner part from solidifying firmly, and causes instead the formation of loose flakes. The flakes and lumps are separated by sifting. Both varieties, when ground, have a reddish-yellow color. Litharge melts at 954° C. (Honsell [1]); it is a good conductor of electricity when molten. It is volatilized at a white heat. It is only slightly soluble in water (1 part in 12,000 parts), but readily so in nitric acid and acetic acid.

Litharge is a strong base and quickly corrodes acid furnace material, with which it forms a silicate. It is an excellent flux, forming fusible compounds with oxides that are infusible alone. They do not always enter into chemical combination with it, but often are simply held in igneous solution by an excess of litharge. Thus fusible mixtures are formed with lime, baryta, magnesia, and alumina. The following table shows the proportion of litharge required to form fusible compounds with the principal metallic oxides:

1 part of	Cu_2O	CuO	ZnO	Fe_3O_4	Fe_2O_3	MnO	SnO_2	Sb_2O_3	SbO_2	As_2O_3	As_2O_5
Requires parts of	1.5	1.8	8	4	10	10	12–13	fusible in all proportions.	5	0.4–0.8	0.25–1

Litharge, being easily reduced to the metallic state, forms an important oxidizing agent. This is seen by its behavior with S, Te, As, Sb, Sn, Bi, Cu, Zn, Fe. They become wholly or partly oxidized, and the oxides either volatilized or scorified by the surplus of litharge, a corresponding amount of lead, which combines with any unoxidized part, having been reduced.

§ 6. **Lead Silicates.**—Lead oxide and silica begin to combine at a temperature where the oxide becomes soft. In fact, it is dis-

---

[1] *Berg- u. Hüttenmännische Zeitung*, 1866, p. 106.

advantageous to raise the temperature quickly if a silicate is to be formed. This can be seen in slag-roasting a galena ore to which fine sand has been added. If the time at which the roasted ore is pasty be shortened and the fusion urged, uncombined silica will be found with combined when the roast is decomposed in nitro-hydrochloric acid. All silicates that do not contain more silica than is required to form the tri-silicate, $Pb_2Si_3O_8$ ($2PbO.3SiO_2$), are fusible at a low temperature, forming a transparent vitreous mass; the singulo-silicate, Pb_2SiO_4 ($2PbO.SiO_2$), is as fluid as water. If the proportion of silica be raised above that of the tri-silicate, the compound becomes less fusible; thus $2PbO9.SiO_2$ gives a porcelain-like mixture, and $PbO.18SiO_2$ fritts only to a porous mass. All fusible lead silicates are yellow; they become darker in proportion to the quantity of lead they contain. They change their color, if they are contaminated with other metallic oxides, as can be seen if lead is slagged in a scorifier; *e.g.*, iron colors brown; copper, green; manganese, purple-black; nickel, brownish-yellow; cobalt, blue; tellurium, yellowish-red, the colors growing dark in proportion to the oxide added.

The lead from silicates is not readily liberated by the ordinary reducing agents. Sulphur decomposes the singulo-silicate to some extent, but it has less effect on the bi-silicate; iron sulphide throws down some lead, a double silicate of lead and iron being the result; carbon reduces from a bi-silicate part of the lead. In order to extract all the lead, it must be first set free from its combination with silica by a basic flux; thus metallic iron decomposes all fusible lead silicates at a bright-red heat, provided enough is added to form a singulo-silicate.

The singulo-silicate and bi-silicate of lead are readily decomposed by nitric acid, the tri-silicate is not completely decomposed; the more acid the silicate, the less soluble it is.

§ 7. **Lead Sulphide,** *PbS.*—This occurs native as galenite. It is formed by heating lead and sulphur, or lead oxide with an excess of sulphur, or by reducing lead sulphate with carbonaceous matter. The artificial sulphide has the same properties as the mineral.

The existence of subsulphides of lead (Pb_2S,Pb_4S) is denied by Percy, who shows that lead sulphide and lead can be melted together in all proportions, and that the properties of the resulting compound will resemble galena or lead according to the predominance of one or the other compound. Also, if such an apparent

subsulphide be heated carefully, comparatively pure lead will eliquate and a residue of hard crystalline sulphide remain behind.

Galena is not as fusible as lead, but it is very fluid when melted, and penetrates the fire-brick of the furnaces in which it is
treated; often a net-work of small veins of bright crystalline galena
is found in furnace-linings. When melted, galena begins to volatilize without being decomposed, if free oxygen be excluded. On the
walls of lead blast-furnaces crystals of sublimed galena are of
common occurrence. Lead sulphide is isomorphous with metallic
sulphides, as Ag_2S, Cu_2S, ZnS, FeS. Such mixtures of sulphides
are found in lead matte and copper matte obtained in smelting
sulphide ores. With the electro-negative sulphides of antimony
and arsenic it forms sulpho-salts. Quite a number of these occur
as minerals;[1] others can be artificially prepared in the dry and wet
ways. Galena as well as matte is a good conductor of electricity.[2]
Iron decomposes galena better than any other metal. For instance,
copper that has a greater affinity for sulphur than iron decomposes
lead sulphide only partially, as it alloys too readily with the liberated lead, and the cuprous sulphide formed combines with the remaining lead sulphide to matte. Zinc decomposes galena partly,
but the zinc sulphide formed is so refractory that a separation of
the liberated lead is not effected; the result is a black, porous mass
containing particles of lead and galena. The reaction that takes
place between iron and galena is generally expressed thus:

$$PbS + Fe = FeS + Pb,$$

and forms the basis of what is called the precipitation or iron-
reduction process. In reality, however, the iron sulphide retains
some undecomposed lead sulphide, and Nolte's[3] formula,

$$3PbS + 3Fe = 2Pb + (PbS + Fe_2S + FeS),$$

might be truer to the actual facts, although he presupposes the existence of an iron subsulphide. In decomposing galena in furnace-work
by means of iron, enough has to be present so as to have 1Fe for
1PbS. If less is added, the resulting matte remains too rich in lead;
if an excess is given, it is wasted. It may in fact be a disadvantage
in decomposing argentiferous galena, as on account of the affinity the silver sulphide has for iron sulphide, more silver will go into
the matte than can be accounted for by the amount of lead pres-

[1] See Dana, "System of Mineralogy," New York, 1880, p. 84.
[2] Kiliani, *Berg- u. Hüttenmännische Zeitung*, 1888, pp. 287, 306, 378.
[3] *Berg- und Hüttenmännische Zeitung*, 1860, p. 165.

ent. In addition to having a correct amount of iron present to decompose galena, the temperature is of great importance; the higher it is, within reasonable limits, the better will be the decomposition. A basic ferrous silicate ($4FeO.SiO_2$) will decompose galena readily; the singulo-silicate ($2FeO.SiO_2$) shows little effect. In practice, the amount of iron that is in excess of that required to flux the silica will be available for the decomposition of the lead sulphide.

Lime as well as baryta has a decomposing action on galena. If air has access, the following reaction takes place (Rivot [1]):

$$4PbS + 4CaO = 3CaS + CaSO_4 + 4Pb.$$

If the air be excluded and carbon present, the following occurs (Berthier [2]):

$$2PbS + CaO + C = Pb + (PbS.CaS) + CO.$$

§ 8. **Lead Sulphate,** *PbSO₄.*—This occurs as anglesite. It is formed in roasting lead sulphide (§ 9). Of all metallic sulphates it is the only one that is not decomposed upon ignition at a bright-red heat; it softens at a white heat and loses some of its sulphuric acid, forming a basic salt. Silica readily decomposes the sulphate, forming a lead silicate, while the sulphuric acid is driven off, being split into sulphurous acid and oxygen. In this way lead sulphate, obtained in roasting a siliceous galena, is decomposed. The operation goes by the name of slag-roasting. Lead sulphate is a poor conductor of electricity (Kiliani [3]). Carbon, if present in sufficient quantity, reduces the sulphate completely to sulphide at a dark-red heat:

$$PbSO_4 + 2C = PbS + 2CO_2.$$

If there is not enough carbon present, only part of the sulphate will be reduced :

$$2PbSO_4 + C = PbSO_4 + PbS + 2CO_2, \text{ and}$$
$$4PbSO_4 + 2C = 3PbSO_4 + PbS + 2CO_2.$$

At a cherry-red heat the resulting sulphate and sulphide will react upon each other, as shown in the next paragraph.

Lead sulphate is only slightly soluble in water and dilute sulphuric acid, more so in nitric acid and solutions of nitrates; it is soluble to some extent in sodium hyposulphite, the solubility increasing with the concentration and the temperature of the solvent (Stetefeldt [4]).

[1] "Traité de métallurgie," Paris, 1872, vol. ii., p. 42.

[2] "Traité des essais par la voie sèche," Liège, 1887, vol. ii., p. 580.

[3] *Berg- u. Hüttenmännische Zeitung,* 1883, p. 237.

[4] The Lixiviation of Silver Ores with Hyposulphite Solutions," New York, 1889, p. 81.

§ 9. **Roasting of Lead Sulphide; Reactions between Lead Sulphide, Lead Sulphate, and Lead Oxide.**—The roasting of galena and making the products react upon still undecomposed sulphide at an elevated temperature is of special interest, as one important lead-smelting process, the roasting and reaction process, also called the air-reduction process, is based upon it. If galena is ground fine and roasted [1] carefully at a low temperature, it will at first be converted into oxide (perhaps only into suboxide) and sulphur dioxide. Lead sulphide does not oxidize readily, and the dioxide will therefore form slowly. As we require a low temperature, only part of the dioxide combines with the oxygen of the air and forms the trioxide by contact, and this combines with the lead oxide, forming sulphate. If lead suboxide was present, the sulphur trioxide converts this first to oxide. Rammelsberg [2] suggests that some lead sulphide is directly oxidized to lead sulphate without passing through the stage of oxide. An experiment with pure galena gave to Plattner approximately the proportion $5PbO : 2PbSO_4$; on roasting a galenite from Bleiberg (Carinthia) that contained a small amount of blende and pyrite this changed to $PbO : PbSO_4$; with 50 per cent. pyrite added the relation was $2PbO : 3PbSO_4$.

This shows that the relation of lead oxide and sulphate in roasted galena depends on the presence of other sulphides. It is generally accepted that slow roasting at a low temperature produces more sulphate than if the operation be carried on quickly at a higher temperature, but, according to Rammelsburg,[3] this is not definitely settled.

If lead sulphide is heated to a strong red heat with lead oxide or sulphate, the following reactions will take place :

(1) $PbS + 2PbO = Pb_3 + SO_2.$
(2) $PbS + 3PbO = Pb_3 + PbO + SO_2.$
(3) $2PbS + 2PbO = Pb_3 + PbS + SO_2.$
(4) $PbS + PbSO_4 = Pb_2 + 2SO_2.$
(5) $PbS + 2PbSO_4 = Pb + 2PbO + 3SO_2.$
(6) $PbS + 3PbSO_4 = 4PbO + 4SO_2.$

These equations show that with correct proportions (1 and 4) all the lead is reduced by the sulphur, which combines with the

---

[1] Plattner, C. F., "Die metallurgischen Röstprocesse, theoretisch betrachtet," Freiberg, 1856, p. 145.

[2] Percy-Rammelsberg, "Die Metallurgie des Bleies," Brunswick, 1872, p. 39.

[3] *Op. cit.,* p. 40.

oxygen to dioxide ; if we have an excess of lead oxide (2) or of sulphide (3), it remains unaltered. We find something similar the case with equations (5) and (6). With too much lead sulphate we retain some or all the lead as oxide ; with a surplus of sulphide it would remain unchanged.

§ 10. **Lead Carbonate,** $Pb\,CO_3$.—This occurs as cerussite. It is a poor conductor of electricity.[1] The white lead of commerce is a basic carbonate. Lead carbonate is readily decomposed at a very low temperature into oxide and carbonic acid.

§ 11. **The Lead of Commerce** : *its Impurities and their Effect.*—In the market we find three kinds of lead : undesilverized lead, desilverized lead, and antimonial lead. The first comes from the non-argentiferous ores of the Mississippi Valley, the second from the refining works which desilverize argentiferous lead (Part III.). The third is a by-product of the second. The two soft leads are manufactured into sheet-lead and lead pipe ; they are used for making alloys, for corroding, and for other chemical purposes that require a good grade. The hard lead is used in making type-metal, bearings, etc.

In the subjoined table are given the analyses of the principal American brands of lead ; some well-known European makes have been added for the sake of comparison.

	Hartz Mountains.			Pribram.	Pennsylvania Lead Co.	Consol. Kansas City Smelting and Refining Co.		Missouri Ore-Hearth.
	Reduced from Litharge	Pattinson Process.	Parkes Process.	Luce-Rozan Process.	Parkes Process.	Parkes Process.		
Cu,	0.0600	0.0150	0.00080	0.0024	0.00007	0.00022	trace.	....
Ag,	0.0028	0.0022	0.00085	0.0018	0.00042	0.00020	0.0004	....
Bi,	trace.	0.0006	....	0.0023	....	0.00808	....	....
Cd,	....	....	0.00068	....	....	trace.	....	....
Sn,	....	....	....	....	....	....	....	....
As,	....	....	....	....	....	trace.	trace.	....
Sb,	0.1340	0.0100	0.00796	0.0056	0.00051	0.00127	0.0004	....
Ni,	0.0050	0.0010	....	0.0011	....	trace.	....	....
Co,	....	....	....	....	....	....	....	....
Fe,	0.0030	0.0040	0.00100	0.0017	trace.	0.00178	0.0006	....
Zn,	0.0040	0.0010	0.00092	0.0010	0.00038	0.00075	0.0018	....
Mn,	....	....	....	....	....	0.00021	....	....
S,	....	....	....	....	0.00018	....	....	....
Pb,	99.7912	99.9662	99.98480	99.9841	99.99844	99.99249	99.9963	....
Reference.	a.	a.	b.	c.	d.	e.	e.	f.

a. Zeitschrift für Berg-, Hütten- und Salinen-Wesen in Preussen, xviii., p. 205. *b.* Private Notes, 1890. *c. Oesterreichische Zeitschrift für Berg- und Hütten-Wesen*, 1890, p. 497. *d. Trans. A. I. M. E.*, iii., p. 822. *e. Eng. and Mining Journal*, July 14, 1888. *f.* See § 49.

[1] Kiliani, *Berg- u. Hüttenmännische Zeitung*, 1888, p. 287.

The analyses show that the soft lead in the market is of great purity; the total of the impurities does not exceed 0.02 per cent. These are copper, silver, bismuth, cadmium, tin, antimony (arsenic), nickel (cobalt), iron, and zinc (manganese).

1. *Copper.*—Copper and lead do not readily form homogeneous alloys. In order to obtain these, the two metals have to be melted together beyond the fusing-point of copper and then chilled quickly. If such an alloy be heated gradually to the melting-point of lead, it is possible to separate part of the lead from the copper. This will retain as minimum 0.08 per cent. copper.[1] The copper retaining some lead will remain behind as a porous mass. In order to remove the copper from the lead an addition of zinc is necessary, which will extract it to the practical limit (Parkes Process, § 94). The percentage of copper in commercial lead does not interfere with the rolling and other mechanical processes; it appears to protect the lead against the action of sulphuric acid. Antimony has a similar effect.[2] If used for corroding or for making flint-glass the percentage of copper ought not to exceed 0.0014 per cent. (Hampe[3]).

2. *Silver.*—What is said about copper in regard to mechanical treatment holds good for silver. Small quantities of silver protect lead against sulphuric acid. Baker[4] says that 1.70 ounces of silver per ton give white lead a reddish tinge, while this is not the case with 0.15 ounces per ton. Landsberg[5] gives as minimum 1.03 ounces for corroding lead.

3. *Bismuth.*—A percentage of 0.118–0.352 bismuth makes lead hard,[6] somewhat crystalline, and more readily fusible (Hampe). According to Napier[7] and Bauer,[8] 0.10 per cent. bismuth protects lead somewhat from sulphuric acid at 20° C., but not at 100° C. On white lead, from 0.0045 to 0.0075 per cent. bismuth has no effect (Hampe). Endemann[9] states that bismuth favors the corrosion of

---

[1] See Softening of Base Bullion, § 97.

[2] See below, 6.

[3] *Zeitschrift für Berg-, Hütten- u. Salinen-Wesen in Preussen*, xviii., p. 209.

[4] Dingler, *Polytechnisches Journal*, clxxiii., p. 119.

[5] *Wagner's Jahresberichte*, 1875, p. 596.

[6] Plattner, *Berg- u. Hüttenmännische Zeitung*, 1889, p. 116.

[7] *Chemical News*, 1880, p. 314; *School of Mines Quarterly*, vii., p. 97.

[8] *Berichte der deutschen chem. Gesellschaft*, 1875, p. 40; *School of Mines Quarterly*, vii., p. 117.

[9] *American Chemist*, 1876, vi., p. 457, and *Wagner's Jahresberichte*, 1877, p. 422.

lead, a small black residue remaining behind containing metallic bismuth. The only means of removing bismuth from lead is by the Pattinson crystallization process (§ 91).

4. *Cadmium.*—This occurs only in traces. It protects lead from the action of sulphuric acid.

5. *Tin.*—Tin makes lead light gray, hard, and increases its fusibility. It is uncommon in market leads. Lead containing it is more affected by sulphuric acid than pure lead (Napier, Bauer[1]). The effect in corroding has not been studied. It is removed by heating the lead to a bright-red heat with access of air; part of the tin collecting on the surface as oxide is first drawn off as a powder, and the rest as a slag consisting of stannic oxide and lead oxide. (Softening Base Bullion, § 97.)

6. *Antimony (Arsenic).*—Even small quantities of antimony give lead a grayish-white color, and make it harder and less malleable than ordinary lead. A bar of lead containing some antimony will show, especially in the centre, an uneven, moss-like surface. Hampe[2] finds that 0.005 per cent. antimony does not harden lead; Heeren[3] states that 0.25 per cent. makes lead hard, but that it is still malleable. Lead with 0.1 per cent. antimony is not so easily attacked by cold sulphuric acid as pure lead, but more easily by hot acid. For corroding, lead may not contain over 0.005 per cent. antimony (Hampe,[2] Landsberg[2]). Antimony and arsenic are removed, if the lead is heated to a bright-red heat, with access of air, as antimoniate and arseniate of lead [$Pb_3.2Sb(As)O_4$], slagged by an excess of litharge. In fact, if tin, arsenic, and antimony are present they will be oxidized in the order named and can be worked up separately. (Softening of Base Bullion, § 97.)

7. *Nickel (Cobalt).*—These hardly occur in market lead. Berthier[4] produced a malleable alloy containing from 0.4 to 0.5 per cent. nickel. Mrázek[5] says that from 1 to 2 per cent. antimony favors the entrance of nickel and cobalt into the lead, but they rise to the surface when the furnace lead is melted down slowly, and can then be easily skimmed off.

8. *Iron.*—Alloys of lead and iron form under special conditions, but market lead contains in maximo 0.07 per cent. iron

[1] See above, p. 22, notes 7 and 8.
[2] *Loc. cit.*
[3] Percy-Rammelsberg, "Die Metallurgie des Bleies," p. 49.
[4] "Traité des essais par la voie sèche," Liége, 1847, ii., p. 595.
[5] *Berg- u. Hüttenmännische Zeitung*, 1864, p. 315.

(Reich [1]), which does not have any effect on the softness and malleability. Corroding lead ought not to contain over 0.003 per cent. iron (Landsberg [2]).

9. *Zinc.*—Zinc and lead can be melted together in varying proportions, but they separate again in part on cooling. The amount of zinc that lead will retain depends, according to Rössler and Edelmann,[3] on the temperature of the lead. The subjoined table illustrates this.

Degrees Centigrade.	400	500	600	700
Per cent. of zinc retained	0.6–0.8	0.9–1.3	1.5–2.8	3.0

Zinc gives lead a silvery color and makes it so hard that it cannot be rolled ; cold and hot sulphuric acid attack it readily. Corroding lead should not contain over 0.003 per cent. (Landsberg [2]). Zinc is removed from lead by heating it to a bright-red heat, and oxidizing it by admitting air, introducing steam, etc. (See Dezincification of Desilverized Lead, § 105.)

10. *Manganese.*—This is present only in very small amounts and has no practical importance.

---

[1] *Berg- u. Hüttenmännische Zeitung*, 1860, pp. 28, 284.

[2] *Loc. cit.*

[3] *Berg- u. Hüttenmännische Zeitung*, 1890, p. 245 ; *Engineering and Mining Journal*, November 15, 1890.

CHAPTER III.

LEAD ORES.

§ 12. **Introductory Remarks.**—Quite a number of minerals contain lead, but only two or three are found in sufficient quantity to constitute lead ore. The lead occurs either as a sulphide or as a salt (sulphate, carbonate, etc.). Both are made more or less impure by other metallic compounds and vein matter. The ores are divided into two classes : Sulphide Ores (galenite) and Oxidized Ores (anglesite, cerussite, etc.), commonly called Carbonate Ores.

§ 13. **Galenite,** *PbS;* 86.6 *Pb*, 13.4 *S.*—This mineral is found well crystallized in cubes, sometimes also in isometric octahedrons. Crystals are not so often found isolated as in irregular bunches. It occurs also in coarsely crystalline to fine granular varieties ; crypto-crystalline galena is rare. Galena is found in most of the geological formations, but especially in the Silurian, Carboniferous and Triassic. It occurs in sedimentary rocks, such as limestone, dolomite, and sandstone, forming regular beds or impregnating the country rock to a greater or less degree, and in fissure and metamorphic veins, where it is contaminated with vein matter, such as quartz, earthy carbonates, barite, clay-slate, granite, gneiss, etc. These are generally removed by a mechanical washing process, and the dressed mineral is then treated by the smelter. The mine which contains probably as little galena as any that is worked is the one of Mechernich, Rhenish Prussia, where small nodules of galena of the size of a pea occur in a soft Triassic sandstone, the grains of which, about the size of millet, are slightly cemented by a clay or lime cement. The ore contains only 2.5 per cent. of galena, and this runs only about six ounces silver per ton.

The following table shows the rock and formation in which some well-known galena deposits occur and the tenor of the ore before and after dressing :

Locality.	Nature of Rock.	Geological Formation.	Raw Ore; per cent. lead	Dressed Ore:		Reference.
				per cent. lead.	oz. silver per ton.	
Mineral Point, Wis. .	Dolomite	Lower Silurian . . .	—	—	3.0	1
Rockville, Wis.	"	" . . .	—	—	0.3	1
Granby, Mo.	"	Sub-Carboniferous	—	84	1.25	1
St. Joe, Mo.	"	Lower Silurian . . .	7	70	—	2
North of England . . .	Limestone.	Carboniferous. . . .	8.5	70–77	8	3
Bleiberg, Carinthia. .	"	Triassic	8	71	0.05	4
Pribram, Bohemia. . .	Graywack.	Lower Silurian . . .	20	37–38	76.5	5
Freiberg, Saxony	Gneiss.	Archæan	3	18–70	17–88	6
Tarnowitz, Silesia . . .	Dolomite	Triassic	6	75.5	13.5	7
Upper Hartz, Prussia	Graywack slate	Sub-Carboniferous	9	64	25	8
Mechernich, Prussia .	Sandstone	Triassic	2	56–60	3–4	9

1. " Geological Survey of Wisconsin," 1873–1879, iv., p. 382.
2. Desloge, *Trans. A. I. M. E.*, xviii., p. 262.
3. Hunt, " British Mining," London, 1884, p. 899; and Phillips "Elements of Metallurgy," Philadelphia, 1887, p. 566.
4. *Oesterreichische Zeitschrift für Berg- und Hüttenwesen*, 1890, p. 286,
5. *Oesterreichische Zeitschrift für Berg- und Hüttenwesen*, 1888, p. 567; and *Oesterreichisches Jahrbuch*, xxxix., p. 10.
6. " Freiberg's Berg- und Hütten-wesen," Freiberg, 1883.
7. *Zeitschrift für Berg-, Hütten- und Salinen-Wesen in Preussen*, xxxii., p. 392.
8. *Zeitschrift für Berg-, Hütten- und Salinen-Wesen in Preussen*, xxx., p. 131 and Private Notes, 1890.
9. " Bergbau und Hüttenbetrieb von Mechernich," Köln, 1886.

Galena often occurs in a very pure state, but it is more generally contaminated with other metallic sulphides. These are either pyrite, arsenopyrite, chalcopyrite, blende, bournonite, etc., as associated minerals; or silver, copper, zinc, iron, nickel, etc., forming isomorphous compounds with the lead sulphide. The associated minerals can usually be mechanically separated from the galena, but not always. Sometimes the admixture is too intimate, causing trouble and loss in the metallurgical treatment.

Galena is almost always argentiferous. The silver is rarely present in the native state; commonly it appears as isomorphous silver sulphide, or as a finely disseminated silver mineral. The difference of form is important in connection with concentration [1]. If the silver occurs as isomorphous sulphide, replacing part of the lead sulphide, the loss in concentration will correspond to the percentage of lead in the tailings; if as associated mineral (*e.g.*, tetrahedrite), it will be very great, as this mineral, being very brittle, is readily crushed to a fine powder, and, being also lighter than galena, is carried off on the water. If the dark scum that is often seen floating on the water of jiggs, where argentiferous ga-

---

[1] Raymond, *Engineering and Mining Journal*, February 11, 1882.

lena is concentrated, be assayed, the main source of loss in silver appears. The tenor of silver in galena ores varies a great deal. The galenite from Bleiberg, Carinthia, with 0.05 ounces silver per ton, represents probably the lowest amount, and occasional specimens from Idaho and Schemnitz, Hungary, with 2,042 ounces, the highest. The amount of silver contained in galena is often said to be dependent on the enclosing rock; thus crystalline rocks would be favorable to a high percentage of silver, and unchanged sedimentary strata unfavorable. While many occurrences of galena appear to sustain this rule, others again disprove it, so that it is not of general application.

It has often been said, and may sometimes still be heard, that coarse-grained galena is poor in silver, while fine granular varieties give higher assays, but Malaguti and Durocher[1] disproved it forty years ago. The minerals oftenest associated with galena—such as pyrite, blende, etc.—do not generally contain as much silver as the galena.

Percy[2] states that gold is as invariably present in galena as silver, but it does not often occur in appreciable quantities.

§ 14. **Anglesite,** $PbSO_4$; PbO, 73.6; SO_3, 26.4; Pb, 68.3. **Cerussite,** $PbCO_3$; PbO, 83.5; CO_2, 16.5; Pb, 77.5.—Carbonate ores, using that term in a general sense as embracing all oxidized lead ores, occur often in the form of a sand or an earth, bearing the name of sand or soft carbonates. In other cases the particles of carbonate are cemented together by clay, iron, manganese, or silicious matter, forming compact amorphous or crystalline lumps called hard carbonates.

The minerals are seldom found as originally deposited; the form and composition are more apt to have been caused by chemical changes that have taken place since the galena from which they resulted was deposited. The sulphate usually formed by the oxidation of the sulphide is in most cases the compound from which the carbonate has resulted by the action of circulating waters holding alkaline or earthy carbonates in solution.

Anglesite as an ore is rare, as it is not often that galena is exposed only to the oxidizing action of air. It often occurs, however, with cerussite, and represents the transition between the sulphide and the carbonate. This is further illustrated by pieces

---

[1] *Annales des mines*, 4th series, 1850, xvii., p. 85.
[2] "Metallurgy of Lead," p. 96.

of galena found in carbonate ores, which contain in the interior anglesite and whose surface is changed to cerussite. Thus these two minerals will be always found near the outcrop of galena deposits. To what extent the decomposition has progressed depends on local circumstances.

The richness in lead of carbonate ores varies a great deal. If galena alone has been oxidized, the ores are likely to be rich ; if the decomposing action attacked also the country rock, this may so contaminate the ore as to reduce the percentage of lead below the limit where it pays to treat the ore. The grade of carbonate ore is not so often raised by wet concentration as that of galena ores, as the losses in lead, and especially in silver, which passes off in the slimes, are almost unavoidable. In some instances the carbonate ore is first leached with sodium hyposulphite to remove as much silver as possible, and then the lead is concentrated in the wet way to a rich product. This used to be done at the Old Telegraph[1] Mine, Ut. Another way is to use Krom's[2] system of dry concentration, the result being a high grade smelting ore, and tailings and dust, to be treated in the wet way. The following table shows some very pure carbonate ores from Missouri resulting from corresponding pure galena :

Locality.	Per cent. lead.	Ounces silver per ton.	Chemist.	Reference.
Red Mountain, Col.	17	128	Kedzie	*Trans. A. I. M. E.* xvi., p. 581.
Leadville, Col	21	65	Rolker	" " xiv., 287.
" "	38	25	Ricketts....	" " xiv., 387.
Granby, Mo	65	..	Williams...	" " v., 315.
South West, Mo ..	72	..	Chauvenet .	Broadhead, " Geol. Survey of Mo.," 1874, p. 710.

The associated minerals undergo a process of oxidation with the galena and are generally found again in part in the carbonate

[1] " Tenth Census of the United States," 1880, vol. xiii., p. 415.

[2] Krom: " Commissioner Raymond's Report," 1876, p. 419 ; *Trans. A. I. M. E.*, xiv., p. 497; *Engineering and Mining Journal*, 1886, Aug. 14, Sept. 25, Oct. 23.—Stetefeldt : *Eng. and Mining Journ.*, Oct. 28, 1876 ; May 2, 1885 ; *Trans. A. I. M. E.*, xv., p. 355.—Furman : *School of Mines Quarterly*, iii., p. 127.—Newberry : *Ibid.*, iv., p. 1.—Heard : *Eng. and Mining Journ.*, 1886, July 3, Sept. 11.—Dry Recorder : *Eng. and Mining Journ.*, 1886, Sept. 4.—Hollister : *Trans. A. I. M. E.*, xvi., p. 1.—Sickel : *Berg- u. Hüttenm. Ztg.*, 1885, p. 313.—Blömecke : *Berg- u. Hüttenm. Ztg.*, 1886, pp. 485, 501, 514.

ore, although, being more soluble than the lead sulphate, they may be carried away entirely. The silver in oxidized ores is present mostly in the form of chloride, although it also occurs in its original form as sulphide and antimonide. There is less likelihood of a uniform ratio between lead and silver in a carbonate ore than there is in the sulphide ore, as lead sulphate and carbonate behave differently with solvents from silver sulphide, chloride, and antimonide. Thus enrichment and impoverishment both in lead and silver can be easily accounted for.

The gold in carbonate ores occurs probably in the native state.

§ 15. **Other Lead Minerals.**—The following six oxidized lead minerals occur often in carbonate deposits, but not in sufficient amounts to constitute an ore :

Pyromorphite, $PbCl_2 + 3Pb_3P_2O_8$; 76.4 Pb.—Calcium fluoride often replaces in part the lead chloride ; calcium, the lead combined with phosphoric acid, and arsenic acid, the phosphoric acid.

Mimetite, $PbCl_2 + 3Pb_3As_2O_8$; 69.6 Pb. In the lead arseniate the lead is sometimes in part replaced by calcium, and the arsenic usually in part by phosphoric acid.

Vanadinite, $PbCl_2 + 3Pb_3V_2O_8$; 65.0 Pb.

Crocoite, $PbCrO_4$; 63.9 Pb.

Wulfenite, $PbMO_4$; 57.0 Pb.

Stolzite, $PbWoO_4$; 45.4 Pb.

There might still be mentioned about twenty lead-bearing sulpharsenites, sulphantimonites, and sulphbismuthites which occur in lead deposits, but they are only mineralogical curiosities.[1]

---

[1] Dana, "System of Mineralogy," New York, 1880, pp. 84 to 109.

CHAPTER IV.

DISTRIBUTION OF LEAD ORES.

§ 16. **Lead Ores of the United States.**—Lead ores occur in many parts of the world. (Table of World's Production, § 3). The mines of Spain, Germany, and England furnish the bulk of the European product, but it is not the present purpose to deal with that branch of the subject.

The occurrence of lead ores in the United States is best discussed under four heads :

    I.   The Atlantic coast.
    II.   The Mississippi Valley.
    III. The Rocky Mountains.
    IV. The Pacific.

I.—LEAD ORES OF THE ATLANTIC COAST.[1]

§ 17. **Lead Ores of the Atlantic Coast.**—The lead ores of the Atlantic coast occur in New York, New England, and Virginia. They were worked in former times, and are practically abandoned now.

The Archæan gneiss of New York is traversed by veins of galena, which, being free from zinc and iron, have a gangue consisting mainly of calcspar. The Rossie mines were worked at intervals for over fifty years.

In the New England States, galena, more or less argentiferous and associated with other metallic sulphides, occurs irregularly distributed in segregated veins in highly metamorphosed palæozoic rocks.

Virginia[1] has some deposits of galena and blende with carbonate and silicate of zinc ; the lead, however, is subordinate to the zinc.

[1] J. D. Whitney, "Metallic Wealth of the United States," Philadelphia, 1854, p. 382.

[2] *Trans. A. I. M. E.*, viii., p. 340.

II.—Lead Ores of the Mississippi Valley.

This heading covers two divisions : the lead region of the Upper Mississippi Valley and that of Missouri. They have many features in common and are often discussed together ; there are, however, so many differences in the mode of occurrence of the galena, in the associated minerals, and even in the geological horizon, that it is advisable to keep them separate.

§ 18. **The Upper Mississippi Valley.**[1]—In the southwestern part of Wisconsin are the important lead deposits of the Upper Mississippi Valley, which extend a very small way into the adjoining States of Iowa and Illinois. They are principally centred around certain districts (diggings), such as Mineral Point and Platteville, Wis.; Galena, Ill., and Dubuque, Ia. About ninety per cent. of the lead produced comes from Wisconsin.

The ore is a non-argentiferous galena; it occurs in wholly undisturbed dolomitic limestone of the Trenton period in vertical crevices, flat crevices, or as an impregnation. The vertical crevices (gash veins) are either thin seams in the rock, a few inches thick by several hundred feet long, extending downward for from twenty to forty (occasionally one hundred) feet, filled up solid with coarse galena; or more commonly they expand in an irregular way, inclosing particles of rock, which are cemented together by galena. If disintegration has taken place, caves are formed in which the galena is found distributed among loose masses of rock, calcareous sand, clay, ochre, etc. The horizontal crevices (flat openings, flat sheets) are either thin seams of compact ore along the bedding planes of the rock, or, if the rock has been disintegrated, have been enlarged, and the galena is found in the same form as in the caves of vertical crevices. Combinations of vertical and horizontal crevices (flats and pitches) increase the size of the deposits. Impregnations of certain strata of dolomite occur, but they are not frequent.

The galena (mineral) from the upper beds is pure and rich. As depth is gained, the associated minerals—pyrite (sulphur), blende (black jack)—increase in quantity and often predominate over the galena. Chalcopyrite (copper) is scarce; it is found more

---

[1] J. D. Whitney, "Metallic Wealth of the United States," p. 408.— "Geology of Wisconsin," 1878–1879, vol. ii., pp. 645–752, by M. Strong (also in *Engineering and Mining Journal*, 1878, July 6, ff.); vol. iv., pp. 877–568, by T. C. Chamberlin.—*Trans. A. I. M. E.*, viii., p. 498, Irving.

with pyrite and blende than with galena. Secondary minerals are not of frequent occurrence. Calcite and barite occur in the lower beds. The absence of nickel, cobalt, and arsenic is to be noted.

§ 19. **The Mines of Missouri.**[1]—The ore is a coarsely crystalline galena, practically free from silver.

In the Southeastern District, represented by the Saint Joe and La Motte mines, galena occurs disseminated through strata of dolomitic limestone of the Lower Silurian period, lying almost horizontally. The thickness of the lead-bearing "Third Magnesian Limestone" varies usually from two to six feet, although it sometimes goes much higher. The ore as mined at Saint Joe[2] runs about 7 per cent. lead and is concentrated to a product of 70 per cent. lead. Associated with the galena occurs pyrite containing nickel and cobalt. Chalcopyrite, as well as nickel and cobalt sulphides with traces of arsenic, is found more at Mine La Motte than at the Saint Joe mines. The absence of blende is to be noted. The pyrite-bearing galena ores are concentrated (separately from the pure galena) to a product called sulphide, which has the following composition :

Locality.	Lead.	Iron.	Nickel, Cobalt.	Sulphur.	Insol.	Chemist.
Saint Joe, 1884	21.86	16.21	0.61	....	....	Setz.
Mine La Motte, 1881.	17.87	....	4.77	....	....	Neill.
" " "	13.34	44.84	4.07	20.37	3.58	"

The dolomitic limestone in which the ores occur contains about three per cent. silica; a marked feature is the presence of barite in places where the usually disseminated mineral has been concentrated to small sheets.

The ores of the Central Lead Region, lying between the Osage and Missouri Rivers, also occur in dolomite of the Lower Silurian period; they resemble those of Wisconsin, but are of no special importance.

In the Southwestern Region, which reaches into Kansas, lead and zinc ores are worked very extensively at present, especially

---

[1] Whitney, *Op. cit.*, p. 417.—Broadhead, "Geological Survey of Missouri," 1874.—*Trans. A. I. M. E.*, iii., p. 116, Gage; v., p. 100, Broadhead.—*School of Mines Quarterly*, ix., p. 74, Kemp.

[2] *Trans. A. I. M. E.*, xvii., p. 659, Munroe; xviii., 263, Desloge; *School of Mines Quarterly*, ix., p. 74, Kemp.

around Joplin. The galena occurs in dolomitic limestone of the Sub-Carboniferous period, containing layers of chert and bituminous matter. When broken, it often emits a bituminous odor. It occurs in single or loosely aggregated crystals, also in crystalline masses of small dimensions imbedded in the limestone and in the beds of chert, the fragments of which are cemented together by a bluish-gray, clayey mass. Cadmium bearing blende occurs plentifully in two forms, a coarsely crystalline and a granular variety; pyrite is subordinate. To be noted is the absence of chalcopyrite and barite. Magnesite and calcite are found with the galena and often occur in crystalline masses in the dolomite.

§ 20. **Other Occurrences.**—The lead deposits of Southwestern Texas in the Quitman Mountains are as yet of no special importance.

III.—Silver-Lead Ores of the Rocky Mountains.[1]

To this division belong the occurrences of argentiferous lead ores in Colorado, South Dakota, Montana, and New Mexico.

§ 21. **Colorado.**—The deposits of argentiferous lead ores of Colorado are more important than any others of the country. They are found in veins of Archæan rock and of eruptive rock of the San Juan region, and, as sedimentary deposits, in Palæozoic limestone, which is penetrated by igneous rock. The latter deposits are the great metal-producers of the State.

Boulder County.—On the eastern slope of the Rockies the ores occur as veins in Archæan gneiss, which, on account of the indistinctness of the bedding, Emmons calls granite-gneiss. These are not real fissure veins but alterations of the country rock along certain planes. The mines are noted principally for the occurrence of rich telluride minerals, but in the Caribou district is found rich galena with silver minerals, such as stephanite and proustite. The Caribou mine contains a massive mixture of rich argentiferous galena, chalcopyrite, and blende, occurring in gneiss near a dyke of eruptive diabase.

A shipment of ore made in 1891 contained :

Pb	44.1 %
Zn	2.5 %
Fe	9.1 %
SiO_2	8.6 %
Ag	72.0 oz. per ton.

[1] S. F. Emmons, "Tenth Census of the United States," 1880, xiii., pp. 60–104.

The San Juan Region.[1]—This embraces the southwestern part of Colorado, *i.e.*, the counties of San Juan, Hinsdale, Ouray, Dolores, La Plata, part of Rio Grande and Conejos.

Characteristic of the region are immense quartz veins traversing older and younger (Triassic) eruptive rocks. The productive ore bodies are found in the older massive and brecciated rocks, but veins occur also in the underlying gneiss and granite. In the neighborhood of Rico (Dolores Co.) and Ouray (Ouray Co.) sedimentary deposits occur in carboniferous limestones. The deposits are mainly argentiferous. The minerals are argentiferous galena, silver-bearing gray copper, ruby silver, and native silver. Bismuth-silver minerals are frequent; small amounts of gold are found; blende occurs in considerable quantities. The gangue is quartz, kaolinite; barite is common; fluorite also occurs. In the bedded deposits of Rico iron and manganese are prominent. Near the outcrop the ores are often completely changed into a sandy carbonate.

Analyses[2] from two characteristic car-load samples of ore from the Red Mountain district by Kedzie are subjoined.

H_2O (by ignition)	6.75	11.94
Insol.	41.63	8.65
Pb	18.40	....
Fe_2O_3	17.12	64.80
Al_2O_3	6.08	13.05
MnO	0.59	1.98
CaO	1.70	....
ZnO	1.23	....
SO_3	3.87	....
CO_2	2.89	....
Total	99.76	....
Ag, ounces	128.00	68.00
Au, ounces	0.22	0.10

Custer County.[3]—Two deposits of this region, the Bassic and Bull Domingo mines, differ from other Colorado occurrences.

[1] Emmons, *Op. cit.*, p. 12.—*Trans. A. I. M. E.*, xi., p. 165; xv., p. 218; xvii., p. 261, Comstock; xvi , p. 804, Emmons; xvi, p 570, Kedzie; xviii., p. 139, Schwarz.—*Proceedings Colorado Scientific Society*, March, 1888, Hills; Abstract in *Eng. and Min. Journal*, 1888, June 9, Emmons.

[2] Kedzie, *Trans. A. I. M. E.*, xvi., p. 581.

[3] Emmons, *Op. cit.*, p. 80.—Grabiel, *Trans. A. I. M. E.*, xi., p. 118.

The ore, which is found in large bodies without any definite boundary, forms concentric layers from ⅛ to ¼ inch in thickness upon spherical fragments of eruptive country rock. They are located near Rosita and Silver Cliff respectively. At the Bassic mine the pieces of wall rock vary from ⅜ to 24 inches; the sizes most common have diameters ranging from 4 to 12 inches; the rock proper shows only traces of precious metal. The metallic coating surrounds the pebbles and bowlders and fills the interestices between them. It forms concentric layers. The first one consists of sulphides of zinc, antimony, and lead, assays 60 ounces in silver and from 1 to 3 ounces in gold, and is dark. The next one is similar, but lighter in color, and assays higher; the third is blende, rich in silver and gold. Chalcopyrite often follows with some pyrite.

At the Bull Domingo mine the occurrence is similar. Around a barren nucleus of hornblendic gneiss is deposited a coating of argentiferous galenite, followed by siderite, but the ore contains no gold.

The Terrible mine[1] at Ilse resembles the other two in some respects. A porphyry dyke, 127 feet wide, traverses the granite country rock for a considerable distance, 87 feet being impregnated with crystals of cerussite. As mined the ore carries from 10 to 12 per cent. of lead and 1 ounce of silver to the ton. It is concentrated and gives an enriched product asssaying 70 per cent. lead and 1.5 ounces silver.

Lake County.—Of the sedimentary deposits those of Leadville[2] are the most important, producing about sixty per cent. of the argentiferous lead of Colorado.

The ore deposits of Leadville occur principally at the contact of the dolomitic "Blue Limestone" of the Lower Carboniferous, with an intrusive sheet of "White or Leadville porphyry" that covers it. They are, however, not confined to the contact, but extend into the limestone below. In some cases there is a gradual transition from ore to unaltered limestone; in others the ore forms pockets and caves in the limestone; again, the ore has entirely replaced the limestone and is included between two sheets of porphyry. It is exceptional when ore occurs unconnected with the

---

[1] Private Communication of Messrs. Taylor & Brunton, December 1, 1890.

[2] S. F. Emmons, "Geology and Mining Industry of Leadville, Col.," Monograph XII., U. S. Geol. Survey, Washington, 1886, pp. 375–582.

contact surface in the form of irregular masses or running across the formation.

The principal ore is argentiferous galena with its secondary products of decomposition—anglesite, cerussite, pyromorphite. Silver occurs in the altered ore principally as chloride and chloro-bromide, rarely as chloro-iodide and native, sometimes as sulphuret. It has been found that nodules of galena are richer in silver than cerussite. Small amounts of gold in the native state have been found in limestone, but it is usually associated with pyrite in porphyry. The accessory minerals are blende and its secondary products, carbonate and silicate of zinc. Arsenic is found as sulphide and in combination with iron as arseniate, antimony as sulphide, copper as carbonate and silicate, molybdenum and vanadium as wulfenite and vanadate of lead and zinc, bismuth as sulphide and sulpho-carbonate. Iron and manganese occur as oxides. The gangue of the ore consists of silica (as chert, quartz, and combined with iron and manganese) and various clays charged with iron and manganese. Barite is not uncommon, but irregularly distributed. Siderite, pyrite, and gypsum are subordinate. A decomposed product, Chinese talc, consisting of a mixture of silicate and sulphate of alumina, occurs at the contact of the white porphyry and the limestone.

The following analysis of carbonate ore by Fluegger[1] represents an average sample of a thousand tons of ore coming from every producing mine. It shows the character of the ore mined at that time (1881 ?).

CO_2	5.58	FeO.	0.89	MgO	3.04	Au	trace
PbO.	25.77	Fe_2O_3	24.86	As	0.01	Cu	trace
Ag.	0.31	MnO_2	4.03	Sb	0.02	Zn	trace
SiO_2	22.59	Al_2O_3	3.99	$(KNa)_2$	0.98	H_2O	5.58
S.	0.90	CaO.	2.36	Cl	0.09	Total	101.00

The composition of the sample in its main features is therefore in round figures in per cents.: Pb, 24; Ag, 90.5 ounces; (FeMn)O, 21.8; Al_2O_3, 3.4; (CaMg)O, 6.6; SiO_2, 22.6. Guyard,[2] comparing this analysis with a number of others of that time,[3] says the figures of lead, iron, and silica agree with the general composition of the smelting charges, but the silver is too high, the common relation being one ounce of silver to six pounds of lead, while Fluegger's analysis would correspond to one ounce to every five pounds.

[1] *Engineering and Mining Journal*, January 8, 1881.
[2] Emmons, *Op. cit.*, p. 620.
[3] Emmons, *Op. cit.*, pp. 621-625.

The varying proportion of silver and lead in different districts is shown by the following figures of Rolker [1] :

Locality.	Mine.	Dry weight, tons.	Ounces silver per ton.	Per cent. of lead.	Relation of Ag: Pb.
Fryer Hill......	Chrysolite............	10,561	65.45	21.45	1 : 6.5
Carbonate Hill .	Evening Star........	6,815	39.00	16.10	1 : 8
Carbonate Hill .	Morning Star	4,794	25.00	38.45	1 : 32
Iron Hill.......	Iron & Silver Mg. Co.	152,457	15.00	18.60	1 : 26

The table shows that Fryer Hill ores run higher in silver than either Carbonate Hill or Iron Hill ores, and that the latter two give a lower-grade base bullion.

These figures refer to the time when mining was carried on entirely in parts of the ore bodies that had been oxidized by surface waters. The ores, just as they came from the mine, could be economically smelted in the blast furnace. Now every plant has a number of roasting furnaces, thereby showing how, as the exploration extends deeper, the unaltered minerals, galena, blende, and pyrite are encountered. Unfortunately, at the same time, less silver is found. This transition from carbonate to sulphide ore is often very rapid. On Iron Hill, according to Blow,[2] the ore changes suddenly from a body of fine oxidized smelting ore to a close-grained sulphide ore consisting principally of zinc and iron sulphides with sulphide of lead in small quantities. The following analyses show the character of some of the sulphuret ores of Leadville :

	Galena.	Mixture.	Galena.	Mixture.	Lead Ore.	Zinc Ore.
	Minnie Mine.	Minnie Mine.	Moyer Shaft.	Moyer Shaft.	Col. Sellers Mine.	Col. Sellers Mine.
Pb..........	72.65	50.86	44.0	15.0	27.40	10.70
Zn..........	5.66	12.86	18.0	24.0	25.00	24.50
Fe..........	1.60	9.30	11.0	16.0	6.00	16.60
S	15.66	24.50	30.0	40.0	35.00	40.00
Ag., ounces.	41.50	11.50	14.0	11.0	26.90	54.30
Au., ounces.	trace	trace.	....	....	trace	trace
Insol........	4.12	1.88	....	....	3.00	3.40
H$_2$O........	....	....	....	....	2.00	2.10
References..	1	1	2	2	3	3

References: (1) *Trans. A. I. M. E.*, xiv., p. 189, Freeland ; (2) xiv., p. 288, Rolker; (3) xviii., p. 173, Blow.

[1] *Trans. A. I. M. E.*, xiv., p. 287.
[2] *Trans. A. I. M. E.*, xviii., p. 170.

The decrease in silver and lead and the increase in zinc with increasing depth have forced the question of concentration upon the mining companies. According to Ihlseng[1] several large and some small concentrating plants are enriching ores that run from 6 to 8 ounces in silver and from 6 to 10 per cent. in lead, saving 60 per cent. of the silver and 70 per cent. of the lead, 7 tons being concentrated to 1 ton.

Taylor and Brunton[2] report that the Colonel Sellers mill, constructed by them, concentrates from 75 to 90 tons of sulphide ores, very free from gangue, and containing less than 10 per cent. lead and 9 or 10 ounces silver, into heads running 55 to 60 per cent. lead and tailings under 2 per cent. lead at a cost of 70 cents per ton. The following analyses by Kellar show the character of the products obtained from coarse-grained ore.

	Crude Ore.	Blende.	Galena.	Pyrite.
SiO_2	0.960	1.090	0.120	0.190
Pb..................	16.185	8.525	79.958	9.592
Fe	22.951	10.864	2.303	39.431
Zn	19.246	47.522	1.734	3.043
Mn.................	1.664	1.477	0.154	0.406
Cu	trace.	....	....	trace.
As	trace.	....	....	trace.
S...................	39.240	37.040	15.760	43.196
Ag, ounces	7.5	8.0	19.8	5.2

Park County.—The deposits of this county have been developed in Archæan and Palæozoic rocks. The Mesozoic porphyries form dykes in Archæan and intrusive sheets in Palæozoic strata. Prominent are the deposits of Mount Lincoln and Mount Bross. The ores are argentiferous galena with sulphate and carbonate of lead and chloride of silver; pyrite, more or less decomposed, and associated with oxides of manganese, is also found, coloring the clay of the gangue; barite is of common occurrence. The deposits of Mount Lincoln form irregular bodies in the blue limestone, and are found generally near its upper surface; those of Mount Bross occur rather in the mass of the limestone near the dykes of porphyry. The character of the ore resembles, therefore, that of Leadville.

[1] "Report of State School of Mines," Golden, Col., 1887, p. 44.
[2] *Engineering and Mining Journal*, May 8, 1886.

Pitkin County.—The Aspen[1] deposits of silver-bearing ores have lately come into great prominence, very little ore having been shipped before the advent of the railroad in 1887. They occur in the same geological horizon as the Leadville deposits, viz., the Lower Carboniferous limestone, but they are not found in immediate contact with the eruptive rock. They occur principally between two varieties of limestone, the blue (a pure carbonate of lime) and the brown (dolomite), which is traversed by numberless small veins containing iron salts. The brown limestone breaks up by oxidation of the iron into small pieces, and is called "short lime." The ore is a limestone impregnated along certain fissures with fine-grained argentiferous galena. The main contents of silver comes, however, from associated silver minerals, such as polybasite and stephanite. Near the outcrop the ore is somewhat decomposed, consisting of barite, carbonate of copper, oxides of iron and manganese, calcite, dice-shaped fragments of dolomite, and galena.

The character of the ore is shown by the following analyses furnished by Taylor and Brunton :[2]

$BaSO_4$.... 0.5	5.5	9.1	11.7	13.0	15.8	24.2	26.0	26.2	29.5	30.0	36.9	40.0
SiO_2...... 6.5	15.4	9.2	20.0	19.7	27.9	17.8	40.0	44.9	14.3	7.0	22.3	19.0
CaO...... 7.6	27.5	29.8	19.0	15.7	14.2	13.7	3.0	1.8	7.8	11.1	12.9	12.5
Fe........ 9.0	5.4	5.5	4.0	6.8	6.5	1.5	9.4	10.6	2.5	7.3	4.8	6.5
Pb........19.8	....	....	....	1.8	....	4.7	8.0	....	16.0	1.4	....	....
Ag. ounces 7.5	18.0	24.0	15.0	26.4	28.0	31.0	25.0	32.0	69.0	26.0	32.0	37.0
H_2O...... 9.75	3.0	5.5	5.8	7.0	9.0	5.0	9.0	10.0	3.0	9.0	10.8	3.0

Summit County.—In the Ten-Mile district the ores occur in Upper Carboniferous limestone and in the sandstone above. The typical mine is the Robinson. The ore is a rich argentiferous galena associated with pyrite and blende. It occurs near the surface of the limestone, in some cases extends into it, and in others penetrates it entirely.

§ 22. **South Dakota.**[3]—The occurrence of lead ores is confined to two small camps in the Black Hills, Galena and Carbonate. Argentiferous galena and carbonate ore occur in shoots in calcareous Potsdam rock, which overlies the upturned Archæan slates. These shoots are especially frequent where the porphyry cuts

[1] S. F. Emmons, "Proceedings of the Colorado Scientific Society;" Henrich, *Trans. A. I. M. E.*, xvii., p. 156.
[2] Letter of December 1, 1890.
[3] Carpenter, *Trans. A. I. M. E.*, xvii., p. 582.

through the rocks and where it overlies them. Associated with galena, pyrite and blende are found. The ore has to be closely sorted, so as to run 20 ounces in silver and 50 per cent. in lead.

§ 23. **Montana.**—The argentiferous lead ores of Montana are unimportant in comparison with the great copper deposits. They occur as metamorphic veins in crystalline rocks or as irregular ore bodies in limestones. According to Lindgreen,[1] in the neighborhood of Helena occur veins from one to ten feet in width, carrying in a gangue of quartz, galena, blende, chalcopyrite, pyrite and arsenopyrite. The galena assays up to 90 ounces in silver, the blende from 10 to 12 ounces in silver, the pyrite from 0.2–0.3 ounces in gold, the arsenopyrite 1 ounce in gold ; the chalcopyrite carries more silver.

The smelting works at Glendale, Helena, and Great Falls obtain a large proportion of their ores from Idaho, especially from the Cœur d'Alene District.

§ 24. **New Mexico.**—Although New Mexico produces much dry-silver ore, lead-silver ores are not of frequent occurrence. The principal ones are found in the Magdalena Mountains.[2] The deposits occur in limestone and porphyry and are from 4 to 40 feet wide. The principal mines are the Kelly, the Juanita and the Graphic. The ore averages 25 per cent. lead and runs very low in silver—8 ounces to the ton.

IV.—SILVER-LEAD ORES OF THE PACIFIC.[3]

Argentiferous lead ores under this head occur in Nevada, Utah, Idaho, Arizona and California.

§ 25. **Nevada.**[4]—The production of argentiferous lead in Nevada has greatly diminished in the last twelve years. Although 31,063 tons of lead were produced in 1878, in 1890 the production had sunk to 1,994 tons. The ore comes principally from two

---

[1] "U. S. Geol. Survey: Mineral Resources of the U. S.," 1883–84, p. 422.

[2] Silliman, *Trans. A. I. M. E.*, x., p. 425; Prince, *Eng. and Mining Journal*, 1890, November 20.

[3] G. F. Becker, ' Tenth Census of the United States," 1880, vol. xiii., pp. 5–59.

[4] J. S. Curtis, "Silver Lead Deposits of Eureka, Nev.," Monograph vii., "U. S. Geol. Survey," Washington, 1884.—A. Hague, "Geology of Eureka Mining District," monograph, "U. S. Geol. Survey," in preparation.

mines, the Richmond mine and the Eureka Consolidated mine, of the Eureka District. It occurs in irregular chambers in the Prospect Mountain Limestone of Ruby Hill, which is a compact magnesian limestone of the lower horizons of the Cambrian. In the upper part of the larger chambers it is in a loose state; in the lower ones it is more compact. It is an argentiferous and auriferous carbonate containing both anglesite and cerussite and a very small amount of galena; mimetite and wulfenite are often found. A remarkable feature is that at a depth of 1,300 feet the ore is still oxidized, and the regular sulphide from which it originated has not yet been reached. The associated minerals are pyrite, arsenopyrite and blende, with their oxidized products. The gangue accompanying the lead ore is principally limonite, which colors it in different shades of yellow, red and brown; lime, magnesia, alumina and silica occur only in subordinate quantities. The character of the ore is shown by the following analysis made by F. Claudet[1] in 1878:

PbO	85.65	Mn_2O_3	0.13	SiO_2	2.95	H_2O+CO_2	10.90
CuO	0.15	As_2O_5	6.84	Al_2O_3	0.64	Ag+Au	0.10
FeO	34.89	Sb	0.25	CaO	1.14		
ZnO	2.87	SO_3	4.18	MgO	0.41	Total,	100.52

27.55 ounces silver per ton; 1.59 ounces gold per ton.

The average tenor of lead, silver and gold is lower than the analysis shows, viz.: 15 per cent. lead, 23 ounces silver, 0.72 ounces gold.

To be noticed is the large amount of arsenic, which causes the formation of the unwelcome speise in smelting. Molybdenum has not been determined. Silver is present as chloride and sulphide; gold, probably finely divided, in the native state.

§ 26. **Utah.**[2]—The argentiferous lead deposits of Utah form more or less regular bodies in limestones or at the contact of limestone and eruptive rock. The ores are carbonate with anglesite strongly prevailing; the secondary minerals reach down to considerable depth. Small amounts of galena occur. A very few of the mines have been worked down to the sulphuret ores. Others which produced only siliceous ores low in lead are running with greater depth into bodies richer in lead.

In Beaver County is the celebrated Hornsilver mine that closed

[1] Curtis, *Op. cit.*, p. 60.
[2] D. B. Huntley, "Tenth Census of the United States," 1880, vol. xiii., pp. 405–489; Hollister, *Trans. A. I. M. E.*, xvi., p. 1.

down in 1885. Prospecting has opened up low-grade sulphide ores. The deposit forms a chimney between limestone and rhyolite. Anglesite and cerussite are the lead minerals ; galena is found in small quantities ; silver occurs as chloride, sulphide and sulpharsenide ; the gangue is calcite, quartz, barite and ferruginous clay. The subjoined analyses[1] by S. B. Newberry, made in 1879, show the character of the ore :

SiO_2	15.17	47.95
$BaSO_4$	0.49	2.71
$PbSO_4$	74.51	28.80
Fe_2O_3	4.80	....
Al_2O_3	1.71	12.55
Sb_2S_3	0.87	....
As_2S_3	1.12	....
$(CaMg)O$	0.50	....
CO_2	0.62	....
Ag, ounces	78.83	....

The average tenor of lead and silver in 1882 and 1883[2] was 37.80 and 36.83 per cent. lead and 34.2 and 27.15 ounces silver, which is considerably lower than the analysis of 1879.

The ores of the Tintic district (Juab Co.) are rather milling ores (Eureka Hill and Bullion mines), but they are smelted at Salt Lake City with lead ores.

In Salt Lake County are the mines of Bingham and the Little and Big Cottonwood Cañons. The Old Telegraph[3] mines of Upper Bingham Cañon are among the oldest of the territory. The ore forms an irregularly mineralized zone between limestone and quartzite and sometimes porphyry. The limestone and porphyry are more or less decomposed near the deposits. The lead ore is carbonate containing more or less silica, ferruginous clay, and some galena. The siliceous ore is an oxidized siliceous pyrite, the quartz having become ochreous, spongy and brittle ; the pyrite yellow, red or black ochre.

The following analysis made by Wuth[4] in 1876 shows the character of the ore :

$PbCO_3$	50.43	SiO_2	12.47	H_2O	0.19
PbS	15.02	Al_2O_3	3.01	Ag, ounces	21.14
Fe_2O_3	3.78	$CaCO_3$	3.64	Sb	trace
CuS	0.67	$MgCO_3$	0.26	As	trace
FeS	7.37	$CaSO_4$	3.04	Co	trace

[1] Huntley, *Op. cit.*, p. 466.
[2] "U. S. Geol. Survey : Min. Resources," 1883–84, p. 417.
[3] Lavagnino, *Trans. A. I. M. E.*, xvi., p. 25.
[4] Huntley, *Op. cit.*, p. 413.

The first-class shipping ore in 1888 contained between 30 and 50 per cent. lead, 10 and 12 ounces silver and 0.048 ounces gold per ton ; the second-class ore is concentrated and runs a little higher in lead than the first-class. The ore of the Brooklyn mine, lying east of the Old Telegraph mine, is similar in character ; first-class ore, 40–50 per cent. lead, 10 ounces silver per ton. Twenty tons of low-grade ore are concentrated into 4 to 5 tons of shipping ore. The Cottonwood Cañons contain the celebrated Emma and Flagstaff mines, which stopped producing about twelve years ago.

A number of other mines might be mentioned, but those described represent sufficiently the character of Utah silver-lead ores.

While a good deal of Utah ore goes east to be smelted, much ore from Idaho comes down to Salt Lake City to be treated there. Just as Denver and Pueblo form the main smelting centres east of the Rocky Mountains, so Salt Lake City is *the* ore market of the West.

§ 27. **Idaho.**—The argentiferous lead ores of Idaho come principally from two districts, viz., the Wood River in the southeast and the Cœur d'Alene in the north.

In the Wood River region limestone, slate and granite are the chief rocks, rich argentiferous galena with its secondary minerals the principal ore. The deposits seem to occur chiefly in limestones, but little or nothing has been published about them. Kirchhoff[1] says that the ore from the Hailey and Bellevue average 60 per cent. lead and from 80 to 100 ounces silver. Blake[2] writes that the Minnie Moore and the Queen of the Hills furnish concentrates assaying 65 and 62 per cent. lead, 80 and 65 ounces silver per ton.

Some ore is smelted near Ketchum, but large amounts are shipped to Montana, Colorado, Utah and California.

The Cœur d'Alene region.—The rocks are, according to Clayton,[3] quartzite, magnesian, shale and schist, which have been much folded. The deposits form fissure veins varying in width from a few inches to twenty feet. The ore, a galena with its products of

[1] "United States Geol. Survey: Mineral Resources of the U. S.," 1887, p. 100.

[2] *Engineering and Mining Journal*, November 29, 1890.

[3] *Engineering and Mining Journal*, November 28 and December 14, 1890.

decomposition, is generally concentrated before it is shipped [1] to a product assaying from 60 to 65 per cent. lead, and from 25 to 30 ounces silver per ton.

§ 28. **Arizona and California.**—Very little lead is mined in Arizona or California. The smelting industry of Arizona is represented by the Benson and Tombstone works, but most of the ore goes to outside smelters. The lead produced in California comes principally from the Selby smelting works at San Francisco, the principal one of the Pacific coast.

[1] "United States Geological Survey: Mineral Resources," 1887, p. 108 ; Clement, *Engineering and Mining Journal*, July 25, 1891.

CHAPTER V.

RECEIVING,[1] SAMPLING,[1] ASSAYING AND PURCHASING OF ORES, FLUXES AND FUELS.

§ 29. **Receiving and Weighing of Ores.**—Ores arrive at the smelting works either in a loose condition or in sacks. The gross weight is taken on platform scales before unloading from the cars. The ore is marked, when unloaded, by a shingle 2 by 12 inches, with the running lot number chalked on it. The entry is made in the receiving-book, a form of which is given below.

Ores contain a varying percentage of moisture ; those arriving direct from the mine are often quite wet or, in winter, frozen solid. After being weighed, the ores should be at once unloaded and the moisture-sample taken. If the ore is in sacks, these are emptied and weighed immediately, before they can become lighter by exposure to the air. They are then dried and tied up in bundles. Generally speaking, it is best not to receive an ore at all until the operations of weighing and unloading it, and taking out and weighing the moisture-sample, can be performed in quick succession. If this is not possible, the weights will have to be taken after unloading, thus causing an extra handling.

§ 30. **The Moisture-Sample.**—It is difficult to obtain an absolutely correct moisture-sample, but a fair average may be arrived at which will satisfy the seller as well as the buyer. A correct moisture-figure is necessary in making up the charges for the blast furnace, as the analytical data from which they are figured refer to dry ore, and a correct allowance must be made for the moisture.

If the ore arrives loose, the surface which has been exposed to the air will be drier than the middle part, and this again less

[1] Austin : *Engineering and Mining Journal*, 1882, July 22, August 5, 26, Sept. 16. Low : *Ibid.*, Sept. 24, 1881 ; Feb. 11, 1882. Reed : *School of Mines Quarterly*, iii., p. 253 ; vi., 351. Brunton : *Trans. A. I. M. E.*, xiii., p. 639. Glenn : *Ibid.*, xx. Hodges : *Engineering and Mining Journal*, 1891, Sept. 5. Bridgman : *Trans. A. I. M. E.*, xx. Johnson : *Engineering and Mining Journal*, 1892, Jan. 16, 23.

dry than the real average; if in sacks, it may show a greater irregularity, as more surfaces are exposed to the drying influences of the air. With ore of uniform size, arriving loose, the sample is obtained by taking a little from the shovel at regular intervals while the ore is being unloaded. Or, if it is discharged through a chute into a bin below, it is better to take the sample in small intervals from the end of the chute. When the same kind of ore arrives in sacks, a small sample is taken from the dryer top of every sack and a larger one from the centre, or the sack-sample is taken at once after emptying from what seems to be an average part. This last way sounds rather vague, but a person accustomed to emptying sacks will be able to take the correct sample. If the ore is not of uniform size, it has to be crushed in order to get a correct assay-sample. In that case, the moisture is best taken at the discharge of the crusher. When the crushing cannot immediately follow the unloading, and the ore has to be temporarily stored in a bin to be sampled down later on, the lumps may either be broken up and sampled with the fine part, or screened out and sampled separately, taking a correct proportion of coarse and fine in making up the final moisture-sample.

In whatever way the sample may be taken, it is put into a deep tin vessel with a closely-fitting lid, so as to prevent the first part from drying while the other is being taken. The moisture-sample must be taken at once to the assay office. A certain amount, varying from one to two pounds, is then weighed out twice for two moisture-determinations. These can be made when convenient, the weighed sample being transferred to a paper bag to await its turn.

Before the moisture is determined the weighed sample is reduced to a uniform small size. This is done on the grinding-plate, the ore being crushed there to the size of a pea. If it is too moist to be crushed it is first dried somewhat. The final drying is performed in a shallow tin or granite-ware pan on an oil-stove, a sand-bath, on top of the boiler, or by any means that permits the controlling of the temperature. It is necessary to stir the sample at intervals with a spatula or spoon so as to dry it completely. A cool beaker-cover held a moment over the drying sample will show by the absence or presence of condensed vapor whether the drying is finished or not.

The following form of receiving-book for ores shows the entries that are necessary to give the requisite information:

| Years. | Date of Receipt. | Shipper's | | Company's Lot Number. | Gross Weight, Tons. | Tare, Lbs. | Net Weight, Tons. | Moisture, per cent. | Dry Weight, Tons. | Placed | | Remarks. |
		Name.	Mark.							In Bin No. ...	On Bed No. ...	

§ 31. **The Assay-Sample.**—From a sample weighing a few ounces the metal value and the chemical composition of an ore are determined. The sample must therefore represent its average character. The most thorough way would be to crush the entire ore to a uniform small size, mix it, and then sample it down by one of the general methods to be discussed later on. If the ore is a sulphide, and has to be roasted before it is smelted, this preliminary fine crushing is necessary ; if, however, as is generally the case, the ore is treated raw, this crushing not only entails increased cost, but it is disadvantageous, as fine ore disturbs the uniform running of the blast furnace, increases the amount of flue dust, and retards the smelting. The best ore for the blast furnace is between the size of a man's fist and that of an egg ; it will be desirable to reduce it as little as possible below this. When, however, the ore is rich and the metal-bearing part is irregularly distributed in the gangue, as is mostly the case, it is necessary to crush finer before reducing to a smaller quantity. In sampling, the sample will grow finer as the bulk is reduced, this occurring more rapidly with high-grade than with low-grade ore. Reed[1] has, by an interesting calculation, arrived at certain definite maximum sizes which ores of a varying tenor in silver and gold may have when they are reduced from a hundred-ton lot down to the assay sample. His tabulated results, slightly condensed, are subjoined.

| Quantity of ore. Reducing | Value of Silver in Ounces per Ton. | | | Size of Ore. |
	Highest : 300. Average : 50.	Highest: 3000 Average: 75.	Highest : 10,000. Average: 500.	
100 tons to 10 tons........	Cocoanut .	Fist.......	Fist	Maximum permissible size of ore for given grade.
10 tons to 1 ton	Orange ...	Egg	Walnut...	
1 ton to 200 pounds.......	Walnut...	Chestnut .	Chestnut. .	
200 pounds to 5 pounds...	Pea.......	Wheat. ...	Wheat. ...	
5 pounds to bottle-sample.	20-mesh...	25-mesh ..	50-mesh...	

[1] *Op. cit.*, vol. vi., p. 357.

The breaking to walnut size is done in a crusher, from this down to the size of wheat by crushing in rolls; smaller sizes are obtained by grinding in a mill or on a plate.

While in practice these quantities and sizes are not rigidly adhered to, they are at least approximated. For example: it is not desirable to make one sample of a lot of moderately rich ore weighing over fifty tons. A hundred-ton shipment would be preferably divided into two lots and sampled separately, or be sampled by the carload (14 tons). Uniform low-grade ores can be sampled in 200 and 250 ton lots. With very rich ores the entire lot is taken as a sample. An ore must run very much lower than 50 ounces per ton; if the seller will be satisfied with a cocoanut size, he will insist on at least fist size, and prefer egg to walnut size. A 50-mesh sieve is generally considered to be too coarse for the final sample that is filled into the bottles; the 80-mesh and often the 100 and the 120-mesh sieves, especially the 80-mesh, are in common use.

Before sampling carbonate ores that are dry, they should be moistened with a hose, so that the fine dust, which often represents the richest part of the ore, may adhere to the lumps. If this is not done, it will be partly lost by being blown away when the ore is handled, or it may drop through the ore-heap to the floor and be partly lost there. In any case, the dust is liable to be unevenly distributed, and thus errors may occur. For instance, a lot of ore sampled wet gave 400 ounces silver to the ton, sampled dry only 390 ounces.

Sampling is done either by hand or machine. There are four ways of doing it by hand: by quartering, by fractional selection, with the split shovel and by channelling. Machine-sampling is of two kinds: that which divides the stream of ore into two unequal parts and takes the smaller one continuously off as the sample, and that which takes intermittently at certain intervals the whole of the stream as sample.

a. *Quartering.*—By this method the ore is reduced one-half in quantity at a time. It presupposes a thorough mixing of the ore, which is done by coning. The ore coming from the chute of a crusher or of rolls is wheeled away and dumped in a circle large enough for the two samplers to be able to stand inside. They work as partners, walking around the circle while forming the cone, remaining always diametrically opposite each other while they pile up the ore. In doing this, care must be taken that every shovelful is thrown directly upon the top of the forming cone (Figure 1), so that it may run and spread evenly down the sides. With experi-

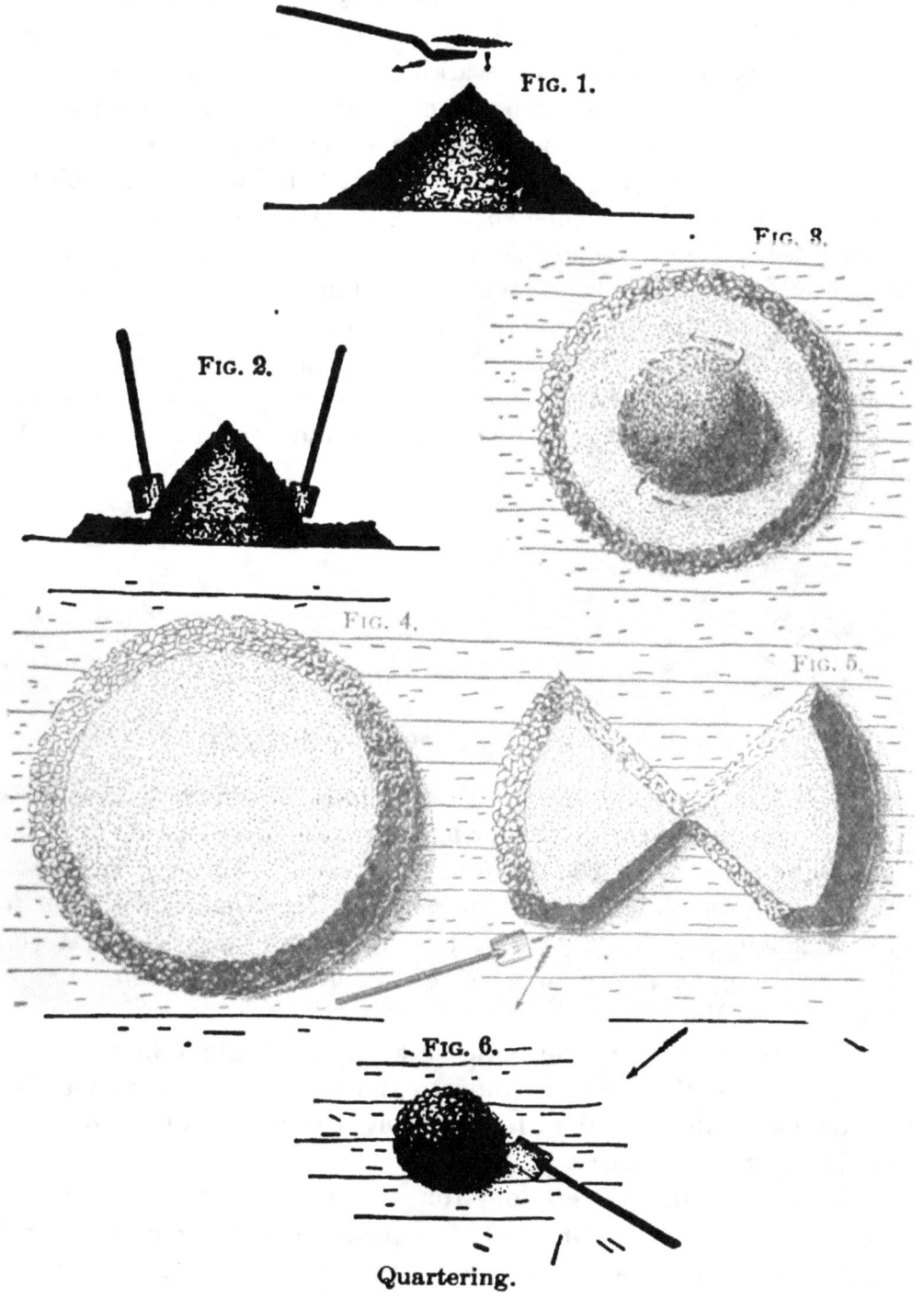

FIG. 1.

FIG. 3.

FIG. 2.

FIG. 4.

FIG. 5.

FIG. 6.

Quartering.

enced men the circumference of this cone will be a true circle. When the ore is exhausted the floor is swept carefully and the sweepings deposited on the top of the cone—not swept simply towards it—forming a ring of fine ore. A 50-ton cone is from 10 to 12 feet high. The shovels used are long-handled and round-pointed. The ore being thus mixed, the cone is pulled down, the men working in opposite pairs (Figures 2 and 3). They begin near the top of the cone, and, walking round it, work it down from centre to periphery until it becomes gradually transformed into a truncated cone from 6 to 18 feet in diameter and from 6 to 12 inches in height (Figure 4). A short-handled, square-pointed shovel is desirable for this work. With the sharpened edge of a straight lath two diametrical lines are drawn across the cone at right angles to each other. The ore is thus divided into four equal parts (hence the name quartering). Two opposite sections are removed to the bin (Figure 5), while from the other two a new cone is formed (Figure 6), and the process of quartering repeated. When the pile has become so small in quantity that it has to be

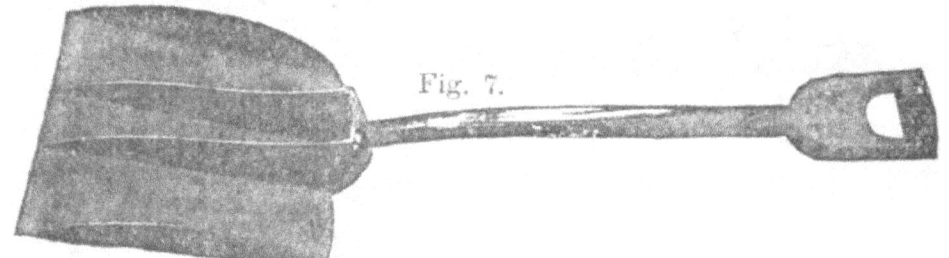

BRUNTON'S QUARTERING SHOVEL.

crushed (see Reed's table) to attain a uniform sample, this is done by crusher or by rolls. The whole process is repeated until the sample is reduced to 100 or 50 or less pounds, forming the finishing sample, which is treated separately. Care should be taken to avoid losing any dusty part of the ore or working other material into the sample, the danger of error increasing as the sample becomes smaller.

Quartering was formerly used as a starting method for the entire mass of the ore as it came to the smelter ; it has generally become a finishing method for reducing a part only of the ore. It is always very satisfactory.

Two men will quarter a one-ton sample down to a few pounds in about two hours. Brunton[1] has patented[2] a quartering shovel by whose use the operation is very much shortenened. It is (Fig-

[1] *Engineering and Mining Journal*, 1891, June 20.
[2] Patent No. 454,120, June 16, 1891.

ure 7) a flat-bottomed steel shovel, 10 inches wide, turned up at the sides and divided into three compartments. The centre compartment is closed at the back and has the width of one-quarter of the shovel. Thus, when the shovel has been pushed into a heap of finely crushed ore and filled, it is raised, and a "sharp rotary motion to the right" will discharge the ore from the outside compartments to one side, forming the rejected ore pile, and leave the sample in the central compartment to be emptied on the other side. In a speed test a sample of one ton of ore was cut down with this shovel to 100 pounds in 15 minutes by one man.[1]

b. *Fractional Selection.*—This is a starting method. It consists in taking, while unloading, every fourth, fifth or tenth shovelful of ore for the first sample. This is then cut down in the same way. If the shovelling is done every time from the floor, as it should be, the ore without previous mixing will give a correct sample.

The same method is used with sacks in making the first sample if the ore is known to be uniform in character, as otherwise the chances of error are too great.

As in quartering, when the sample becomes small it has to be crushed. Fractional selection is rarely, if ever, used for finishing a sample.

c. *Split-shovel.*—The tool used resembles a fork with a long handle, the prongs being replaced by from four to six troughs made of sheet iron. The width of the troughs, which is equal to the distance between them, depends on the size of the ore to be treated, and the largest piece of ore in the heap ought not to exceed one-fourth the width of the trough. Thus if the ore is half an inch, a 2-inch trough is required. The depth of a trough is such that a piece of ore striking the bottom will not rebound and fly out—say from 2 to 4 inches deep for a 2-inch trough. The length of the trough varies from 15 to 18 inches; the handle is that of a long-handled shovel. In sampling, the split-shovel is placed on the floor by one man, while the other, the sampler, facing the ends of the troughs, delivers the ore from a square-pointed shovel in a thin stream in the direction of the troughs. One-half of the ore is caught in the troughs, one-half passes into the spaces between. Re-sampling over the fork is continued until the troughs

[1] The writer is informed by disinterested parties that the results with the Brunton shovel check well with their own results, obtained from regular quartering.

are full, when it is lifted out and the contents discharged on a separate heap, forming the reduced sample. The whole heap is passed over the fork in this manner, and the first reduced sample cut down similarly until it is necessary to crush finer, when it is passed over another split-shovel with smaller-sized troughs—say from ¾ to 1 inch wide.

That sampling with this tool requires even less previous mixing of the ore than with fractional selection is clear. It is also evident that sampling with the split-shovel can be used as a starting method only with pretty fine material. Smelters make the objection to the split-shovel that an undue amount of fine ore is liable to be caught by the troughs. This gives an incorrect sample if the fine ore assays differently from the coarse ore. The split-shovel is used at some sampling works for the entire sampling, but this is not common. It is rather the tool for finishing a sample, and is even used in the laboratory in a modified form as assayer riffle.

This method works as fast as quartering, if not faster.

d. *Channelling.*—This consists in spreading out the crushed and thoroughly mixed ore in a square, about 4 inches thick, and then taking the sample out in parallel grooves—say one foot apart—across the square, first one way, then the other. The sample is reduced by repeating the same operations. The greater the number of the grooves (channels) drawn, the less thorough need be the preliminary mixing of the ore.

This method, used formerly at Batopilas and Silver Islet,[1] is hardly used in any silver-lead smelting works now as a starting method. It is still in use as a finishing method, where the mixing of a small amount of fine ore is easily effected.

e. *Continuous Mechanical Sampling.*—Quite a number of machines[2] for continuous sampling have been invented and are in use. The drawback to the method is, however, that a falling stream of ore is never uniformly mixed. In order to counterbalance this disadvantage, a preliminary fine crushing and sizing becomes necessary to obtain a satisfactory sample. Why fine ore is undesirable for the blast-furnace has already been discussed.

The Pipe Ore-Sampler, as represented in Figure 8[3] is an ap-

---

[1] Lowe: *Engineering and Mining Journal*, September 24, 1881.

[2] Egleston: *Engineering*, December 15, 1876. Reed: *School of Mines Quarterly*, ili., p. 253.

[3] Taken from a drawing of Messrs. Fraser and Chalmers, Chicago, Ill.

paratus of this class which is much used at present in lead-silver works. The finely crushed and sized ore is fed into the hopper *a*, ending in a small funnel *b*. Through this it passes over the divider *c*, which cuts it into two equal parts. One-half is discharged into the ore-bin *d*, while the other descends in the pipe and is thoroughly mixed by passing through the funnel *b'*, and then being scattered towards the sides of the pipe before it is brought

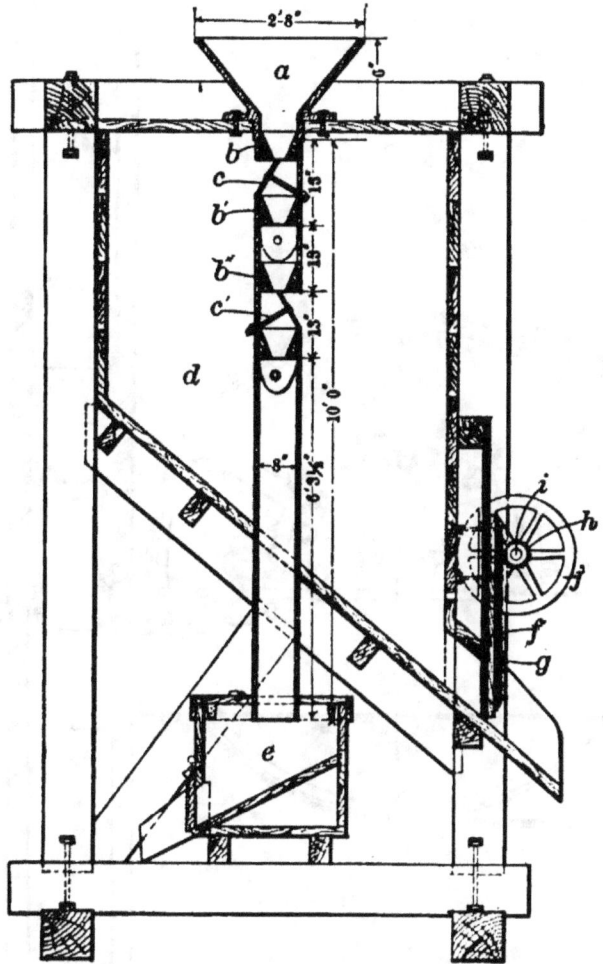

FIG. 8.—PIPE ORE-SAMPLER.

together again in a small stream by the funnel *b'*, which delivers it to the second divider *c'*. The rejected ore falls into the bin *d*, while the sample is collected in the sample-bin *e*, to be put again through the sampler until it has been sufficiently reduced in size. The ore-bin is closed by a wooden gate *f*, running in heavy iron guides. This is raised and lowered with rack and pinion (*g*, *h*), the pinion-shaft *i* being turned by a hand-wheel *j* that is keyed to it. The sample-bin *e* can be locked, if desired. The main disadvantage of the apparatus is that it is difficult to inspect and to clean.

f. *Intermittent Mechanical Sampling.*—This method gives better results, as the entire falling stream of the ore is deflected at certain intervals to cut out the sample. The amount of ore taken

FIG. 9.

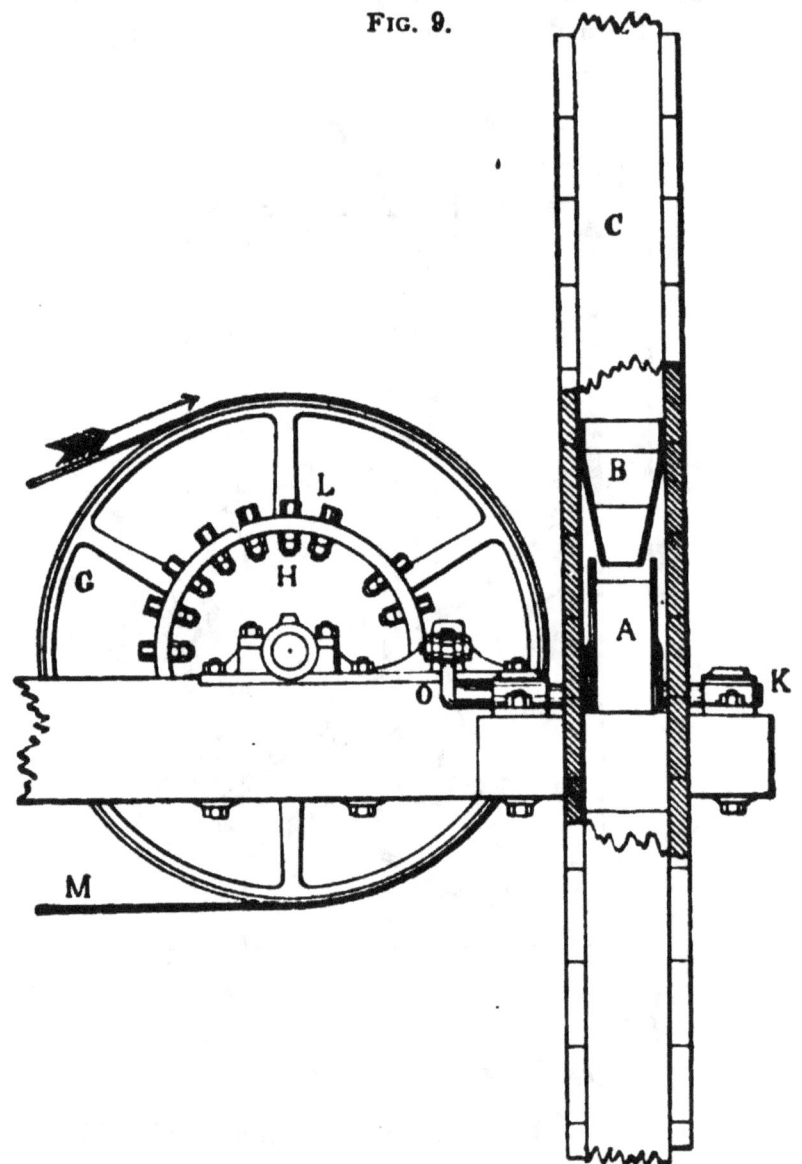

out for the sample is regulated by the number of deflections that take place and the length of time they last. The advantage is that the bad effect of the irregular distribution in the stream of ore is neutralized, and thus fine crushing and screening made unnecessary. Rittinger [1] describes an apparatus of this kind. Two

[1] " Lehrbuch der Aufbereitungskunde," Berlin, 1867, p. 583.

modern machines by Brunton and Bridgman may serve as examples :

The principle embodied in the Brunton[1] sampler is shown by Figures 9 and 10. The ore coming down through spout C and the funnel B is divided by the deflecting chute A into two parts,

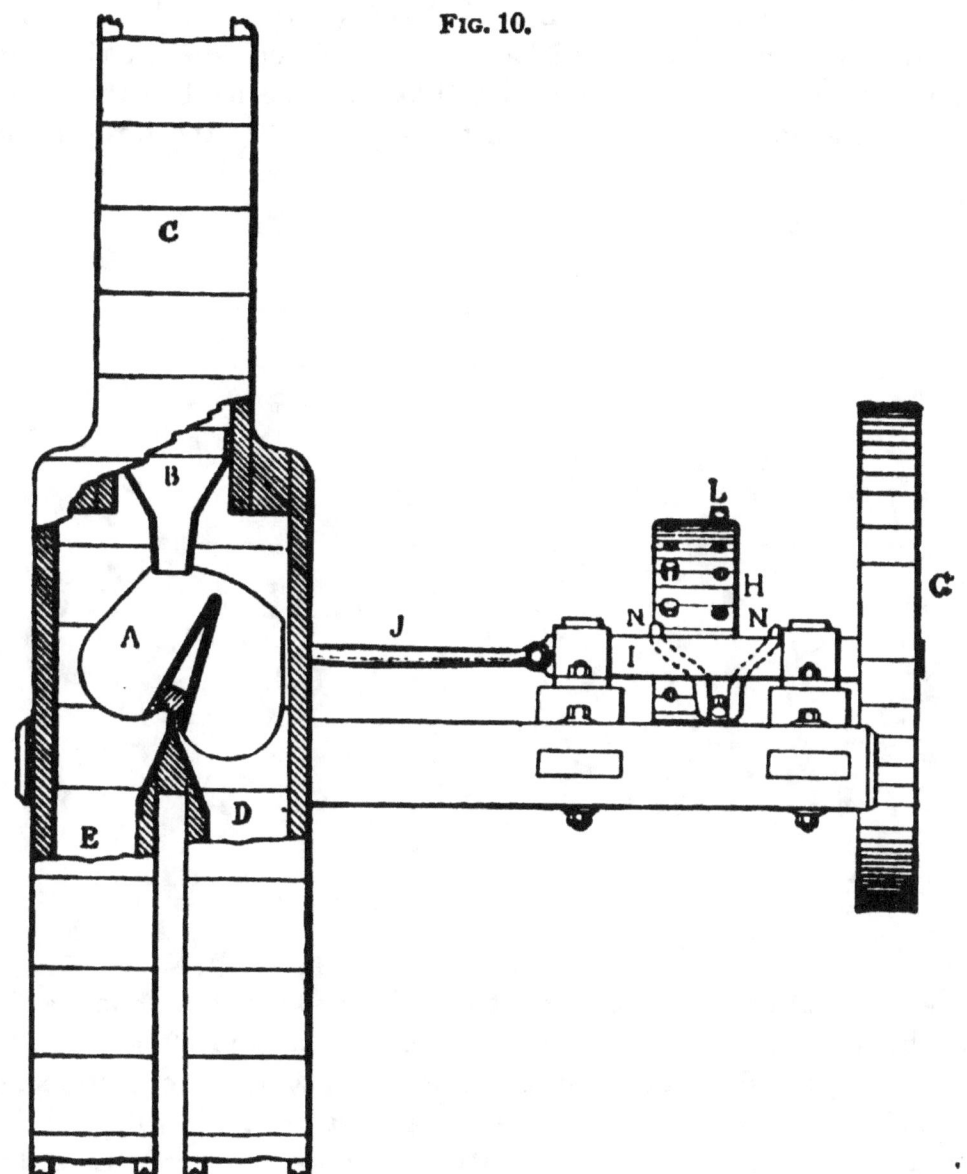

FIG. 10.

which are discharged separately by the spouts D and E. The movement of the chute A is effected in the following way : to the arms of the pulley G, driven by the belt M, is attached a smaller wheel H, the face of which is perforated by two rows of, say, twenty holes. Into these are inserted alternatingly on one side.

[1] *Trans. A. I. M. E.*, xiii., p. 639.

and the other a number of pins *L*, and then fastened like bolts on the inside face by means of jam-nuts. When the wheel *H* revolves, the pins will move the guides *N, N* alternately to the right and to the left, and these the driving-bar *I*, with the pitman *J*, which gives the deflecting chute a reciprocating motion.

The proportion of ore that is taken out as a sample depends on the disposition of the pins *L*. If these are inserted half in the right-hand row of holes and half in the left, on opposite sides of the wheel-face, then the deflecting chute will be held to the right during one-half of a revolution of the wheel and to the left during

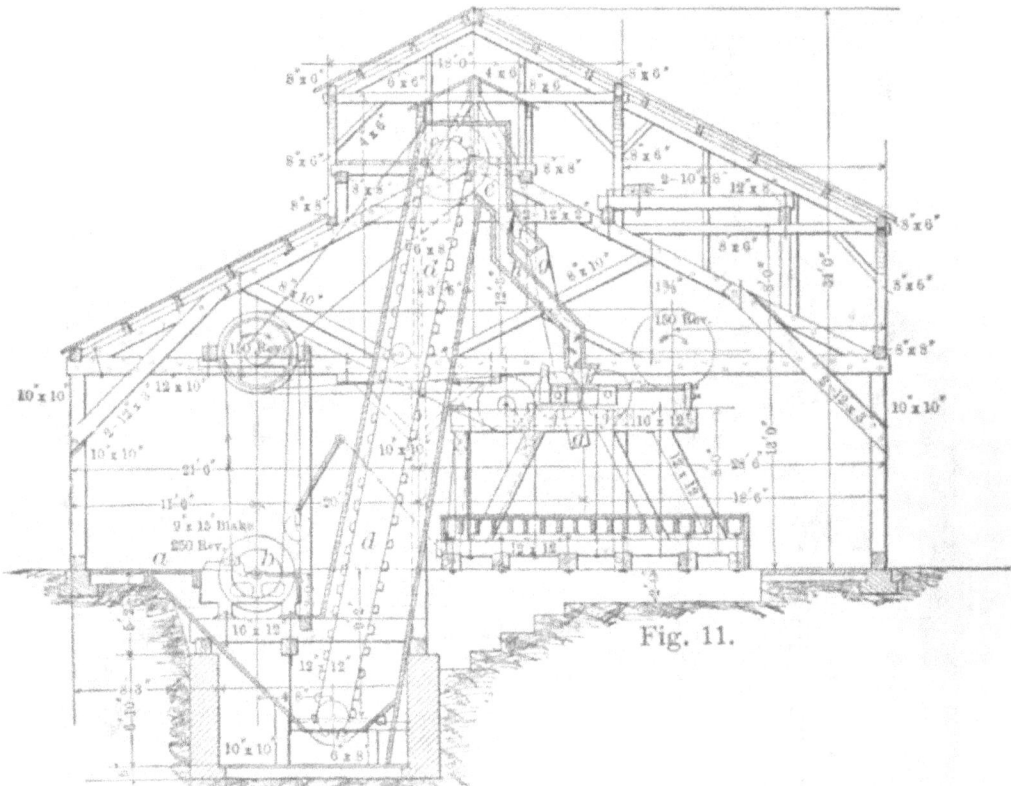

Fig. 11.

the other half, and thus the ore-stream divided into two equal parts. If the division of pins is 10 per cent. to one side and 90 per cent. to the other, the deflecting chute will remain $\frac{1}{10}$ of a revolution on one side and take out $\frac{1}{10}$ of the ore as a sample. It will thus be seen that by a suitable distribution of pins any percentage desired of a falling stream of ore can be recovered as a sample. The number of revolutions wheel *H* makes per minute must be large if the sample is to be a correct one. The further reduction of the first sample can be effected, after crushing fine with rolls, by a smaller machine of similar construction, down to the finishing sample.

FIG. 12.

The Brunton machine gives satisfaction on account of its simplicity, the accuracy and speed of its operations, and its doing away with the necessity of fine-crushing the entire lot of ore that is being sampled.

Figure 11 represents a cross-section through a sampling-mill using the Brunton sampler. The ore is discharged on the screen-plate a, and the coarse ore crushed in the Blake crusher b. Both fine ore and crushed coarse ore drop into the bin c, whence they are delivered by a belt-elevator d at the top of the building to the funnel e. The deflecting chute f divides the stream of ore into two parts; the rejected ore passes down through the chute g, which discharges it into an ore-car to be transferred to a bed or bin. The sample is conducted by the chute h to the hopper i of the rolls j, and the finely-crushed sample is then further reduced by quartering.

The Bridgman Sampler.[1]—Type A, represented by Figures 12, 13 and 14, approximately drawn to scale, may serve as an example of the machine. It occupies a floor-space 3 by 4 feet, and is 7 feet 6 inches high. It consists of three hollow truncated cones, I., II., III. (called apportioners), driven by the pulley F, and three stationary concentric receptacles, R_1, R_2 and H. The first two discharge the original and duplicate samples they receive through the spouts T_1, T_2 into the sample-boxes Z_1, Z_2; the third discharges the rejected ore that falls into it through the chute S.

Apportioners I. and III. (Figure 14) move in the same direction, making respectively 5 and 45 revolutions per minute; apportioner II., moving in the opposite direction, makes 15 revolutions. Apportioner I. (Figure 14) consists of two concentrical rings having eight compartments, L_1 to L_8. To each of these an adjustable spout is attached, one in the direction of M—1, a second in that of M—2, and six in that of M—D. When the apportioner is in motion, M—1 describes the circular path 1—1, M—2 the path 2—2, and the rest the path W—W. The intermediate (II.) and lowest (III.) apportioners have the same construction: a hollow truncated cone with the central discharge opening W, and four vertical chutes N_1—N_1 and N_2—N_2, each of which represents one-eighth of the path covered by the spouts M—1 and M—2.

If crushed ore is fed through the chute F it will fall equally into all the spouts; thus, with every revolution of apportioner I., one-

---

[1] *Trans. A. I. M. E.*, xx.

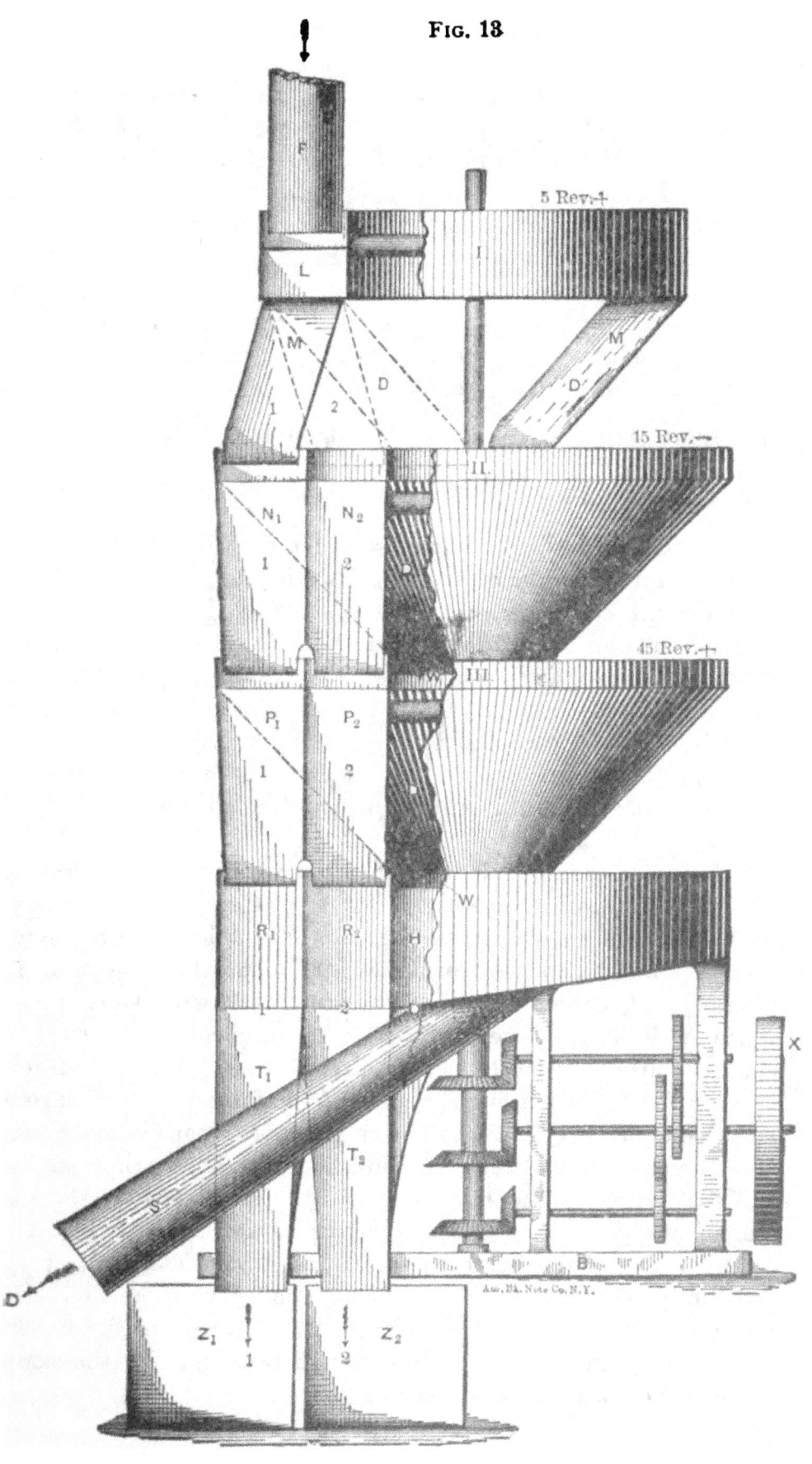

FIG. 13

eighth of the ore passing through M—1 goes to make up the original sample, another eighth passing through M—2, the duplicate sample, and the remaining six the rejected ore, which passes off through W, W, H (Figure 14) and spout S (Figure 13). The samples caught by M—1 and M—2 now undergo separately the same operations. To follow the one in M—1 : it is intercepted by apportioner II., which cuts out and delivers to D three-fourths of it, the remaining fourth passing through chute N—1 to receive the same treatment by the apportioner III. Thus the amount a', cut out by M—1 with every revolution of apportioner I. and forming one-eighth of the ore fed during the revolution, is reduced to $\dfrac{a'}{4}$ by apportioner II., and to $\dfrac{a'}{16}$ by the apportioner III., when it drops into the sample-box Z^1. Thus a^2 cut out by M—2, undergoing the same operation, is reduced in like proportion and collected in box Z^2.

The machine differs from all others in taking out duplicate samples and cutting them down as far as possible without re-crushing. It does its best work with ore that is not larger than 1-inch cube. It has a great capacity, varying from 15 to 25 tons per hour, is worked cheaply, and can be easily cleaned.

Type B, described and illustrated in the reference given, is a smaller machine of similar construction furnishing only one sample.

The general arrangement of a sampling-mill using the Bridgman sampler is shown in Figures 15 and 16, the scale being 8 feet to 1 inch. The ore arriving in the box car F is unloaded after having been weighed on the platform scales X. At the same time it is sampled by fractional selection, the sample being delivered through chute S (located between the bins D) into car Q, which is discharged into hopper B^1. From this the sample passes through crusher C, thence through hopper B^2 and over rolls R into the large sampler H (type A). The rejected ore drops into the bin of elevator E, which discharges into car M. This brings the discarded portion to the bin D, where the main part of the lot is stored. Original and duplicate samples in boxes 1 and 2 are elevated by a chain-tackle to the tilting frame T, and discharged separately into hopper B^3 of the fine rolls O, which deliver the fine ore to the small sampler (type B). This gives the finishing sample, weighing a few pounds, which now goes to the sample-grinder and thence to the assay office.

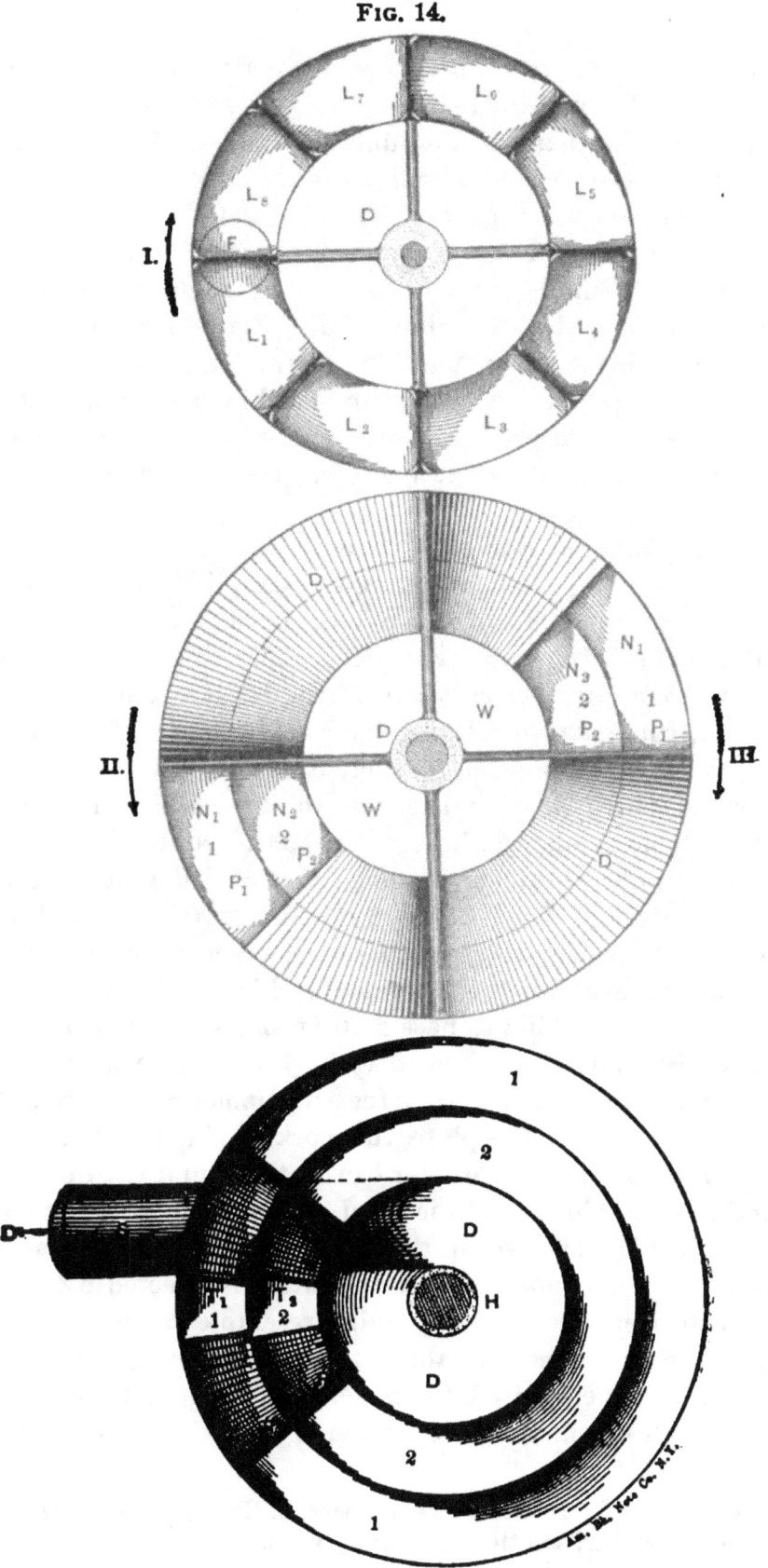

F𝗂𝗀. 14.

g. Comparison of Methods.—With proper care an accurate sam plecan be obtained byany of these methods. Fractional selection and mechanical sampling are the best for handling large quantities of ore ; small amounts are effectively sampled by quartering, fractional selection, and by the use of the split-shovel. The best finishing methods are quartering, using the split-shovel, and channelling. Finally, to compare hand sampling and machine sampling, it may be said that, while in the former you are dependent on the workman and the method is slow, expensive, and requires a large space to work in, machine sampling has not as yet displaced it, and a smelter will be best served by not depending entirely upon either method, but using both as occasion may require.

h. Finishing the Sample.—This is best done in a well-lighted room, the finishing-room adjoining the sampling floor. The sample must be protected from the wind, from dust, and from any possibility of being tampered with. The seller is permitted to watch the sampling through one of the windows, but not to enter the room, where the head sampler works alone. If the sample is moist, it is dried somewhat, so that it can be ground. Ore that is very moist or frozen when it comes to the works can generally be sampled down to one ton before it needs to be dried.

The sample obtained by one of the different methods described, weighing from 50 to 100 pounds (the particles being, according to Reed's table, for 50-ounce ore not larger than a pea), has to be prepared for the laboratory. If larger it is ground either by hand or by machine to pass a 10-mesh sieve. If this is done by hand, a heavy muller, with head $4\frac{1}{2}$ by 7 inches, is used on a circular cast-iron plate of from 3 to 4 feet in diameter and from 1 to $1\frac{1}{2}$ inches in thickness. The plate, the working side of which is planed, stands free, so that the sampler can walk around it while grinding. Sometimes an oblong plate is used, 18 by 24 inches, with a rim $\frac{1}{4}$ of an inch high on two or three sides and a muller head 4 by 6 inches. More commonly the comminution is effected by a machine, the so-called sample-grinder,[1] which resembles a coffee-mill. The sample is now reduced to about ten pounds by quartering, using the assayer's riffle, rarely by channelling. In quartering a funnel is very handy to form the cone, which is then stirred down with the

---

[1] The makes of Fraser and Chalmers, in Chicago, and of Hendrie and Bolthoff, in Denver, are those generally used.

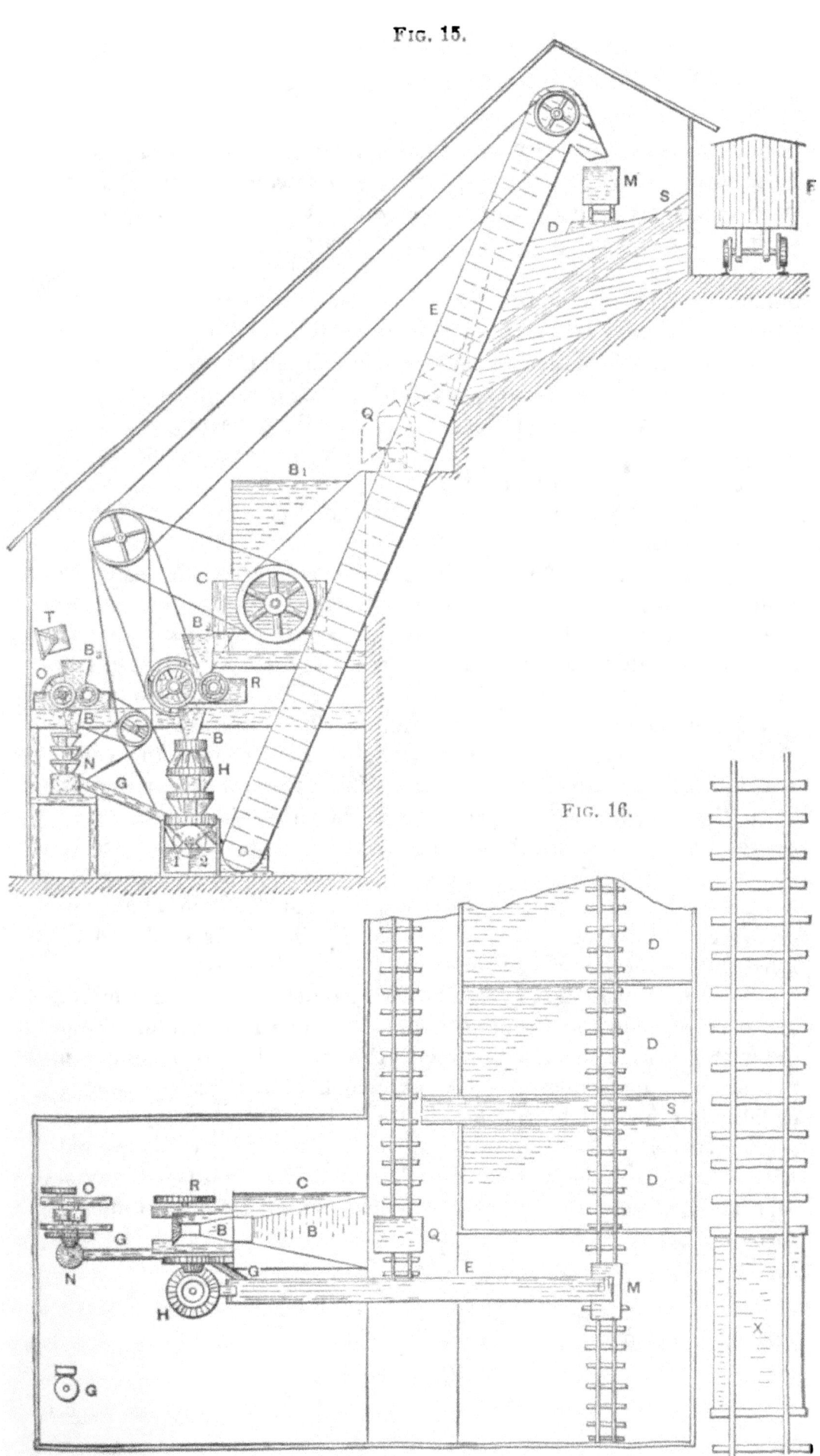

Fig. 15.

Fig. 16.

palm of the hand, with a 2 by ½-inch lath from 12 to 18 inches long held horizontally, or with a spoon held vertically. The assayer-riffle resembles a small split-shovel without a handle. The other tools are a small scoop and a flat paint-brush.

The Bridgman laboratory sampler[1] (Figures 17 and 18) is a very handy machine to simplify the quartering, occupying a space 14 inches square and being about 14 inches high. It has as its main parts a divider D (which is set in motion by hand, by clockwork, or by any convenient power), a funnel 1, and a receptacle 2, with discharges O_1 and O_2 respectively. The ore filling the hopper F runs in a continuous stream on to the divider, which gives eight cuts to the revolution: four of these, forming the sample, are delivered into the funnel; the other four, the discarded portion, are collected in the receptacle.

From the 50 or 100 pounds either a single sample may be made, or two, called original and duplicate. In the latter case the two halves obtained of the ore are treated separately; if the riffle is used, that part which has been caught by the troughs is dumped alternately to one side and the other, forming two heaps, which are then worked up separately. The duplicate sample is useful in two ways: first, to verify the work; secondly, to expose any tampering with the samples. When these are completely dried they are ground to pass a 20-mesh sieve, then well mixed by rolling on an oil-cloth or thin rubber cloth, and cut down to about one pound. This is again ground on the plate to pass a 60-mesh (or more commonly an 80-mesh) sieve, and is then again well mixed by rolling on glazed paper, after which it is ready to be filled into the large-necked four-ounce sample bottles.

To simplify this last operation Bridgman[2] has devised a mixer and divider, shown in Figures 19 and 20. The mixer, a large funnel with movable cover, is filled with the ore, well shaken, and then passed to and fro over the divider, which discharges the sample evenly into the four bottles beneath it.

Sometimes scales of metal in the ore are too large to be ground with the rest to pass the sieve. They are then weighed, assayed separately, and the average result calculated by the following formula:

$$\frac{S.a + P.a'}{S + P} = a''$$

[1] *Trans. A. I. M. E.*, xx.
[2] *Trans. A. I. M. E.*, xx.

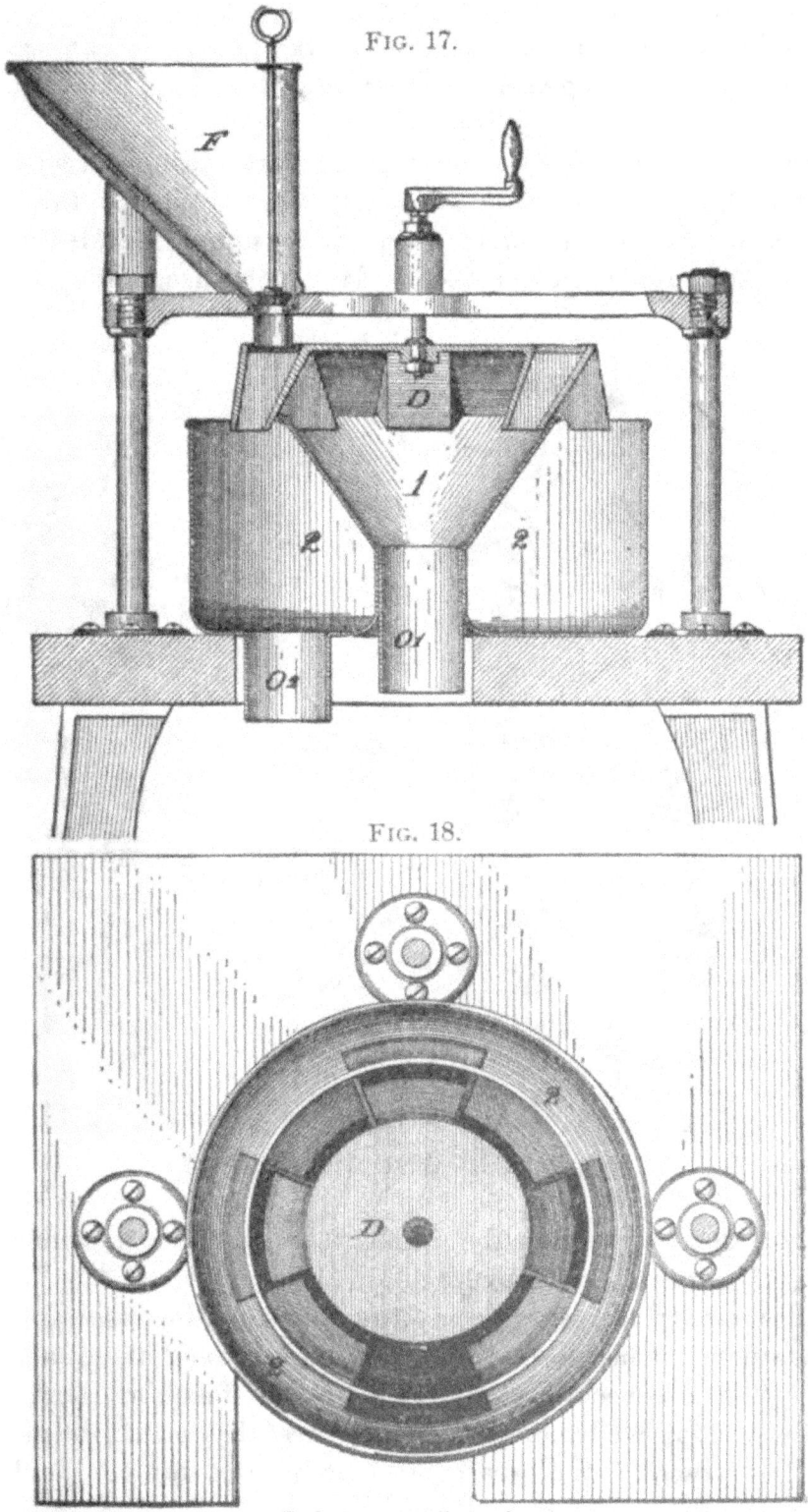

FIG. 17.

FIG. 18.

Laboratory Sampler.

S = weight of scales ; P = weight of pulp ; a and a', the respective assays in ounces per ton ; a', the average assay.   The scales are best assayed all together, as dividing them is liable to cause error.

Each sample is divided into three parts, filling three bottles : one for the seller, one for the smelter, and one for the control. Triplicate assays are made from each sample, and thus both sample and assay are checked.   If much discrepancy is found

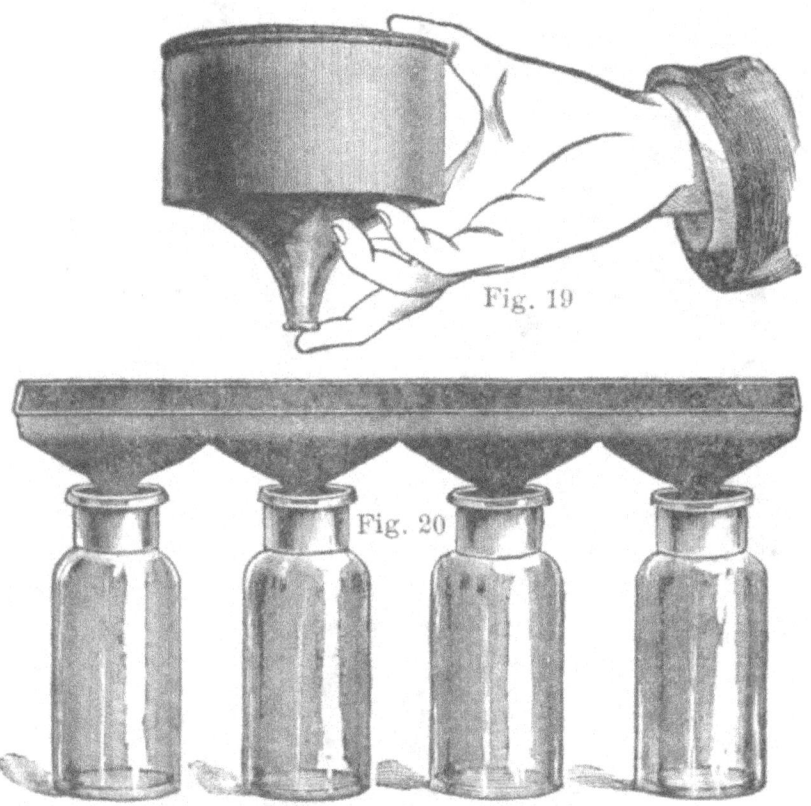

MIXER AND DIVIDER.

between the original and the duplicate, re-sampling becomes necessary ; if only little, it is averaged.[1]

It is hardly necessary to say that the sampling floor and all the apparatus must be thoroughly swept and cleaned after each operation ; this is especially important in the finishing-room, as the smaller the sample becomes the more carefully must it be protected. The last traces from previously comminuted ore can be best re-

[1] "Leadville, Smelter's Regulations :" *Engineering and Mining Journal*, 1883, October 27.

moved by using a small quantity of the new sample, taking care to return this to the general bulk.

The charges made for sampling vary from $1.50 to $3 per ton, according to the grade and character of the ore. Rich silver ores, consisting mainly of gangue and silver mineral, and gold ores with free gold, often require triplicate samples, and always closer work than uniform low-grade ores. The last are sampled, if in large quantities, at $1 a ton; and with concentrates requiring no crushing the price is often as low as 85 cents.

§ 32. **General Arrangement of Sampling Department.** —The sampling department should be arranged so as to avoid all unnecessary handling of the ore.

For hand sampling the simplest way for small works is to have the ore-bins below the main sampling floor, so that the rejected ore can fall at once into the bins and remain there until required for the blast-furnace. The distance must be sufficient for the ore to be discharged from the centre into any of the bins. The receiving floor need only be raised enough above the sampling floor that the ore discharged by the crusher shall fall into a wheelbarrow placed below, the mouth of the crusher being on a level with the receiving floor and near the middle of the edge. A distance of $4\frac{1}{2}$ feet between receiving and sampling floors is sufficient for a No. 5 Blake crusher. The receiving floor is about a third as large as the sampling floor. The floors are made three inches thick and consist of two layers of planking, the upper being boards 1 inch thick, running across the heavier planks, and parallel with the mouth of the crusher to facilitate the shovelling. In front of the crusher the floor is protected by a sheet-iron plate, say $\frac{1}{8}$ inch in thickness, fastened by countersunk screws.

Another arrangement, found in larger works, is to place the sampling-room opposite the ore-bins on the other side of the track. This is laid so low that the floor of the car is on the same plane as the sampling-room and the runways on top of the ore-bins. The sampling-room is a simple, oblong building with stalls on the four sides to keep the samples. It has only one floor, on which are placed crusher and rolls. The finishing-room is partitioned off from the main room. While the ore is being unloaded from the car, the sample is taken by fractional selection and wheeled to one of the stalls on the sampling floor, the bulk of the ore passing straight to the bins.

In machine sampling, which is used only in the largest works, the sampling mill is never directly connected with the smelter building. Its arrangement will depend on the character of the machine and the configuration of the ground.

A lot of ore, after being discharged into a bin or on a bed, is spread out evenly before another lot is received. Each subsequent lot will thus form a distinct layer over the preceding one.

A special book is kept for entering the contents of the different bins and beds, the averages being calculated according to the formula in § 20.

RECORD OF BINS AND BEDS.

Year.	Bin. No.	Bed	Date.			Contains Lots No.	Total Weight, Tons.		Average p. c. of moisture.	Average					
			Closed	Taken for Smelting.	Smelted.		Wet.	Dry.		Oz. per ton.		Per cent.			
										Au.	Ag.	Pb.	Cu.	Zn.	S.

Assay.								Remarks.
Per cent.								
SiO₂.	FeO.	MnO.	CaO.	MgO.	BaO.	Al₂O₃.		

Assay.								Remarks.
Per cent.								
SiO_2.	FeO.	MnO.	CaO.	MgO.	BaO.	Al_2O_3.		

§ 33. **Receiving and Sampling of Fluxes and Fuels.**— Fluxes and fuels on arriving at the smelter are weighed on platform scales and unloaded as near the feed floor as possible. It is necessary to keep large quantities of fluxes always on hand. As they take up a great deal of room and are not injured by the weather, they are generally left outside the building. The fuels are placed in sheds for protection. The breaking of fluxes is done by hand or in a coarse-set crusher. The latter method makes more fines. The most desirable size for iron ore and limestone is that of a man's fist; for quartz, that of an egg.

In sampling fluxes and fuel, a so-called grab-sample gives sufficiently near results. This consists in taking out bits or handfuls here and there from the top, middle, and bottom of the heap until about 50 pounds of iron ore, 30 of limestone, or 10 of coke have been taken. From the grab-sample the moisture-sample is taken, and the remainder crushed fine and cut down to be analyzed in the same manner as a regular ore-sample.

§ 34. **The Assaying of Lead-Silver Ores.**—This is for two purposes : first, to ascertain the amount of metal contained in the ore ; second, to determine its chemical composition so far as this relates to the making up of the smelting charge.

Assays are made in the dry way to ascertain the amount of lead, silver, and gold an ore contains ; all the other elements of importance are determined in the wet way.

For the dry assay two kinds of furnaces are required, the crucible and the muffle furnace, the two being usually placed side by side. Drawings of furnaces are given in all books on assaying, and are unnecessary here.[1] A few comments may, however, be in place. With crucible furnaces the grate should be of wrought-iron bars that can be shaken and drawn out from a horizontal slit above the ash-pit. Portable muffle furnaces are used in small works and in places where bituminous coal is not obtainable. One of the best models in the market is "Brown's portable assay furnace." In large works permanent furnaces built of common brick and lined with fire-brick are the rule. The use of bituminous coal is preferable to that of carbonized fuel or anthracite, as it is easier to regulate the temperature and the bottom of the muffle remains hotter. In building such a furnace special attention is to be given to having the passages of the correct width, if the fuel is to surround the muffle, or to giving the inside wall of the furnace the form of the muffle, if the flame is to play within it. With coke, anthracite or charcoal the passages ought to be about 4 inches wide at the sides,

---

[1] W. L. Brown, "Manual of Assaying," Chicago, 1889 ; P. de P. Ricketts, "Notes on Assaying," New York, 1879 ; C. Balling, "Probirkunde," Brunswick, 1879 ; "Supplement," Berlin, 1887 ; B. Kerl, "Metallurgische Probirkunde," Leipsic, 1882 ; "Supplement," *ibid.*, 1887 ; B. Kerl, "The Assayer's Manual" (an abridged treatise of the larger work), transl. by W. T. Brannt and W. H. Wahl ; new edition by F. L. Garrison, Philadelphia, 1890 ; A. Riche, "L'art de l'essayeur," Paris, 1888. See also Wood, *School of Mines Quarterly*, xii., p. 186 ; *Engineering and Mining Journal*, 1891, September 26.

so that the fuel need not be broken into small pieces, and 6 inches at the back, as this part chokes up readily. With bituminous coal the flue at the back can have the same width as that at the sides, about 2 inches. The vertical distance from the grate bars to the bottom of the muffle must be at least one-half the length of the muffle. Muffle furnaces are better fired from the back than from the front. The muffles used at large smelting works hold from 12 to 16 2½-inch scorifiers.

Lead assays are made both in muffle furnaces and in crucible furnaces. With the former the models of the Denver Fire Clay Company are commonly used: the 5-gramme (2½ inches high by 2½ inches wide at the top) and 10-gramme crucibles (3 inches high by 2½ inches wide at the top), called by the Battersea Company "Colorado *AA*" and "Colorado *A*." In the crucible furnace, Hessian and Battersea crucibles, 4½ inches high and 3 inches wide at the top, form a convenient size for 10 grammes of ore. The muffle assay is generally preferred to the crucible furnace assay at smelting works. Already, in 1870, Percy[1] proved that a wrought-iron crucible was preferable to a clay crucible, and this has been substantiated over and over again since. The reasons are the shorter time required for the assay and the more intimate contact of the iron with the ore.

Every laboratory contains two or more mixtures for assaying lead ores. They are measured into the crucible first, filling it about two-thirds, and the ore is added afterwards. The table on the opposite page gives a number of charges for special kinds of ores and for slag, in detail, and a few mixtures for general use.

Silver assays are uniformly made by scorification, as it is adapted to all kinds of ores, and gives excellent results[2] unless organic matter[3] is present. In that case the crucible assay is preferable.

In scorifying, 0.1 assay ton of ore is mixed with 1 ton of granulated lead and 1 gramme of glass borax added. A 2½-inch scorifier is used. Results of triplicate assays should check to within 0.5 of an ounce per ton. If the ore is refractory, or if it contains volatile metals, half the usual quantity of ore must be used in the 2½-inch

---

[1] "Lead," p. 104.

[2] *Berg- u. Hütten-männische Zeitung*, 1867, p. 85 (Arents); 1867, p. 102 (Fournet); 1874, p. 68 (Richter and Hübner); 1877, p. 232 (Richter and Hübner).

[3] *Berg- u. Hütten-mannische Zeitung*, 1886, p. 441 (Görz).

INGREDIENTS.	CHARGES IN DETAIL.															MIXTURES					
	Pure Galena.		Galena.				Impure Galena.			Copper-bearing Galena.	Carbonate ore.			Slag.	Carbonate ore.	Ore or Slag.					Slag.
Ore..................grs.	5	10	10	10	10	10	5	5	15	5	10	15	5	5	5	10	10	10	10	10	10
Sodium bicarbonate...."	15	25	20	15	30		15	15		15	15	20	10-15	15	5	2	15	4	2	1	
Potassium carbonate...."		10		10												1		1	1	1	
Borax, dried........."	4			1*	3*		4-6	5	5-10	6	5				0-5	1		1	1	1	
Argol..............."				5	7		4	4	4	4	5	5	6	6	6	½	3	1	1	1	
Black Flux Substitute."		35	35													¼					40**
Sulphur............."										0-1					1	½			½		
Flour..............."				2	5		2	2	1-3	4					1	1	3	1	1	1	
Nails..............No.	3	3	3	3	2	3	3	2		4	5	5	6	6	6	1½	1	1		1	
Silver.............grs.																To suit.					1
Salt................	Cover.										Cover.					With slag 1 gramme. Cover.					
Reference...........	b	a	ac	c	a	c	c	b	b	b	a	c	a	b	b	b	b	b	b	d	c

* Borax glass.   ** Black flux.   a, Brown, Op. cit.   b, Clark, Op. cit.   c, Ricketts, Op. cit.   d, Plattner's flux.

scorifier to leave room for the extra amount of lead and borax, the latter being added from time to time as the case may require.

The following table for scorification assays is made up principally from Kerl's manual, already quoted :

Weight of Argentiferous Substance: 0.1 assay ton.	Grammes Test Lead.	Grammes Glass Borax.	Remarks.
Galena, pure..........................	15–18	up to 0 5	
" with blende or pyrite	24–36	0.6–0.9	
Wall accretions.....................	30–40	0.3–0.6	
Lead matte, rich in iron...........	27–36	0.5–1.0	
" " copper or nickel bearing.	36–48	0.5–1.0	Low temperature with copper; the button may have to be rescorified.
" speise	30–60	0.5–0.7	High temperature; the button may have to be rescorified.
" carbonate	10–15	up to 0.5	
Blende.	30–45	0.3–0.6	High temperature.
Copper ores or matte..............	36	0.3–0.5	Low temperature; the button may have to be rescorified.
Gray copper ores..................	36–48	0.3–0.5	And more borax.
Antimonial ores...................	48	up to 3.0	
Arsenical ores....................	up to 48	up to 1.5	High temperature; often addition of litharge.
Telluride ores	50	0.3	Add litharge; rescorify the button.
Dry silicious ores..................	35–40	up to 0.5	
Dry basic (iron) ores..............	35–40	0.3–1.0	
Hearth accretions.................	25–50	up to 1.5	If necessary, first evaporate to dryness with HNO_3.
Cupel-bottom	20–25	up to 0.75	

The gold of an ore is determined by dissolving the silver buttons obtained from triplicate scorification assays.

The scorification assay is not suitable for slags, because they contain too little silver, viz. : from 1 to 2½ ounces per ton. With a crucible assay, taking 1 assay ton of slag, a silver button of satisfactory size is obtained. The usual way is to prepare the smelting mixture in considerable quantities beforehand. For the assay, it is measured out to fill two-thirds of the crucible. From one-half to two-thirds of the amount is mixed with the slag, the rest is added on top, and this again covered with salt. A nail is then inserted to assist in liberating the lead.

CRUCIBLE MIXTURES.

Litharge................................parts.	1	1	1
Sodium bicarbonate........................... "	0.5	2	1
Potassium carbonate.......................... "	..	..	1
Borax, dried................................. "	0.25	0.5	0.5
Argol....................................... "	0.3	0.3	..
Flour....................................... "	..	..	0.2

The wet assays of other components of the ore, as well as of the fluxes, will be treated under the heads of matte (§ 79) and slag (§81).

§ 35. **Purchasing of Lead-Silver Ores.**[1]—In purchasing lead-silver ores the character of the lead-bearing mineral and the chemical composition of the gangue have to be considered. If the lead mineral is a sulphide, the ore has generally to be roasted; this is not necessary if the lead is present in the form of carbonate. Further, an ore may be either self-fluxing, acid, or basic, *i.e.*, requiring no fluxes, requiring a base (iron, manganese, lime), or requiring an acid (silica) to form a desirable slag. Lead-silver ores are commonly acid, thus the basic ores command a higher price. In purchasing basic ores, the base paid for is iron, with its substitute, manganese. The so-called "base excess" is that amount of available iron and manganese which is obtained by adding the percentage of metallic iron to that of metallic manganese, and deducting the percentage of silica. The impurities in an ore affect its price. They may necessitate a preliminary roasting (sulphur, arsenic), may impair the fusibility of the slag (zinc, magnesia, baryta), or may cause loss of lead and silver by slagging or volatilization (zinc, arsenic, antimony), or finally may render the lead impure (zinc, arsenic, antimony, copper). The price paid for an ore will therefore be in inverse ratio to the percentage of impurities present.

A smelter, especially if so located as to draw its supplies from a number of mining districs, treats not only argentiferous lead ores, but extracts the precious metal also from real silver ores, called dry ores on account of their want of lead, by mixing them with ores that contain more lead than is required for the charge. Smelting can thus often compete with milling on account of the higher percentage of precious metal extracted from the ore.

In bidding for an ore deductions are made for loss and for cost of smelting, which will vary from the causes just mentioned. In the smelting charges are included the cost of shipping and refining the base bullion obtained.

The value of lead is given in units of 20 pounds to the ton of 2,000 pounds avoirdupois. Its price is regulated not simply by the fluctuations of the market, but also largely by the scarcity or abundance of available lead ore at the works. If the lead runs lower than 5 per cent. it is not paid for.

[1] Kirchhoff, "United States Geol. Survey: Mineral Resources of the United States," 1885, p. 251.

The value of silver and gold is given in ounces per ton. The price of silver is regulated by the New York quotation of the day, that of the gold varies slightly according to the amount present, which is generally small. The gold in an ore is not counted if it runs less than 0.1 ounce per ton.

Competition has forced works to pay something for copper contained in sulphurets, if it runs as high as 5 per cent.

There are various ways of arriving at the minimum smelting charge for an ore that is offered for purchase. The following gives very satisfactory results, and can be made to suit all cases. It is based on the cost of smelting what is called a neutral ore, and then debiting and crediting the actual ore for which the smelting-charge is to be fixed, as it differs in composition from the standard. A neutral ore is one in which the insoluble residue is equal to the sum of the iron (Fe) and the lime (CaO) it contains, and in which the lead amounts to 13 per cent. Taking Denver and Pueblo rates, f.o.b. works,[1] the manner of figuring may be shown by two examples:

1. A sulphide ore containing :

Ag.	Pb.	Au.	SiO_2.	Fe.	CaO.	Zn.
75 ozs.	13 %.	0.5 ozs.	25 %.	35 %.	....	10 %.

Debit :

    Smelting....... $4.50
    Roasting....... 2.50
    Fluxing........ 0.00 (there is an iron excess and enough lead).
    Zinc charge.... 5.00 (10 units @ 50 cents).
                    ——————
                    $12.00

Credit :

    Gold.......... $0.50 ($20.00 an ounce received by the refiner
                        and $19.00 an ounce paid to the miner).
    Iron excess ... 1.00 (35 − 25 = 10 units @ 10 cents).
                    ——————
                    $1.50

The minimum charge to be made for treating this ore will be : $12.00 − $1.50 = $10.50 per ton.

[1] Cost of smelting, per ton.................................. $4.50
    " roasting, " 2.50
    Pay for gold, per ounce 19.00
    " lead, " unit 0.10
    Charge for zinc, per unit.... .... 0.50
    " " grade silver (every 50 ounces silver above 100
                    ounces per ton) 1.00
    Pay for iron or lime excess, per unit...................... 0.10
    Charge for silica excess, " " 0.10

The value of the ore is calculated on the following basis :

Lead : 30 cents a unit is paid when the price of lead in New York is $4 for 100 pounds, which is called paying on a four-dollar basis. The pay grows 1 cent for every rise of 5 cents for 100 pounds in the New York market.

Silver : 95 per cent. of the New York quotation for the day.

Gold : $19 an ounce.

The value of the ore for which the smelting charge is calculated will be per ton :

Lead .... 13 units @ 30 cents	=	$3.90
Silver .... 75 ounces @ 95 cents	=	71.25
Gold................... 0.5 ounce @ $19.00	=	9.50
		$84.65
Smelting charge		10.50
Net value of ore per ton	=	$74.15

2. A dry siliceous ore containing:

Ag.	Pb.	Au.	SiO$_2$.	Fe.	CaO.	Zn.
250 ozs.	4 %.	....	60 %.	10 %.	12 %.	6 %.

Debit :

Smelting	$4.50	
Roasting	0.00	
Fluxing	0.90	(13 − 4 = 9 units lead @ 10 cents).
Silica excess....	3.80	(60 − 22 = 38 units silica @ 10 cents).
Zinc charge	3.00	(6 units @ 50 cents).
Grade silver	3.00	(250 − 100 = 150 ounces; $1.00 for every 50 ounces.)
	$15.20	

Credit : none.

The smelting charge will be $15.20 per ton.

The value of the ore per ton is :

Lead.........................4 units, no pay	=	$ 0.00
Silver.................... 250 ounces @ 95 cents	=	237.50
Gold.............................none	=	0.00
		$237.50
Smelting charge		15.20
Net value of ore per ton	=	$222.30

§ 36. Purchasing Non-Argentiferous Ores.[1]—In the Mississippi Valley this is not done by bidding in the market. The

---

[1] Clerc: *Engineering and Mining Journal,* July 4, 1885, p. 4.

" buyer " goes to the different mines and, guided by previous experience in purchasing from the same mine, offers a certain amount for the concentrated galena. The price paid for pure concentrated galena is about $48 per ton.

§ 37. **Purchasing of Fluxes and Fuels.**—In regard to the purchase of fluxes and fuels there are no general standards as there are with ores. The manner of buying and price paid vary according to local conditions.

PART II.

THE METALLURGICAL TREATMENT OF LEAD ORES.

PART II.

THE METALLURGICAL TREATMENT OF LEAD ORES.[1]

§ 38. **Classification of Methods.**—Lead ores are treated exclusively in the dry way. If free from silver, the resulting lead goes to the market, after it has been purified by liquating and poling. If the ore is argentiferous, the silver passes for the most part into the lead (base bullion), which has then to be desilverized (Part III.). Wet methods[2] and the electrolytic[3] extraction have been tried, but so far without success. The smelting of lead ores is carried on in furnaces of various forms. They may be classified under three heads :

THE REVERBERATORY FURNACE.

THE HEARTH FURNACE.

THE SHAFT FURNACE.

[1] Percy, "Metallurgy of Lead," London, 1870.—Phillips-Bauerman, "Elements of Metallurgy," London, Philadelphia, 1887, pp. 564-658.—Balling, "Metallhüttenkunde," Berlin, 1885, pp. 49-165.—Kerl, "Metallhüttenkunde," Leipsic, 1881, pp. 1-128.—Stölzel, "Die metallurgie," Brunswick, 1863-86, pp. 851-1050.—Rivot, "Traité de métallargie," vol. ii., Paris, 1872 —Grüner, "Sur l'état actuel de la Métallurgie du plomb," *Ann. d. Mines*, 1868, xiii., p. 325 ; also as pamphlet : Dunod, Paris.—Grüner, "Notes additionelles sur l'état actuel de la métallurgie du plomb," *Ann. d. Mines*, 1869, xv., p. 519 ; also as pamphlet : Dunod, Paris.—Cahen, "La métallurgie du plomb," Paris and Liége, 1863.

[2] *Berg- u Hüttenm. Ztg.*, 1880, p. 1. (Schaffner), p. 2. (Maxwell).—*Wagner's Jahresberichte*, 1877, p. 151 (Meunier).

[3] *Berg- u Hüttenm. Ztg.*, 1883, p. 252 (Kiliani), p. 878 (Blast-Miest).

CHAPTER VI.

SMELTING IN THE REVERBERATORY FURNACE.

§ 39. **Introductory Remarks.**—The process carried on in the reverberatory furnace is the Roasting and Reaction Process (§ 9). The Precipitation Process (§ 7) was formerly used with raw sulphide ores at Vienne [1] (France) and at Chicago, [2] and with roasted ores at Par and Point [3] (Cornwall), but has now become obsolete on account of the high cost and loss of metal.

The roasting and reaction process in the reverberatory furnace has the following advantages : the ore is treated in the raw state, the apparatus is inexpensive, inferior raw fuel is used, hardly any fluxes are required, the bulk of the metal in a pure state is quickly extracted at a low temperature with little loss by volatilization, and if the ore is argentiferous, the larger part of the silver follows the lead, and only a small quantity is left in the residue, which is either thrown away or treated at a higher temperature in the blast-furnace. The great disadvantage of the method is that it is very much limited by the character of the ore. To be suited for the reverberatory furnace an ore must be a rich galena or a mixture of galena with carbonate (the former prevailing), that does not contain less than 58 per cent. of lead. It may not contain over from 4 to 5 per cent. silica ; and the non-siliceous associated minerals, such as blende, pyrite, chalcopyrite, calcspar, barite, may be present only in small quantities. The process requires much fuel and many hours of skilled labor per ton of ore treated.

It consists of two operations, one following closely upon the other, and both being repeated several times.

1. *Oxidation.*—The ore, crushed fine enough to pass a 4- or 5-mesh sieve, is spread in a thin layer over the hearth of the furnace and is heated gradually to a low red heat. The roasting is carried on in such a way that only a part of the lead sulphide is

---

[1] Kerl, "Metallhüttenkunde," p. 24.
[2] *Trans. A. I. M. E.*, ii., p. 279 (Jernegan).
[3] Percy, "Lead," p. 257.

converted into oxide and sulphate, the rest remaining undecomposed. The temperature at which this roasting is carried on and the time given to it depend on the character of the ore. Pure galena requires a low temperature to avoid agglommeration ; if pyrite, blende or calcite are present the roasting can be accelerated ; the lower the temperature, the more sulphate will be formed. During the operation the fire on the grate is kept low and open and the charge is raked frequently, to expose as much of the ore as possible to the action of air and heat and to prevent agglomerating, which obstructs oxidation.

2. *Reduction.*—The second operation is that of raising the temperature, so that. the oxygen compounds may react on unchanged sulphide. The resulting lead runs down the inclined hearth and collects in a basin, the sulphurous acid passes off through the flue, and the residue remains on the hearth. The temperature during the reduction period must be low, as the reactions do not take place if the ore is melted. By filling the grate well up with fuel the required temperature is obtained and unconsumed air excluded. The charge is stirred at intervals to bring sulphide and oxide constituents into intimate contact.

As it is not possible to roast a large amount of lead ore uniformly in one operation, the first reaction that takes place on raising the temperature will not extract all the lead. The resulting pasty residue will be rich, consisting mainly of lead sulphide with some oxide, sulphate, silicate, and gangue. The temperature is lowered and air is admitted. A second roasting takes place and is followed by a second reacting. It takes several repetitions of the process to extract the bulk of the lead. With each one the temperature must be slightly raised as the amount of lead diminishes. To counteract the melting of the charge, slacked lime is added, which acts mechanically by rendering the charge less fusible and more spongy. It also assists the process chemically by liberating the lead (§ 7) and by decomposing the sulphide, thus helping the silver in the residue to pass into the lead. Towards the end of the process there will not be enough lead sulphide left to react on the excess of lead sulphate (§ 9) and lead oxide. To reduce these to sulphide and metal and to make the charge more porous, charcoal or coal is mixed into it. Then the roasting and reacting can proceed again. Each successive operation will be of shorter duration than the preceding one and the lead each time a little less rich in silver. The first lead may contain four times as much silver as the last.[1] The products of smelting in the reverberatory furnace are :

1. *Lead,* holding in suspension particles of ore and other solid matter, which are removed by liquating and poling (§ 109). If the ore contains arsenic, antimony, or copper, some of these elements combine with the lead and have to be removed by refining (§ 1c5).

2. *Gray Slag,* a more or less matted mixture of lead, lead sulphide, oxide, sulphate, silicate, gangue, cinders, and lime in varying proportions. Its tenor in lead and silver depends on the character of the ore and on the degree to which the residue has been treated in the furnace. In some cases it is crushed and washed to save only the metallic lead ; in others, especially with silver-bearing ores, it is smelted in the blast furnace.

3. *Fluedust* is composed of particles of unchanged or oxidized ore, volatized lead that has been converted into oxide, carbonate and sulphate, and parts of fuel. If the ore contains blende, oxidized zinc compounds will also be found. The amount of fluedust formed will vary with the temperature at which roasting and reacting operations have been carried on and also with the skill of the furnace-man in manipulating the furnace and the charge. As it consists principally of oxidized compounds, it is worked in with subsequent charges and shortens the time required for roasting. If very impure, *e.g.,* from arsenic and antimony, it is smelted in the blast-furnace (§ 52) with the gray slag. The resulting lead is hard and has to be refined (§ 105).

4. *Hearth Bottom,* consisting of hearth material soaked to some depth with metal. It is worked up in the same manner as the residues.

§ 40. **Influence of Foreign Matter.**—The quantity and quality of lead that can be obtained from a given lead ore will depend largely on the nature and proportion of the other constituents. These may be silica and argillaceous matter, oxides of iron, limestone (dolomite), barite, fluorspar, pyrite, chalcopyrite, blende, antimony, arsenic, silver (gold), and oxide lead ores.

Silica and Argillaceous Matter have an injurious influence in both stages of the process on account of their readiness to combine with lead oxide (§ 6). It has been found by experiment that

---

[1] *Berg- u. Hüttenm. Ztg.,* 1860, p. 859 ; 1863, p. 285 (Fallize) ; 1871, p. 152 (Bouhy). *Zeitschrift für Berg-, Hütten- u. Salinen-Wesen,* xiv., p. 232 (Teichmann).

with more than 5 per cent. of silica an ore cannot be treated by the roasting and reaction process. But even such a small amount as ½ per cent. makes itself felt by coating particles of ore with the silicate that has been formed, thus preventing the action of the air during the first period and obstructing the reactions when the temperature is raised.

Oxides of Iron.—Siderite is sometimes found with galena ores, but most of it can be removed by dressing the ore before smelting. The small amounts which remain with the galena quickly lose their carbonic acid during roasting, and the resulting magnetic or ferric oxide acts as a stiffening ingredient while the lead is being extracted.

Limestone (Dolomite) acts on the whole advantageously during the entire process, as it hinders the fusing of the charge. Any chemical action it may have is so slight that it can be regarded as practically inert matter. It loses some of its carbonic acid and is to a small extent converted into sulphate. Like all indifferent substances it will retard somewhat the roasting by preventing the air from having free access to the particles of galena and obstructing the reactions by interfering with the necessary intimate contact of sulphide and oxide. The highest allowable amount is 12 per cent.

Barite remains practically unchanged in the reverberatory. ·

Fluorspar, the same. If fluorspar and barite are present together, they may increase the fusibility of the charge by combining with lead sulphate.

Pyrite is beneficial in the first stage. It favors the forming of lead sulphate (§ 9) and assists the oxidation of galena through sulphur trioxide liberated by the decomposition of ferrous or ferric sulphate. Small quantities of pyrite have also a favorable effect during the reaction period, the ferric oxide making the charge less fusible. If present to a considerable amount, say from 10 to 12 per cent.,[1] too much iron sulphide remains in the charge and is liable to form double compounds with the lead sulphide. With from 35 to 40 per cent. of pyrite[2] the reverberatory process has to stop.

Chalcopyrite has no special effect during the roasting and behaves in a manner similar to that of iron pyrite in the reaction period, with this addition, that some of the copper enters the lead and impairs the quality.

[1] Bouhy: *Annales des Mines,* 1870, xvii., p. 179.
[2] Bouhy, *Op. cit.,* p. 178.

Blende.—During the roasting period blende is partly converted into oxide and sulphate, and the latter perhaps somewhat decomposed, but much blende remains unaltered. From 4 to 5 per cent.[1] assists the roasting; 10 or 12 per cent.[1] prolongs it greatly and reduces the output of lead; with from 20 to 24 per cent.[2] very little lead is obtained; and from 35 to 40 per cent., as with pyrite, stops the process. The loss in silver with blende-bearing ores is principally a mechanical loss that takes place during the roasting period.

Antimony.—This occurs with some galena ores as a simple or multiple sulphide. It has a deleterious effect even if present in small quantities of 2 or 3 per cent. It is readily fusible and causes caking of the ore. The sulphide and the oxide are both volatile, thus causing loss. The oxide also combines readily with lead oxide, which is retained to a great extent by the gray slag; further sulphide and oxide of antimony react upon each other, giving metallic antimony, which is in part volatilized, causing again loss, and the remaining metal finally combines with the lead, making this hard. Thus antimony is probably the worst metal that can be associated with the lead.

Arsenic.—The pyrite formed with lead ore sometimes contains arsenic. Next to antimony, arsenic is the worst enemy of the process. It causes losses: by volatilization, through the arsenious acid formed in roasting; by slagging, through the combination of arsenic acid with lead oxide; and finally by the reduction of both to metal. This combines with the lead and impairs its quality.

Silver and Gold.—Most of the silver of galena ores passes off into the lead; gold behaves in a similar way.

Oxide Lead Ores, such as anglesite and cerussite, assist the operation, as they reduce the time of roasting.

§ 41. **Classification of Reverberatory Methods.**—The practice of the roasting and reaction process varies in different smelting works. At some the principal stress is laid on extracting as much lead as possible in the reverberatory, while others aim to obtain only the major part of the lead in the reverberatory and to extract the rest from the rich residue by smelting it in the blast-furnace; they thus recover a larger percentage of lead. Then, some establishments roast the ore slowly at a low temperature, which is advisable for the recovery of a large percentage of lead, while others hasten roasting by raising the temperature quickly to

[1] Bouhy, *Op. cit.*, p. 178.
[2] Rivot, " Métallurgie du plomb," p. 46.

the permissible limit, the aim being to save labor and time, which is done at the expense of the lead. As regards the form and size of furnaces and the position of the lead-well there are also characteristic differences.

The reverberatory furnace practice may be classed under three main heads :

The Carinthian Method.
The English Method.
The Silesian Method.

§ 42. **The Carinthian Method.**—The characteristics of this method are the smallness of the charge, the slow roasting, so that for every part of lead sulphide one part of sulphate and at least two of oxide are formed, the low temperature at which all the operations are carried on, and the aim to extract all the lead in the reverberatory. The hearth is inclined towards the flue and the lead is collected outside of the furnace.

A. *Lead-Smelting at Raibl,*[1] *Carinthia.*—The ore that is worked is galena (partly coarse, with from 72 to 75 per cent. of lead, and partly fine, with from 67 to 73 per cent. of lead); in exceptional cases the lead contents go as low as 58 per cent. The other constituents of the ore are anglesite, cerussite, wulfenite, blende, willemite, calcite, fluorite, and abestos. As seen from the assay these other constituents form only a very small quantity. The following analyses made by Phillips[2] in 1845 show the composition of low-grade coarse and fine ore :

	Coarse.	Fine.
PbS	76.6	76.0
$PbCO_3$	4.0	4.0
ZnS	13.2	13.0
Sb_2S_3	0.2	0.2
$CaCO_3$	4.6	5.0
Insoluble	0.4	0.4
	99.0	98.6

The drawings (Figures 21–24) of the furnace show an inclined hearth with only one working-door, *g*, below the flue *d*. On the same side is the door *b* leading to the fireplace. The grate, slightly more inclined than the hearth, is parallel to the long axis of the furnace. It is built of stone and has six transverse openings. The fire-bridge

[1] Thum : *Berg- u. Hüttenmännische Zeitung.* 1863, p. 196; "Official Report :" *Oesterreichische Zeitschrift,* 1889, p. 297.
[2] *Annales des Mines,* 1845, viii., p. 293.

is at *c*; the opening *f* carries off any lead fumes into the main flue. The hearth terminates at its lower end in a cast-iron gutter, over which the lead runs into the mould. Figure 22 shows the support for the working tools. The furnace is built of sandstone and ordinary red brick; the working bottom, which is renewed every four or five weeks, is made by tamping down firmly fire-clay, probably a mixture of

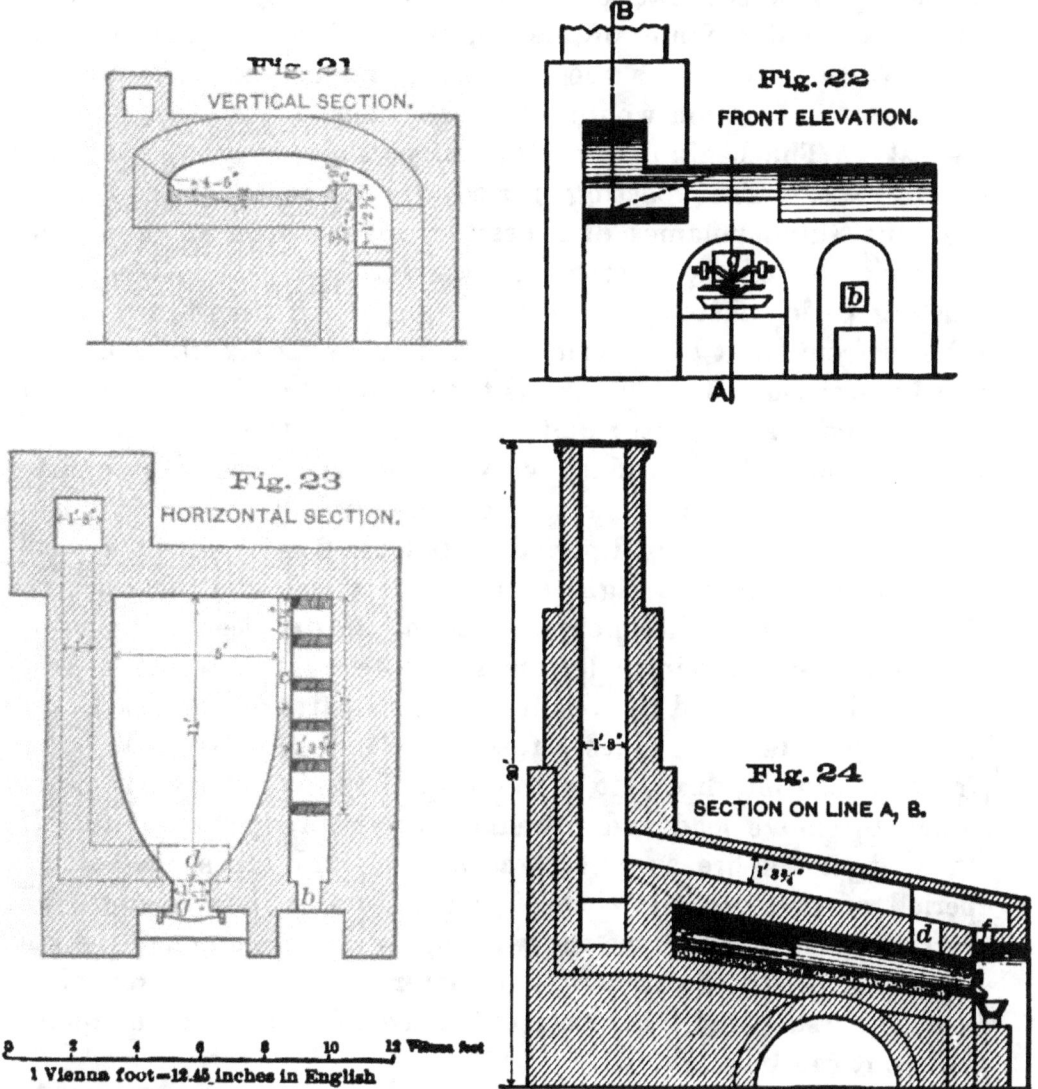

raw and burnt clay. It is made impervious to lead by glazing with gray slag. The heating up of the furnace, leading to the fritting of the slag, is done slowly to prevent the cracking of the tamping. Furnaces are usually built in pairs, being placed side by side. They last from five to six years. The fuel used is cordwood. Dimensions are given in the figures and in § 45.

The mode of operation is as follows : the furnace, barely red-hot from a previous charge, is repaired, if necessary, and the charge of about 400 pounds introduced through the working-door *g*, and spread out over the upper part of the hearth near the bridge to a thickness of from 1⅛ to 1½ inches.   No fresh fire is made, the heat of the furnace and the glowing fuel from the previous charge furnishing sufficient heat for the first slow roasting.   The ore, containing a small amount of blende, is raked every quarter of an hour. With pure galena the raking is not repeated so frequently, as the quick oxidation would liberate sufficient heat to make the ore cake.   The beginning of this is recognized by its adhering to the rake.   The roasting period has come to an end when the blue sulphur-flames disappear, drops of lead are seen near the front end of the hearth and the ore has a sandy feel.   The roasting period lasts three hours, during which time the charge is rabbled from eight to nine times and from five to six sticks of cordwood are consumed.   The fire is then urged, to raise the temperature.   This is kept pretty uniform throughout the reaction, which now sets in.   The attendant works his charge once every quarter or half hour, and raises the temperature slightly when the flow of lead ceases.   This second period lasts from five to six hours, consumes from sixteen to eighteen sticks of wood, and furnishes the first half of the lead, which, on account of its freedom from impurities, is called virgin lead.   The attendant now stops firing until he has collected the residue from all parts of the hearth into one heap.   He then takes a few shovelfuls of ashes and breeze from the ash-pit, throws it on the heap of residue, and works it in in order to remove lead and to reduce lead oxide and sulphate.   He then urges the fire as quickly as he can, and the so-called third period of the process, that of slag reduction, has begun.   The further manipulations are the working of the residue and the stirring-in of breeze of charcoal, until after three hours the rest of the lead, the "slag-lead," has been extracted.   This has to be liquated before it can be marketed.

The practice is varied in some works by raking out the residue after the ashes and breeze have been stirred in and introducing a new charge.   This is worked in the usual way.   The residue from this second charge is not withdrawn, but that from the first added and both worked together for slag-lead.   In this case the reduction of the slag occupies from seven to eight hours.

The final residue is withdrawn from the furnace and sorted into gray slag with 4 per cent. of lead, which is thrown away, and a product to be crushed and concentrated ; the heads which assay from 50 to 60 per cent. of lead, going back to the furnace in one of the subsequent charges of the residue.

There is one furnace-man, working twenty-four hours, who has a helper during the day (12 hours).

Tabulated results are given in § 45. Of the products no satisfactory analyses exist except of the lead.

	Virgin-Lead.		Slag-Lead.	
	a^1	b^2	a^1	b^2
Cu	0.00069	0.00010	0.00075	0.00100
Ag..........	0.00025	0.00008	0.00025	0.00008
Fe	0.00055	0.00700	0.00088	0.00770
Ni..........	....	....	trace.	....
Zn	0.00076	....	0.00082	....
S	0.01476	0.00400	0.01785	0.00130
Sb	trace.	0.05700	0.00703	0.14340
As	trace.	....	0.00721	0.01920

[1] *Oesterreichisches Jahrbuch*, vol. xxii., p. 389.
[2] *Oesterreichisches Jahrbuch*, vol. xxvii., p. 188.

B. *Lead-Smelting at Engis,*[1] *Belgium.*—The method of working at Engis differs in some respects from that at Raibl. The ores are pure. They contain, according to an average of several analyses : PbS, 93.56 ; ZnS, 3.74 ; FeS_2, 2.31 ; $CaCO_3$, 0.35 ; traces of silver, and are free from arsenic and antimony.

The furnace has the ordinary form of a reverberatory. The hearth, widest in the middle, is slightly contracted at the bridge and narrows down considerably at the flue. The furnace has two doors, one at the side and one beneath the flue, below which there is a small kettle with a separate fireplace to receive the lead. The fuel used is bituminous coal. The furnace bottom, oval in cross-section, begins at the top of the bridge, where it is 2 feet $8\frac{3}{4}$ inches thick, and is inclined towards the flue, so that no lead is collected in the furnace. At the flue the thickness is 1 foot $5\frac{3}{4}$ inches. This is the only furnace on record having a brasque working-bottom. The brasque consists of two parts by volume of ordinary brick clay and one of

[1] Bouhy : *Annales des Mines*, 1870, xvii., p. 159 ; reprint : *La fabrication du plomb*, Dunod, Paris, 1870 ; also in *Berg- und Hüttenmännische Zeitung*, 1870, p. 381 ; 1871, p. 52 ff.

coke ground fine enough to pass a 4-mesh screen. Old bottoms, containing usually about 2 or 3 per cent. of lead, are ground up and mixed with new brasque. All the brasque needed for one furnace bottom (about 4,000 pounds of clay and 1,600 pounds of coke) is prepared by two men in 24 hours. In addition to the ordinary test for the correct amount of moisture (that the brasque, when squeezed in the hand, shall cohere into a lump, but not have sufficient moisture to adhere to the hand) another one is used, that of throwing with force a ball of brasque against the wall, to which it should adhere. In tamping a layer of brasque 8 inches thick is first spread out evenly on the brick hearth and tamped down to 1 inch ; the second 8-inch layer is not rammed down as firmly, as it is reduced only to $2\frac{3}{8}$ inches ; the subsequent layers are made by using smaller amounts of brasque, as when spread out they are only about 4 or 5 inches thick. The bottom is not impervious to lead, which filters through to a slight extent, collects on t p of the underlying red brick, and also passes into the joints. A bottom lasts about six weeks. It wears off quickly during the first two weeks, but then resists pretty well abrasion by the tools and chemical action.

In starting a furnace with a new bottom the warming lasts six hours, slight cracks that may form being closed with brasque. Then a small charge of 220 pounds of low-grade ore with from 45 to 48 per cent. of lead is spread over the hearth and the temperature raised gradually for five and a quarter hours, the ore being raked from eight to ten times. The brasque becomes superficially soaked with lead and coated with a mixture of various more or less melted lead compounds. There result from this charge 4 pounds of lead, 5 pounds of rich slag-like material, and 130 pounds of matte-like material containing 48 per cent. of lead. The charges are increased in weight and richer ore is worked until, after the third day, the furnace can do economical work.

The method of working is the same as at Raibl, with the exception that all the lead is not extracted in the reverberatory. The residue, forming 12 per cent. of the original charge and assaying from 17 to 20 per cent. of lead, is melted with puddle-cinder in a small shaft-furnace. One furnace-man with a helper works a charge in twelve hours. Tabulated dimensions and results are given in § 45.

The subjoined analyses show the composition of the residue :

SiO$_2$	29.86	28.10
PbO	25.50	32.80
ZnO	25.33	20.80
(FeAl)$_2$O$_3$	15.03	14.70
CaO	3.70	3.20
S	0.38	0.10
	99.80	99.70

C. *Lead-Smelting in the Air-Furnace.*[1]—The roasting and reaction process, as carried out in the so-called air-furnace, is given here as the American improvement of the Raibl furnace.

The galena ore from the Mississippi Valley is concentrated to a high-grade pure product ranging from 70 to 84 per cent. in lead.

The furnace has a peculiar construction. The drawings (Figures 25 and 26) show an inclined hearth *c*, with *e* as charging and working door, and *f* as discharging and cleaning door ; beneath is a small kettle *g*, into which the lead flows as it is set free in the furnace. The chimney is a sheet-iron pipe. Thirteen inches from the front— *i.e.*, the lower or pot end of the furnace—is the fire-bridge *b*, with fireplace *a* at right angles to the axis of the furnace. The bottom of the furnace is a cast-iron plate with 6 inches of gray slag or residue melted upon it. Cord-wood is used as fuel. The cost of a furnace, including shed, is estimated by Broadhead at $550.

The charge, from 1,400 to 1,600 pounds of galena of from pea size to hazelnut size, is introduced through door *e* and spread evenly over the hearth. It is roasted for from one to one and a half hours at a low temperature, the shorter time being sufficient if the galena contains some oxidized ore. During the roasting

---

[1] Williams : "Geological Survey of Missouri : Industrial Report," 1877, pp. 8–101 ; *Trans. A. I. M. E.*, v., p. 314.—Broadhead : "Geol. Survey of Missouri," 1874, p. 492. The furnace has been used in the Mississippi Valley, where it originated, to a considerable extent ; but even ten years ago it had given way largely to the ore-hearth, as this has about three times the capacity, although it saves a smaller amount of lead. The percentage of lead ore worked in 1880 by different furnaces in the Mississippi Valley is shown by the following table, taken from the Tenth Census of the United States, vol, xv., p. 818.

Air-Furnace	10.61	per cent.
Flintshire Furnace	7.64	"
Scotch Hearth	62.47	"
Blast-Furnace	19.28	"

the charge is constantly raked and moved from fire-bridge to flue and from the cooler part of the furnace to the hotter, in order to heat and roast it as uniformly as possible. When this is accomplished, the heat is raised and lead begins to flow. Ashes and breeze are used as stiffening ingredients. The charge is rabbled at intervals. When no more lead appears (after from seven to eight hours), the residue is drawn without any attempt to extract slag-lead in the furnace as at Raibl.

One furnace-man with a helper works in twelve hours a charge which consumes one and one-half cords of wood. About the yield nothing is known, as the contents of ore charged is not taken. Tabulated dimensions and results are given in § 45.

The following two analyses by Williams[1] show the composition of residue and poled lead from the Granby works :

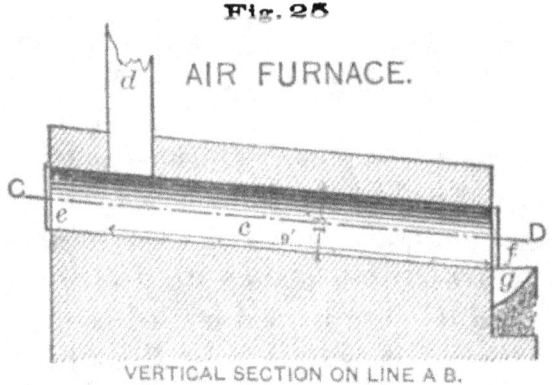

Fig. 25
AIR FURNACE.
VERTICAL SECTION ON LINE A B.

Residue : SiO_2, 21.396 ; CaO, 4.650 ; MgO, 3.948 ; Fe_2O_3, 3.680 ; Al_2O_3, 0.152 ; ZnO, 7.146 ; $PbSO_4$, 2.349 ; PbS, 20.929 ; PbO, 34.914. Total, 99.063. The sample yielded 3.52 per cent. of metallic lead, which makes the total metallic lead 54.82 per cent.

Lead : As, 0.01122 ; Sb, 0.00077 ; Ag, 0.00080 ; Cu, 0.05091 ; Fe, 0.01582 ; Zn, 0.00090 ; Ni, 0.00281 ; Pb (by difference), 99.91777. Total, 100.00000.

§ 43. **The English Method.**—The characteristics of this method are a large charge, a quick roast (with the result that for every part of lead sulphate formed there shall remain two parts of lead sulphide unchanged), a high temperature throughout, and the aim to extract all the lead in the reverberatory. The hearth inclines towards the middle of one of the sides ; the lead collects in the furnace, and is tapped at intervals into an outside kettle.

[1] *Trans. A. I. M. E.,* v., pp. 320–324.

Lead-Smelting at Stiperstones [1] *(Shropshire).*—The ore is a
concentrated galena with some carbonate and blende, assaying
77.5 per cent. of lead, the little gangue being principally carbonate
of lime.

The construction of the furnace is given in detail in Figures
27 to 32. The horizontal section (Figure 27) shows the usual
trapezoidal form of the English reverberatory with its three working doors k on either side, the well b at the front, and the fire-
opening t at the back. It is to be noted that centres of hearth a
and fireplace g are not opposite each other, the latter, with fire-
bridge h, being 10 inches further back. The object of this is to

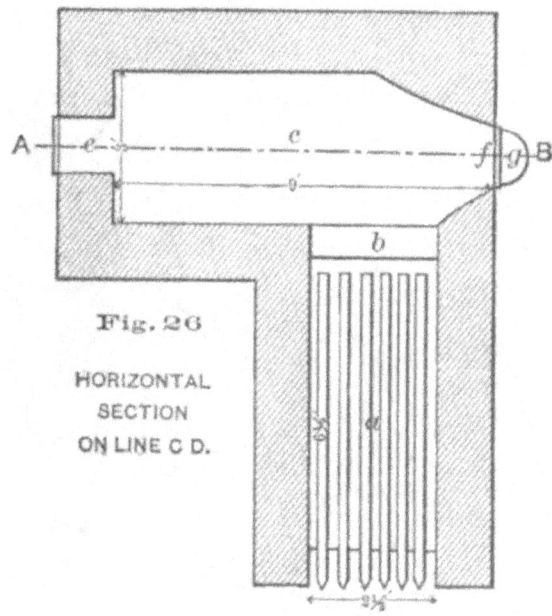

Fig. 26

HORIZONTAL
SECTION
ON LINE C D.

draw away the flame from the well b, for the same reason the
flue d, near the back, is made 1 inch wider than flue d, while both
have the same height. Each has its own damper w and w'. The
cross-section, Figure 29, shows the inclination of the hearth from
the back to the front (24 degrees), where well b is placed. This
has nearly vertical walls and (Figure 27) extends $3\frac{1}{2}$ feet into the
furnace, where it is 2 feet wide, narrowing down to a few inches
in the front. The lead collected in the well is tapped through tapping-hole m into the cast-iron kettle or pot c. The longitudinal

[1] Moissenet : *Annales des mines*, 1862, vol. i., p. 45 ; reprint, " Traite-
ment de la galène au four gallois," Dunod, Paris, 1862 ; also in *Berg-und
Hüttenmännische Zeitung*, 1868, pp. 243, 251, 261, 265.

section (Figure 28) shows the inclination of the hearth, which is 30 degrees from the bridge and 23 degrees from the flue. The roof *P* slopes in a straight line from top of grate *g* to flues *d d'*. The gases from the grate pass over fire-bridge *h* to the hearth and thence through flues *d d'* into flue *e*. In the front elevation (Figure 30) can be seen the chimney *f*, which is 2 feet 6 inches by 2 feet in the clear and 20 feet high from the ground. It does not communicate with the interior of the fireplace. It carries off gases that would enter the smelter building when the fire is fed or stoked or when the doors are open, and also vapor that comes from the ash-pit *u* (Figure 28), which is filled with water.

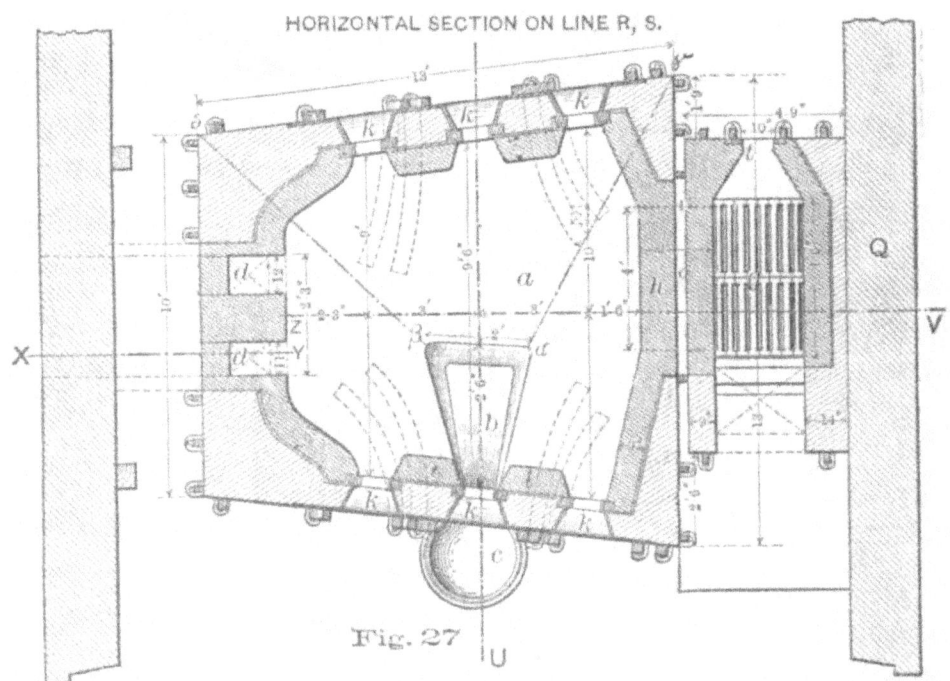

Fig. 27

The following details will elucidate the drawings. In Figures 27, 28 and 30 one end of the furnace is seen to be built against the wall, *Q*, of the building. At the others there is a passage, *D*; 15 feet of free space at front and back give sufficient room for working the furnace.

The furnace is built of red brick and lined with a full course of fire-brick, which is indicated by heavy shading. The foundation is 3 feet deep; the space in the centre between the bent ends (*j j'* in Figure 29) of the lower tie-rods is filled with sand. The upper part of the foundation is solid and slopes from the back and both ends to the front. The edges of the last course of brick are left sharp, as shown by the zigzag line in Figures 28 and 29. In determining the slope, a line 3 feet 6 inches long is drawn from the tap-hole, and the points a and β located at a distance of one foot on each side of its inner extremity. To them, lines are then drawn from the corners γ and δ, starting at an elevation of 2 feet 8 inches. The side-walls are raised slightly over the roof P, and have a total height of 5 feet 8 inches, measured from the floor. The space between the two is filled with sand, R'.

The binding of the furnace is clearly shown in the figures. The buckstays are 6 feet 6 inches long, the tie-rods being slipped over them and tightened with wedges. The upper tie-rods reach over the furnace; the lower ones, $j\,j'$, are turned down one foot at

Fig. 28

LONGITUDINAL SECTION ON LINE X, Y, Z, V.

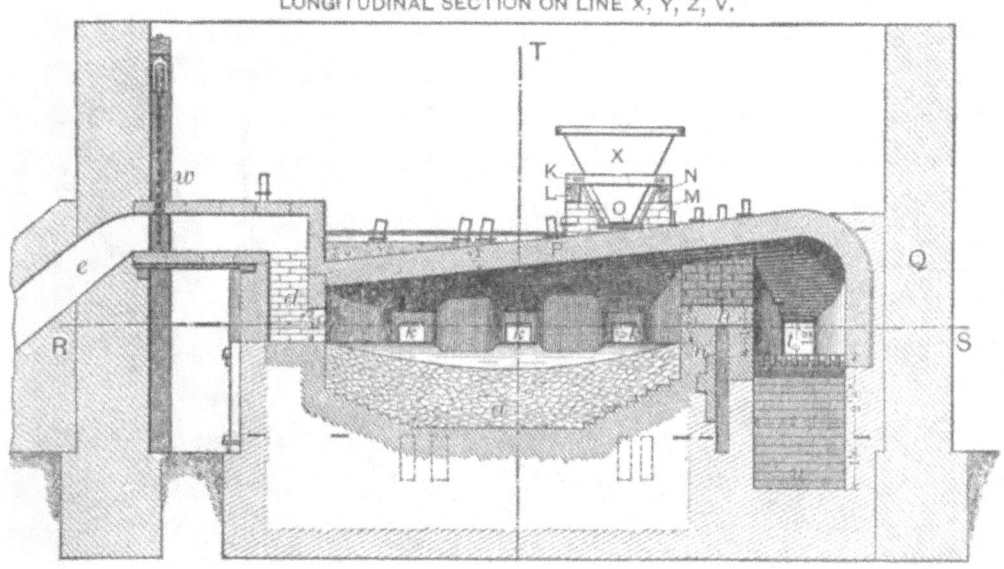

a distance of 4 feet 6 inches. The castings are not shown with sufficient distinctness in the drawings.

The tap-hole plate (Figure 29 and y in Figure 30) is 5 feet long and 20 inches high; it is two inches thick at the bottom, increasing to three, which it reaches at a height of 6 inches and retains for the rest of the distance. Eight inches above the centre of the lower edge is the tap-hole m, and 4 inches above this begins a narrow vertical opening (4 by 9 inches), with a hinged door, p, only 8 inches high, so that when the door is closed a small current of cold air may pass over the molten lead in the well. The tap-hole plate is overlapped or each side by five cast-iron plates (4 feet long and $\frac{1}{2}$ inch thick), placed horizontally (z and z' in Figure

30). They extend from the working opening down to the tap-hole.

The back-plate (5 feet long, 20 inches high and 1 inch thick) is placed just opposite the tap-hole plate. Its main object is to protect the brick from the gray slag, which is drawn out from the back door. To facilitate this, two hooks, *H* (Figure 29), carrying an iron rod (3 feet 6 inches long and 1 inch in diameter) as support for the working-tools are placed inside of neighboring buckstays.

The bridge-plate *n* (Figure 28), counteracting the longitudinal thrust of the hearth, gives strength to the 4-inch air-space *o*, and prevents any leakage of lead. It is 6 feet long and 20 inches high.

Fig. 29

CROSS SECTION ON LINE T, U.

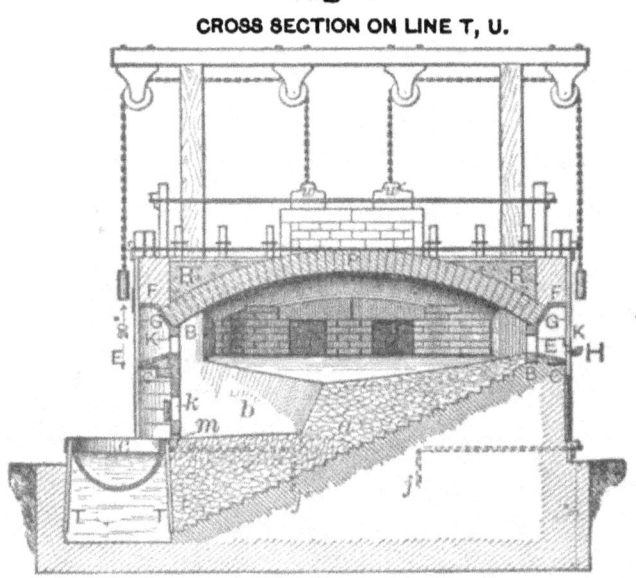

From the upper edge, which is on a level with the grate bars, it is 3 inches thick for a distance of 6 inches, then suddenly increases to double the thickness for another 6 inches, and for the last 8 inches returns to its original thickness.

The working-openings, *k*, have a number of castings. The door-frames (B, Figure 29), enclosing each an opening 10 by 6 inches, are 4 inches square in cross-section. They are protected on the sides by fire-brick, and on the bottom by the hearth. On the level of the grate-bars, or 2 feet 8 inches above the floor, are two horizontal plates, C, 6 inches wide and $\frac{1}{2}$ inch thick, that extend below the six working-openings. On them rest two inclined (3 : 8) plates, E, 8 inches wide and $\frac{1}{2}$ inch thick, which abut against

the bottom of the door-frames. Two plates, F, of the same size as
C, form the upper part of the linings. The skew-backs, G (9 inches
wide and ½ inch thick), resting on these upper plates, F, and the
door-frames, B, support the roof. The sides of the working-openings are lined with ½-inch jamb-plates, as shown in Figure 27. The
cast-iron working-doors, *g*, with handle, *r* (Figure 31), are placed
against the door-frames, B.

The cast-iron kettle, *c*, in front of the tap-hole is 2 feet 6 inches
in diameter and 20 inches deep. It is embedded in clay, which is

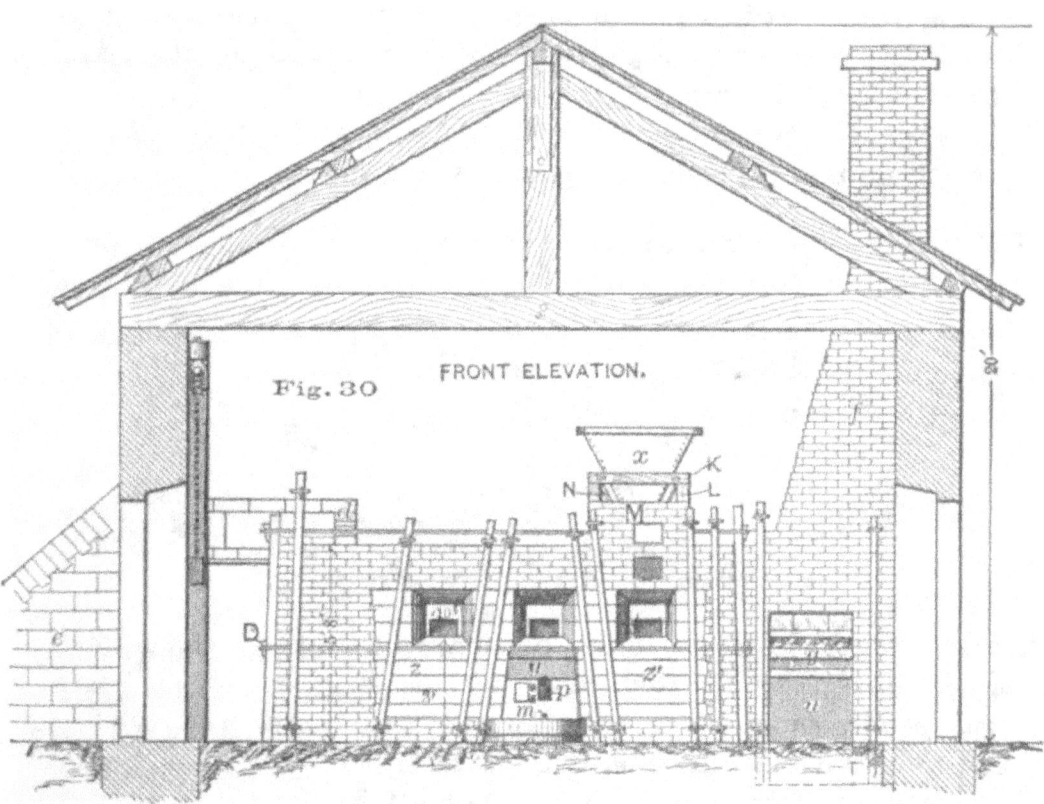

FRONT ELEVATION.

Fig. 30

enclosed on three sides by masonry and on the fourth by the slag-
bottom. Its rim, 8 inches above the floor, is surrounded by an iron
hoop, 4 inches wide and ½ an inch thick. Between it and the kettle
wedges, T, are driven to prevent any lead that may penetrate
through the joints and collect under the kettle from raising it up.

The fire-opening, *t*, 10 by 12 inches, has an iron casing (Figure
27). It is closed by a swinging fire-door (Figure 32), consisting of
fire-brick held together by a wrought-iron frame. The grate
consists of two sets of cast-iron grate-bars of eight each, supported at the ends by cast-iron cross-pieces.

The hopper (*x*, Figures 28 and 30) is a truncated sheet-iron pyramid 36 and 6 inches square at the two ends and 3 feet high. It is suspended in a wooden frame, K (3 inches square), which rests on two iron cross-beams, L, 4 by 6 inches. These are supported by brickwork, 14 inches high, forming the continuation of the side-walls of the furnace. Four iron rods, N (Figure 28), fastened in the wooden frame, K, have at their lower ends a sheet-iron frame, O, on which is placed a movable slide, which closes the discharge of the hopper. The 6-inch opening in the roof of the furnace, over which the mouth of the hopper is placed, is indicated on the hearth in Figure 27. It is 1 foot 9 inches distant from the fire-bridge, and 2 feet 6 inches from the working-opening nearest the bridge at the back of the furnace.

When the furnace is finished, the working-bottom, made of coarsely-broken gray slag and sand, is put in. Part of the mate-

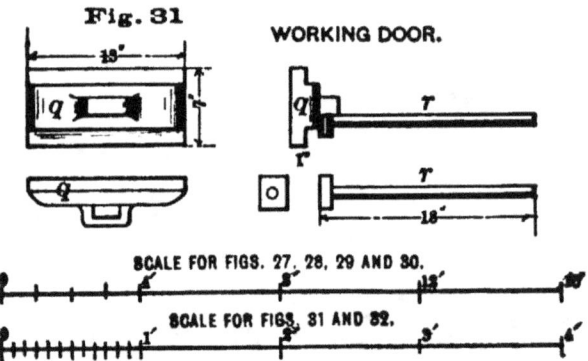

rial is spread out on the brick bottom, which has previously been made red-hot. It is heated till it becomes pasty, and is then patted with paddle and rake. A second part is made to adhere to the first, and so on until the entire bottom has been built up with successive layers of gray slag and sand, and has attained the desired form and thickness. The upper edge of the well is 10 inches below the working-opening. The thickness of the working-bottom increases towards the tap-hole. It is 4 inches at the back doors, 12 at the bridge, 14 at the flues, 30 at the front, and 8 at the back of the well.

The furnace, being built solidly, lasts a long time; the roof of the fireplace requires repairing every two years, that of the hearth every five years. The hearth is repaired after every charge if necessary.

This furnace has the representative form of the English rever-

beratory. Slight variations exist as regards detail. For instance,
the foundation, instead of being built up solid at the back, sometimes has an arched vault [1] at *j* (Figure 3), extending longitudinally,
communicating with the air-space in the fire-bridge, and being acces-
sible at the flue end of the furnace. Another variation is when the
entire hearth is built on cast-iron plates [2] resting on rails supported by a brick pillar. Sometimes the general form of the
hearth differs by having a gentle slope towards the tap-hole. This
makes the sides of the well less steep than those of the drawings.

The tools used in the furnace are shown in Figures 33 to 46.
They are wrought-iron, with the exception of the mould (Figure 45)
and the handles of shovel (Figure 37) and ash-pit hoe (Figure 40).

Figures 33 to 36 : Paddle, rake, old paddle used as chisel, ham-

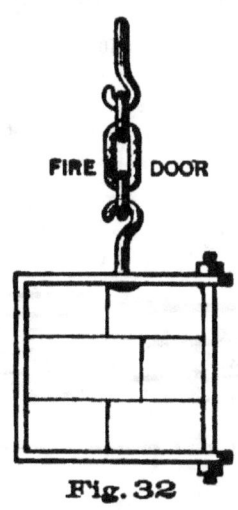

FIRE DOOR

Fig. 32

mer to remove adhering slag, are the tools for working on the
hearth. Figures 37 to 40 : Coal-shovel, hammer to break coal,
fire-poker, ash-pit hoe, are required for firing. Figures 41 to 46 :
Tapping-bar, rectangular skimmer, circular skimmer, ladle, mould,
lead-carrier, are used in handling the lead.

Method of Working. To simplify the description the doors
may be be designated by numbers 1, 2, 3, in front, and 4, 5, 6, at
the back, starting both times from the bridge.

The charge, 2,350 pounds, is let fall from the hopper through
the roof into the furnace, still red-hot from a previous charge, the
dampers having been closed. The lead is left in the well and
covered with lime. The ore is spread over the hearth through

[1] Percy, "Lead," pp. 222–229.
[2] Phillips-Bauerman, *Op. cit.*, p. 591.

doors 1, 4 and 5, with rakes. It decrepitates and gives off vapor of water. Then the dampers are slightly raised and the fire is gradually increased for an hour and a half. During the first two hours (first firing) the ore is turned over four times with padddles through doors 1 and 4, the other doors being kept closed. The paddling requires only a few minutes. When the first firing is nearly accomplished, feeding of fuel is stopped, and doors 1, 2, 4 and 5, fire-door and dampers are thrown open and the charge cooled (first cooling). This lasts for half an hour. The grate is then freed from clinkers, and the charge, nearly 6 inches deep next to the bridge, which has fritted, is broken up and turned over. Any ore that had fallen into the well is raked up on the furnace-bed. The doors are now closed, the dampers lowered, the grate is well filled with coal, and the second firing begins. This lasts from 55

WORKING TOOLS.

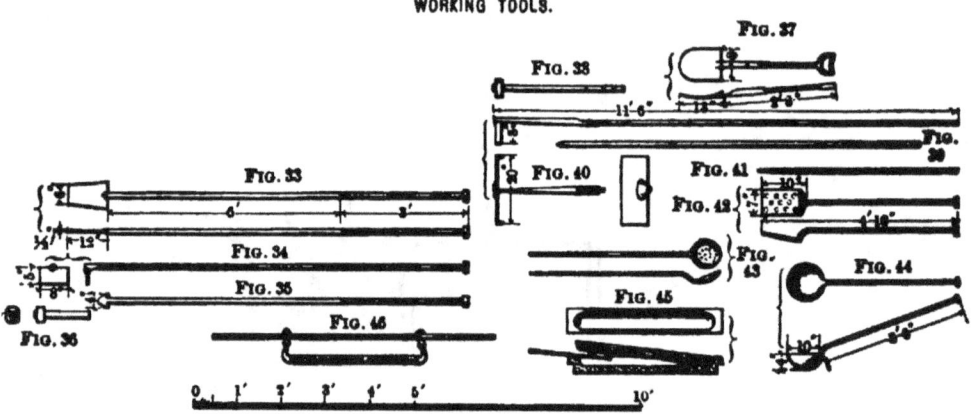

to 60 minutes, and is followed by a second cooling. Some lead now flows into the well. Near the bridge and towards the centre of the furnace, parts of the charge have begun to fuse, while at the flue the ore is only sintered. The charge is worked as in the first cooling, door 6 now being also used. After ten minutes the lead which has accumulated is tapped from the well into the kettle. The tap-hole is closed from the inside by inserting a clay plug fixed to a wooden handle through the little tap-hole door and pressing it in until the clay oozes out in front. Before tapping, fine coal and wood-shavings are put into the kettle to pole the lead. It is then stirred vigorously for six or seven minutes with the rectangular skimmer to bring all the impurities to the surface. These are removed, first with the shovel and then with the rectangular skimmer, and thrown back into the furnace, through doors 1 and 3, on both sides of the well. The charge, now freed from part of its lead, is turned over with paddles through doors 1 and 4. This ends the second cooling, four hours after charging. The doors are now closed, coal is put on the grate, and the dampers are lowered to begin the third firing. Meanwhile the lead in the outside kettle is skimmed and ladled into moulds holding one hundred and twenty pounds. This takes about twenty minutes. During the third firing, which lasts two hours, the ore is turned over several times, care being taken to open the doors as little as possible. The furnace shows a bright-red heat when the third cooling begins. The charge is now worked with paddles for fifteen or twenty minutes, and parts of it that have collected in the well are stiffened by the addition of lime and are raked on the hearth. The residue on the hearth is collected near the bridge and fine coal worked into it. The doors are again closed, the fire is urged for a quarter of an hour, and the residue turned over with the paddle through doors 1 and 4, and finally drawn out through door 5. Any repairing of the hearth that may be necessary takes place now, and the furnace is ready again for a new charge, seven hours from the time when the previous one was first introduced.

Two men work as partners in twelve-hour shifts. Tabulated results are given in § 45.

The method of working near Holywell, North Wales, differs from that at Stiperstones. According to Percy[1] it is as follows.

After dropping the charge of 2,350 pounds on the hearth and spreading it over the upper part of the bed, the doors of the furnace and that of the fire-box are left open for an hour and a half to let the air have free access during the first roasting stage, while the damper is raised just enough for the gases to escape. Working-doors 3 and 6 and the fire-door are now closed and the fire is urged. Lead soon appears. During this heating, which lasts two hours, doors 2 and 5 are closed, but 1 and 4 are kept open. Through these the charge is rabbled at intervals. Towards the end of this first reaction period lead begins to flow. Now doors 1 and 4 are closed, the damper is thrown open, the fire urged for forty minutes, and the charge melted down. The furnace is then cooled for half an hour by throwing open all the doors. What charge remains on the hearth is rabbled, doors 3 and 6 are closed, slacked lime is thrown through door 2 on the charge, which has collected in the well, and worked into it through the tap-hole door.

1 "Lead," p. 232.

The stiffened residue is collected ("set up") near the bridge as well as other parts that have been detached from the hearth. Doors 2 and 5 and the fire-door are now closed, the damper is lowered, and the temperature raised gradually for half an hour, when the damper is entirely thrown open, doors 1 and 4 are closed, and the fire is urged to melt down the residue, which takes twenty minutes. The fire-door and the working-doors 1, 2, 4 and 5 are then thrown open, lime is added through door 2, and worked into the slag to thicken it. The lead is tapped, and the stiff gray slag raked out on the floor through door 5. The lead is poled as at Stiperstones, and the hearth repaired, if necessary. The entire time required to work the charge is six hours.

Two furnace-men and one helper work two charges in a twelve-hour shift. Tabulated results are given in § 45.

The changes that take place in the ore during the process have been examined by Percy,[1] and are given here in their lead contents only.

Lead Compounds.	Hours after Charging.				Gray Slag.
	1½	3½	4½	4¾	
PbS	63.82	53.32	24.76	4.85	0.90
PbO	27.25	31.49	43.12	47.50	48.87
$PbSO_4$	3.82	4.78	6.94	14.02	9.85
Total.Pb,	83.16	78.66	66.22	47.86	52.88

§ 44. **The Silesian Method.** — The characteristics of this method are a large charge, slow roasting, and a low temperature. It is not aimed to extract all the lead in the reverberatory, as this is supplemented by the blast-furnace. The hearth is inclined toward the flue, beneath which the lead is collected and tapped at intervals into an outside kettle.

Lead-Smelting at Tarnowitz,[2] Prussia.—The ore is a mixture of sulphide and oxide lead minerals. Its composition was in 1870, in round numbers :

[1] *Op. cit.,* p. 235.
[2] *Zeitschrift für Berg-, Hütten-, und Salinen-Wesen in Preussen,* vol. xiv., p. 225, Teichmann ; xix., p. 157, Wedding ; xxxii., p. 94, Dobers and Dziegiecki ; xxxiv., p. 292, Dobers and Althans.

PbS	61		40
PbSO₄	11		9
PbCO₃	24		45
(Ca, Mg, Fe, Zn) CO₃	8		5
SiO₂	1		1 and less.

The furnace-charges contained from 70 to 74 per cent. of lead and from 21 to 22 ounces silver per ton. At present the ore runs lower, the percentage of blende having increased considerably.

The construction of the latest furnaces is given in Figures 47 to 50. After the detailed description of the reverberatory at Stiperstones attention need only be called to some of the principal

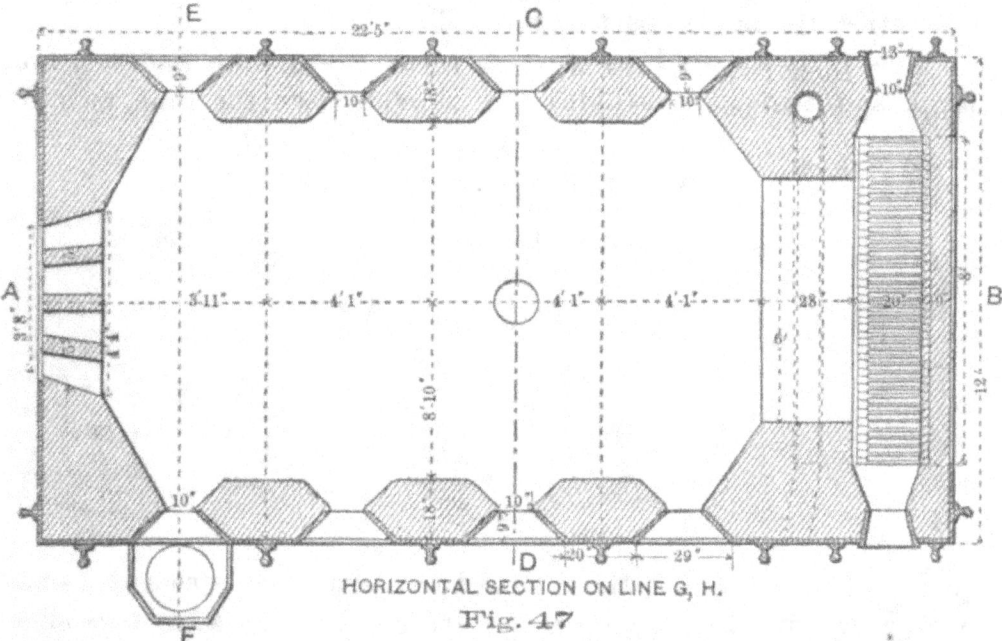

HORIZONTAL SECTION ON LINE G, H.
Fig. 47

features. The horizontal section (Figure 47) shows the rectangular form of the furnace with four working openings on either side, the well being below the door nearest the flue end, the coolest part of the furnace. The grate bars are placed parallel with the axis of the furnace. The fire is stoked at the end and fed from both sides. Four branch flues lead the gases into one main flue. The bridge-wall has an air-passage. The entire furnace is encased in iron plates and bound by buckstays (iron rails) and tie-rods. In the longitudinal section, Figure 48, the roof is seen to form a horizontal line from the fireplace to the beginning of the hearth, sloping thence in a straight line to the flues. The hearth, in Figures 48 to 50, is built up differently from that of any other furnace.

First a layer of sand is tamped down between the furnace-walls so as to have the form of a trough inclining from bridge to flue. At the well the brick bottom, of red-brick set dry, replaces some of the sand. It forms a good support for the brasque bottom, which is followed by the slag bottom, consisting of tap-cinder melted down in the furnace. This covers the brasque bottom, wherever it would otherwise come in contact with working-tools, leaving it exposed only at the well (Figure 49). During the run some residue adheres to this slag bottom and forms the smooth working bottom of the furnace. The hearth is seen in Figure 50 to be trough-shaped from the bridge to the second working-door; from there the front part (Figure 49) slopes down to the level of the rim of the kettle, the lowest part of the well.

The tools required to work a furnace are four paddles, four large and two small rabbles, five shovels (two for lime, two for

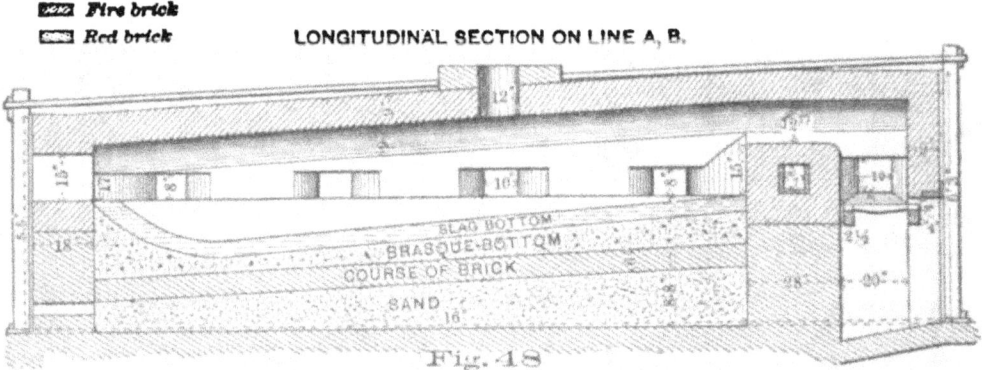

Fig. 48

coal, and one for slag), three steel bars (two large ones and a small one), a tapping-bar, a skimmer, a ladle, a sample-ladle, two slice-bars, a sledge, two hammers, and four hooks for handling the lead.

Method of Working.—The furnace, if new, is heated to a dark-red heat; the damper is then closed and the charge, crushed to pass a 5-mesh sieve, is let fall from the hopper through the opening in the roof and spread out evenly over the hearth by means of rabbles to a thickness of from 3 to 4 inches. The fire is fed with cinders and the temperature never allowed to exceed dark-redness, say 500 or 600 degrees Centigrade. The galena decrepitates, the temperature rises, the ore becomes a dark-red, and after from three-quarters of an hour to an hour the roasting begins. During this time the charge is turned over once with the paddle. The working-doors and fire-door are opened, and the damper raised sufficiently to allow the sulphurous acid and other gases to pass off. The ore roasts on the surface. When the fumes begin to cease, samples are taken to see if a white crust of oxide and sulphate has been formed. This indicates that it is time to renew the surface, which is generally done by paddling or rabbling every twenty or twenty-five minutes, *i.e.*, about eight or nine times during the three or four hours required for roasting. Care is taken to prevent the ore from clotting.

Up to 1885 the roasting usually took about three or three and a half hours, but the gradual increase of blende in the ore brought the required time up to four hours. In 1886 it became necessary to make a change in working the charge. This could be done in two ways. Either the normal time could be prolonged or part of the ore roasted separately. The latter method was chosen, and now one-quarter of the charge, fine concentrates especially rich in blende, is roasted separately in a long-hearth cal-

VERTICAL SECTION ON LINE E, F.

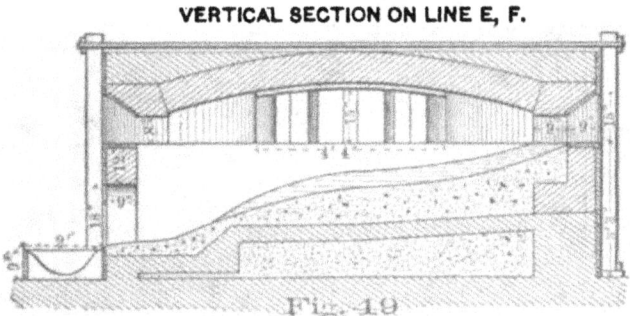

Fig. 49

cining furnace and then added to the ore charge. Towards the end of the roasting period from 330 to 660 pounds of carbonate ore or fluedust, with 45 per cent. of lead, are added to increase the amount of oxides in the charge.

The grate is now cleaned, well filled with good coal, and the damper opened to raise the temperature. The reactions now set in. The roasted ore gradually softens, white fumes are given off, and lead begins to flow. Lime is added to prevent the lead from carrying down to the well particles of ore floating on top of it and to prevent liquefying of the charge. The lime is well worked in and the charge then turned over. The furnace becomes filled with fumes of lead, the damper is raised to remove these, and fusing of the charge is prevented by regulating the fire and the damper and by adding a few shovelfuls of lime at a time. At from an hour to an hour and a quarter after the reaction has set in (three hours with a new furnace) the well is

full of lead. This is tapped into the outside kettle. The dross
floating on the surface is skimmed off and put back into the furnace and the impurities held in suspension by the lead are removed
by stirring in slack coal (poling). This second dross is put aside
and the lead ladled into moulds. During the reaction stage the
working-doors have to be open on account of the necessary rabbling and paddling ; thus air enters, cools the furnace, oxidizes the
charge and the flow of lead begins to cease.

At this stage drosses obtained from melting down base bullion
in the desilverizing plant are added. They consist principally of a
mixture of lead sulphide and lead, assaying 75 per cent. lead and 9
ounces silver per ton. At the same time 1,100 pounds of oxides
low in silver, containing from 75 to 80 per cent. lead oxide and
from 19 to 20 per cent. zinc oxide, are charged. They result from

Fig. 50

VERTICAL SECTION ON LINE C, D.

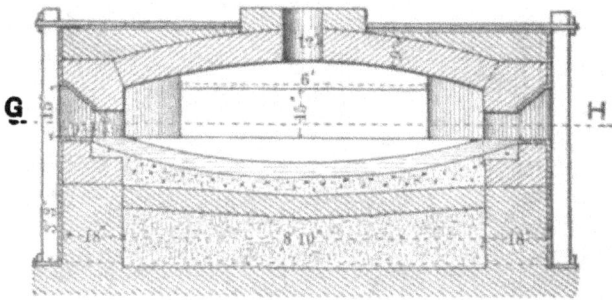

refining lead, desilverized by the Parkes process, by means of
steam (§ 107).

The doors are then closed and fuel is piled up high on the
grate to insure a smoky flame. This reduces lead sulphate to sulphide, and soon the flow of lead begins again. The doors are
opened, the charge is worked, and its liquefying prevented as before. From two and a half to three and a quarter hours after the
first reaction sets in the well is again filled with lead. This is
tapped and lead and dross are treated as before. In this way all together five reactions are caused, the amount of lead obtained becoming gradually smaller and the charge more dry. Before the
last reaction takes places, the mixture of dross and fine coal, obtained from poling the different leads in the kettle, is thrown into
the furnace. It is worked into the charge, which is then covered
with fine coal. The temperature is raised and the last yield of lead
obtained. The damper is now opened fully and the residue drawn
from the back of the furnace into a box filled with water, so as to

prevent fumes of lead from passing into the building. Care has to
be taken to avoid explosions. The hearth is repaired with a mixture
of crushed residue and slacked lime, which is beaten down with the
paddle and rammed into holes with a bent bar. When this is finished, a little lime is spread over the hearth and a new charge dropped from the hopper. The whole reaction period lasts seven hours.

If the hearth should become so incrusted as to raise the charge
too high to be protected by the fire-bridge from the direct action
of the flame, the temperature must be raised so as to soften the accumulated residue that it may be removed. A hearth lasts about
three months.

The results, from one year's (1865) work in a smaller furnace, 11
feet 9 inches by 10 feet 10 inches, with six doors, and with ore low
in blende and without the different additions, were : 10,000 pounds
of ore (assaying 72.97 per cent. lead and 21.76 ounces silver) gave
6,384 pounds base bullion, assaying 32.95 ounces silver ; 1,592
pounds residue (38.80 per cent. lead and 3.93 ounces silver) ; 275
pounds fluedust (50.00 per cent. lead and 0.02 ounces silver) ;
showing that in the reverberatory 87.49 per cent. of all the lead
and 99.9156 per cent. of all the silver was recovered in the form of
base bullion. Of the 12.51 per cent. of lead, forming the difference, 9.31 were recovered in the blast-furnace, so that the actual
total loss amounted only to 3.2 per cent.

The consumption of fuel was 4,600 pounds, or 46 per cent. of
the ore charged, and of lime 100 pounds or 1 per cent. Additional
data are given in the next paragraph.

The following analysis of residue was made by Pietsch in 1865:
PbO, 24.375 ; $PbSO_4$, 13.269 ; $PbSiO_3$, 12.373 ; ZnO, 22.857 ;
FeO, 8.957 ; FeS, 1.823 ; CaO, 11.190 ; C, 4.821 ; Al_2O_3 and MnO,
traces ; silver, 0.015 (= 4.360 ounces per ton). Total, 99.680.

The fluedust, in 1881–1882, forming 2.91 per cent. of the ore
charged, had the following composition, according to Dobers and
Dziegiecki :

PbO..........	66.44..........	66.53..........	65.79
ZnO..........	3.77..........	4.55..........	4.05
Fe_2O_3........	0.50.........	0.80..........	0.85
SO_3..........	27.48..........	27.11..........	27.79
Insol...........	2.14..........	1.20..........	2.30
	100.33	99.69	100.28

§ 45. **Comparison of Reverberatory Methods.**—To facilitate comparison, the main data of the furnaces discussed have been
brought together in the subjoined table:

EXAMPLES OF SMELTING IN REVERBERATORY FURNACES.

Items.	Carinthian Method.					English Method.		Silesian Method.	
								Tarnowitz, Silesia.	
	Raibl, Carinthia.	Engis, Belgium.	Granby, Missouri.	South-west Missouri.	Southeast Missouri.	Holywell, Flintshire.	Stiper-stones, Shropshire.	Old Fur-nace.	New Furnace.
The Furnace.									
Hearth, length	11'	6' 6¾''	9'	9'	10'	12'	9' 9''	11' 9''	16'
" width at bridge	5'	4' 3⅛''	3' 2''	3'	2' 6''	3' 6''	4'	....	6'
" " " middle	5'	4' 4¼''	3' 2''	3'	4'	9' 6''	9' 6''	10' 10''	8' 10''
" " " flue	1'	2' 3½''	3' 2''	3'	2'	4'	3' 3''	Similar to new furnace.	4' 4''
" thickness of bed	6''	1' 5¼''-2' 8¼''	6''	6''	....	2''-10''-26''	4''-18''-30''		4''
" life of bed, weeks	4 5	6	....	....	....	....	....		13
" inclination to Side or Flue	21'' F	17'' F	18'' F	— F	— F	29 S	24 S		27⅛ F
Grate, length	7'	3' 7¼''	5' 6''	6' 6''	3'	4' 6''	4' 6''		8'
" width	1' 3⅜''	1' 6''	2' 6''	2' 6''	2'	2' 6''	2' 6''		1' 8''
" depth below top of bridge	1' 2⅜''	3' 7¼''	....	6''	....	2' 9''	1' 4''		3'
Bridge, width	5⅝''	1' 3¾''	8''-30''	....	21''	24''	24''		28''
" length	3' 7¼''	3' 11¼''	30''	30''	24''	42''	48''		72''
" height above hearth	4-5''	none.	....	6''	....	12''	12''		15''
Roof, height above bridge	9''	8⅝''	14''	....	6''-8''	9''	19'' †		12''
" " " hearth at flue	12''	2' ⅜''	14''	....	6''-8''	16''	20''		17''
Flue (leading to chimney), size	1' x 1' 3¾''	8¾'' x 2' 3½''	8'' x 12''	....	12'' diam.	{ 12'' x 10'' 10'' x 10''	12'' x 12'' 11'' x 12'' }		4(15'' x 20'')
Chimney, inside size	1' 8'' x 1' 8''	1' 3'' x 1'	....	....	12'' diam.	8' x 3' 6''	....		....
" height	20'	....	....	20'	15'	....	....		....
Grate area, ratio to hearth area	1 : 6	1 : 5.25	1 : 2.0	1 : 1.66	1 : 5.4	1 : 8	1 : 6.6		1 : 10
" " " sect'n of chimney	1 : 0.14	1 : 0.47	....	....	1 : 0.13	1 : 0.93	...		....

Work Done.									
Charge, weight in lbs.	370–430	1,375	1,500	1,500	...	2,350	2,350	4,400	8,250
" assay, per cent. lead	70	78	...	...	...	75–90	77.5	70–74	...
" thickness on hearth	1¼"–1¾"	about 3"	...	...	...	1⅝"?	3⅜"–4⅛"	3"–4"	3"–4"
Operation, length in hours	10–12	12	12	9–11	...	8	7	12	12
Men, number in 24 hours	1.5	4	2	2	...	6	4	4	6
Gray slag, amount in lbs.	95	275	...	...	...	288	...	600	2,475
" per cent. of charge	24	12	...	...	...	12	...	15	30
" assay, per cent. lead	4	17–20	55	...	...	55	...	88.8	56
Of lead charged, recovered per ct. In bars	98	92.5	...	...	...	81	88	87.5	50
From gray slag and fluedust	...	2.8	...	...	...	8	...	9.3	...
Total loss in lead	7	5.2	...	...	...	11	12	3.2	...
Outlay for Labor, Fuel and Material per ton of ore of 2,000 lbs. avoirdupois. Labor, hours	90	61	36 aver.	20	...	15	12	11	9
Fuel, charcoal, lbs.	...	...	...	...	...	...	...	...	...
" bit. coal, tons	1.60	1.10	1.10 aver.	0.83	...	0.57–0.76	0.56	0.46	...
" wood, cords	...	...	...	...	...	...	...	...	...
Material, lime, lbs.	...	...	...	...	...	...	...	20	...
Reference	§ 42, A.	§ 42, B.	§ 42, C.	§ 42, C.	*	§ 48	§ 43	§ 44	§ 44

* Kemp: *School of Mines Quarterly,* ix., p. 212.   † Height of roof above grate, 38".

In comparing the amounts of ore treated in twelve hours at the different smelting works, the order in which they are placed in the table shows a steady increase from Raibl to Tarnowitz. The figure for Tarnowitz, 8,250 pounds, requires the explanation that the charge contains a considerable proportion of oxidized ore, which shortens the time required for roasting. If the ore were pure galena, twice the time, or eight hours, would have to be allowed. This would make the amount for twelve hours 6,187 pounds. The large amount of ore treated is due to the size of the furnace. As to the amount of labor required per ton of ore, the table shows that Tarnowitz uses less and Raibl more than any of the other smelting works. With fuel the same is the case. As the wear and tear of a furnace depend on the height of the temperature, the Carinthian and Silesian furnaces will outlast the English ones, other things being equal. If, finally, the amount of lead recovered is considered, it is clear that slow roasting yields more than quick roasting (Carinthian and Silesian *vs.* English), and that recovering in the reverberatory only that amount of lead obtainable at a low temperature and not melting the charge, gives a higher percentage of lead (Tarnowitz and Engis *vs.* Raibl and English methods). The inferences to be drawn are clear. They are stated by Cahen,[1] and Grüner[2] has formulated them at length. He says (Percy's translation[3]): "Rich, pure, non-quartzose ores ought always to be treated by this latter (the roasting and reaction) method. The operation ought to take place in large reverberatory furnaces, with easy access of air, provided with a single fireplace and a receiving basin, internal or external, placed in the least heated region of the furnace. The operation ought always to be conducted slowly, and to consist of two phases very distinct, *roasting* and *reaction*. For roasting, the layer of ore must never exceed from 3.15′ to 3.54′ in thickness. Roasting is to be effected at a low temperature, and ought to proceed as far as the theoretical limit of one equivalent of sulphate, or two equivalents of oxide, for each equivalent of sulphide. After the first firing, which produces lead, and fresh roastings and firings twice or thrice repeated, the rich residues must be withdrawn from the reverberatory furnace, without having recourse to *ressuage* (reduction of rich residue in the same furnace immediately afterwards), but rather by practising this *ressuage* in a blast-furnace."

[1] *Op. cit.*, p. 117.
[2] *Op. cit.*, reprint, p. 511.
[3] "Lead," p. 491.

Why in our own lead districts the reverberatory practice has not made more headway is to be answered in two ways. In the lead-silver districts the ores have on the whole not been of sufficient purity and richness to warrant the use of reverberatory methods. In the Mississippi valley, where the ore is just of the quality required for the process, the question of skilled labor has had some influence, but whether justly so is very doubtful.

CHAPTER VII.

SMELTING IN THE ORE-HEARTH.

§ 46. **Introductory Remarks.**—The process carried on in the ore-hearth is the roasting and reaction process. It resembles that in the reverberatory furnace, with this difference, that oxidation and reduction go on simultaneously, the charge floating on a bath of lead. The lead oxide and sulphate, as soon as formed, react on undecomposed sulphide, and the liberated lead trickles through the charge into the hearth-bottom, overflowing into an outside kettle.

The same conditions are necessary for the hearth-treatment as for the reverberatory, with the exception that the ore should be coarser. The smallest permissible size is that of a pea, and nut-size is desirable. If fine ore is to be treated, it must first be agglomerated in a reverberatory, as in the ore-hearth it would be blown away. It requires power and a blower. Then much lead is volatilized; hence it is not suited for argentiferous ores. The comparative loss at Raibl[1] where ore-hearth and reverberatory work on the same ore, was, in 1888, 10.6 to 7.3 per cent. It requires purer and higher grade ore than the reverberatory, but consumes less fuel. According to Tunner,[2] a furnace similar to the one at Raibl consumed, per 100 pounds of galena, 7.31 cubic feet of wood, while the ore-hearth required only 2.90 cubic feet. The ore-hearth has three times the capacity of the air-furnace. The cost of treatment per ton of ore is the same with a single ore-hearth; with several running side by side it becomes less, as only one set of men is necessary to run the more powerful engine, and this consumes relatively less steam than the smaller one. That the ore-hearth cannot compete with the English or Silesian furnaces as regards capacity and cost, is clear. It has, however, one advantage over all reverberatories, that it is quickly started and stopped without

---

[1] *Oesterreichische Zeitschrift für Berg- u. Hütten-Wesen*, 1888, p. 329.
[2] *Leobener Jahrbuch*, 1852, i., p. 262.

much consumption of fuel or loss in heat, and thus serves its purpose in extracting at intervals from small amounts of non-argentiferous ore the major part of the lead. This is probably the reason why it found such favor in the Mississippi Valley, where small amounts of ore were often treated, and still are, by men who have mined it themselves.

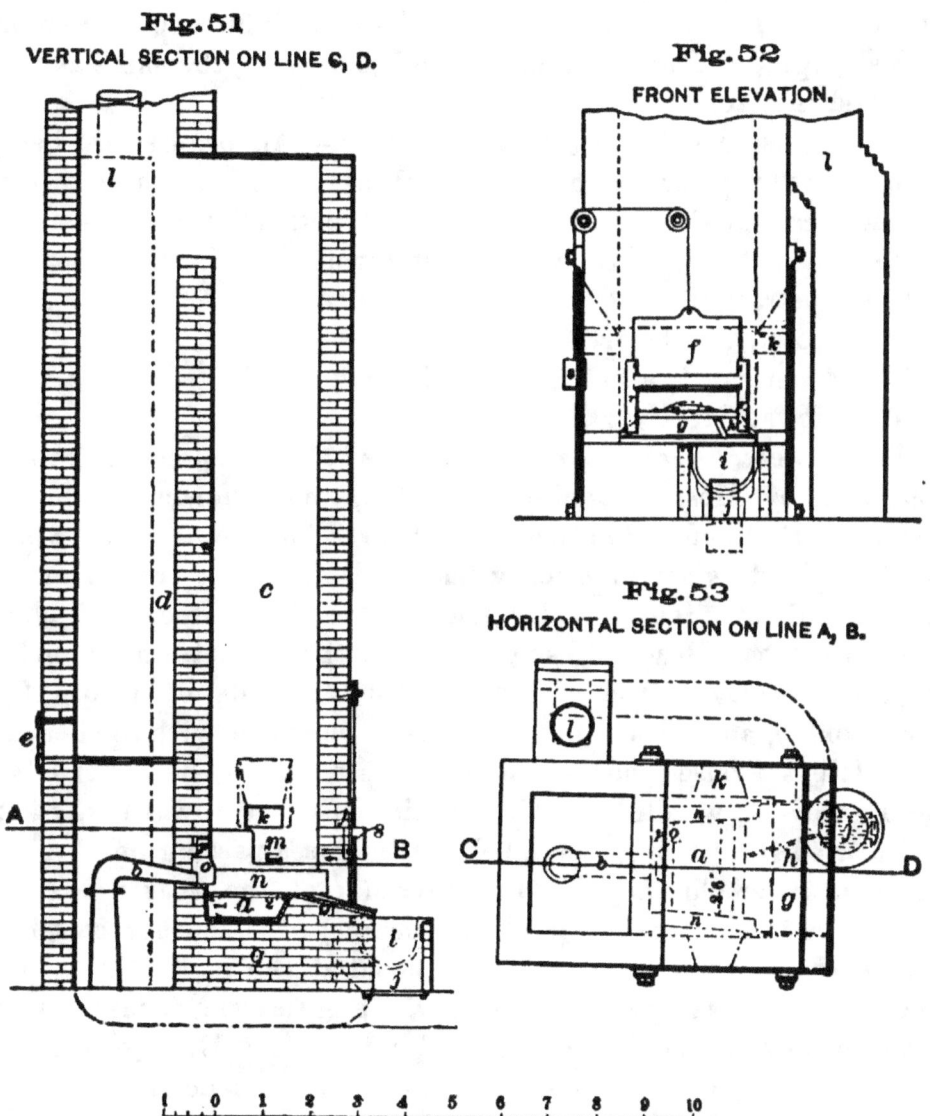

Fig. 51
VERTICAL SECTION ON LINE C, D.

Fig. 52
FRONT ELEVATION.

Fig. 53
HORIZONTAL SECTION ON LINE A, B.

The ore-hearth as worked at Joplin, Mo., is exceptional and will be treated further on (§ 51).

The practice in the different ore-hearths is so very much the same that nothing from a general point of view need be said about it.

The products are similar to those of the reverberatory; there is, however, an intermediary product, a mixture of ore, slag, and fuel, called browse in England, which goes back to the charge in the ore-hearth. The slag contains much lead, and is smelted in a small blast-furnace, the slag-eye furnace.

§ 47. Influence of Foreign Matter.—It has already been said that the ore subjected to hearth treatment must be purer and richer than is necessary for the reverberatory furnace. This is because the foreign matter shows its bad influence in a more marked degree. The chemical action is, however, the same as that described in § 40.

§ 48. Description of Ore-Hearths.—An ore-hearth, being a small low fireplace surrounded by three walls, with a tuyere at the back, cannot show much variety in construction or in manner of working. Three slightly different forms have been chosen by way of illustration :

The Scotch Ore-Hearth,

The American Water-Back Ore-Hearth ; and

The Moffet Ore-Hearth.

a. *The Scotch Ore-Hearth.*[1]—Figures[2] 51–53 represent the furnace used by Messrs. Cookson and Co., near Newcastle, in the North of England. The cast-iron hearth-box or well *a*, which holds the lead, is set in brick-work *q*. It is 2 feet from front to back and about 2 feet 6 inches wide ; it is 1 foot deep and holds about two tons of lead. In some furnaces the depth is only 6 inches and the capacity of the well about 1,340 pounds of lead. The work-stone *g*, an inclined plate, is cast in one piece with the hearth-box. It has a raised border on either side and at the lower edge, and a groove *h*, which leads the overflowing lead towards the kettle *i*, heated from a fireplace *j*, below, the gases passing off through a flue into the chimney *l*. On either side of the hearth-box and resting on it is a cast-iron block (bearer) *n*. Another cast-iron block *o* (back-stone, pipe-stone) is placed at the back. It is perforated for the passage of tuyere *b*, which enters the furnace about 2 inches above the surface of the lead in the well. Upon *o* rests the upper back-stone *p*, also of cast-iron. Thus lead and ore are entirely surrounded by cast-iron. The fore-stone *m* appears to be rather small. The brick shaft *c* carries off the fumes ; at the

[1] Percy, "Lead," p. 278.

[2] "Eighth Annual Report of the Local Government Board," 1878-79, supplement containing the Report of the Medical Officer for 1878, London, 1879, p. 281.

back is a "blind flue or pit" *d*, forming a sort of dust-chamber, to· be cleaned through door *e*, the gases passing off upward. On the side of the shaft is the feed-door *k*, for the introduction of fuel in front of the tuyere and the removal of any slag adhering to it, the charge being fed from the front. The hearth has at the front a shutter *f*, sliding in a grooved frame *r ;* it is raised and lowered by means of counterpoise *s*. Peat, formerly used as fuel, is now replaced by bituminous coal.

It is characteristic for the method of working in the Scotch hearth that the process is non-continuous. After smelting from twelve to fifteen hours the hearth becomes too hot ; it has to cool for about five hours before work can be started again.

VERTICAL SECTION.

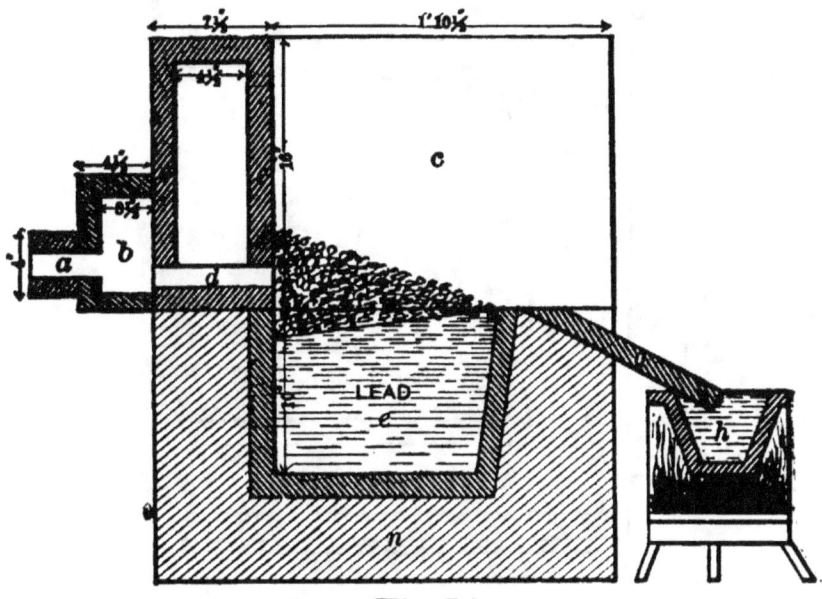

Fig. 54

b. *The American Water-Back Ore-Hearth.*[1]—Figures[2] 54 and 55 show the larger-sized furnace with three tuyere-holes *m* at the back, where the older, smaller form had only one. The hearth-box *e* (filled with lead, the charge of ore, and charcoal floating on top) is here also set in brickwork *n*. It holds about 2,500 pounds of lead. The work-stone *g*, leading to the kettle *h*, forms a separate casting from the hearth-box. The three sides of the furnace are formed by a water-cooled cast-iron jacket, *c c*, $1\frac{1}{2}$ inches thick, called tuyere-plate, resting on the hearth-box. The water enters at *i*, and

[1] Williams : "Industrial Report," "Geological Survey of Missouri," 1877, p. 36.—*Trans. A. I. M. E.*, v., p. 324.

[2] Broadhead : "Geological Survey of Missouri," 1873–1874, p. 492.

passes out at *k*. At the back of the tuyere-plate is the wind-box *b*. The blast enters this at *a*, and passes through three wrought-iron tuyere nozzles *d* (from 1 to 1½ inches in diameter) in tuyere holes *m*, into the hearth, at from 1 to 3 inches above the level of the lead. The hood placed over the furnace to carry off the fumes and gases is not shown in the drawings.

The work in the American ore-hearth is continuous, as distinguished from the Scotch hearth. This is made possible by water-cooling the sides of the furnace which protect the castings and the tuyeres. The fuel used is wood, charcoal, and bituminous coal.

PLAN.

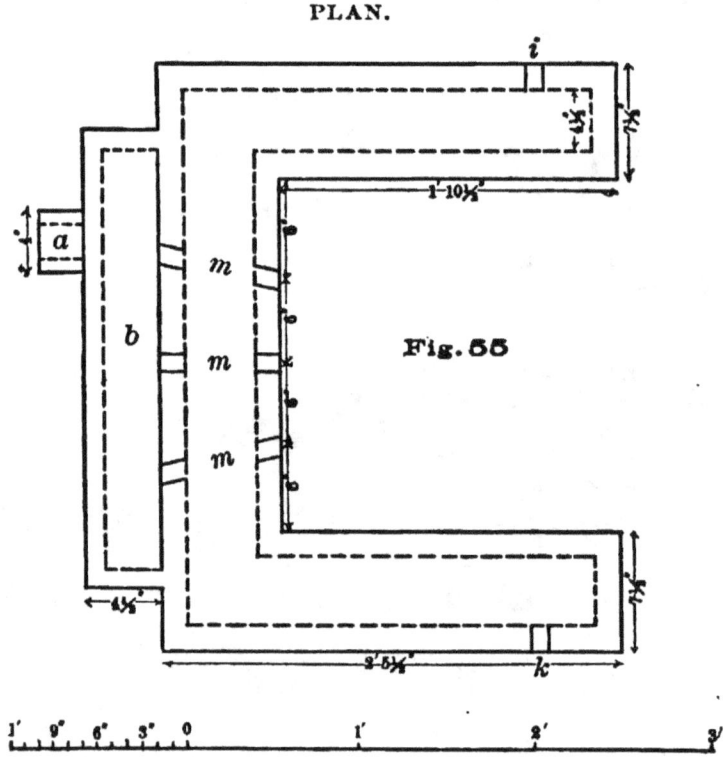

Fig. 55

In the ore-hearth used formerly at Rossie,[1] N. Y., and for some time, also, in the Mississippi Valley, but entirely abolished now, this cooling was effected by letting the blast circulate instead of the water. The resulting hot blast caused much volatilization of lead.

c. *The Moffet Ore-Hearth*[2] (Figures 56 to 59).—In Figures 56 and 57 the entire furnace is seen to rest on four pillars.

---

[1] Percy, "Lead," p. 289.

[2] Dewey, *Trans. A. I. M. E.*, xviii. p. 674; Clerc, *Engineering and Mining Journal*, July 4, 1885; Ramsay, *Scientific American Supplement*, May 14, 1887, No. 593.

The hearth-box (lead basin) is not set in brickwork, and the lead is thus kept cool. Two furnaces are set back to back, the fumes passing off under one hood. The lead runs through a separate spout near the top of the work-stone into a cast-iron kettle 31 inches in diameter and 44 inches deep, not shown in the drawings. It is set in a

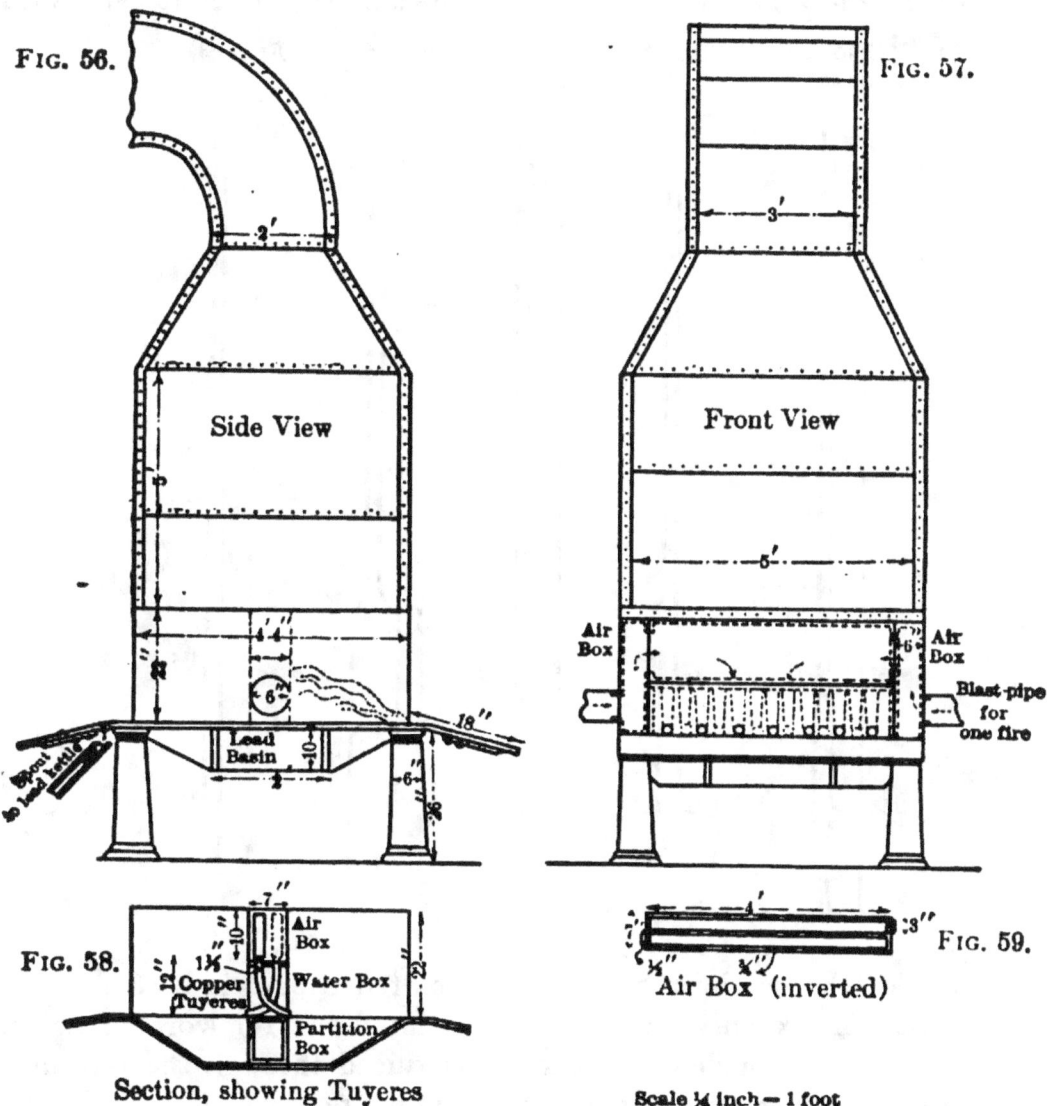

FIG. 56.

Side View

FIG. 57.

Front View

Air Box

Air Box

Blast-pipe for one fire

FIG. 58.

Air Box

Copper Tuyeres

Water Box

Partition Box

Section, showing Tuyeres

Air Box (inverted)

FIG. 59.

Scale ¼ inch = 1 foot

cast-iron cylinder and heated from below, one cord of wood a week being required for the purpose. The two furnaces work independently. On the bottom of the lead basin rests the partition-box, having an opening near the bottom for the lead to enter. It serves as a support for the water-box, which cools the hottest part of the furnace, and upon which rests the air-box, consisting of two separate chambers (Figure 59), where the heated blast passes down through the water-box by means of fourteen 1-inch copper tuyere pipes, seven on either side. The working opening is 15 inches high. A No. 5 Baker blower furnishes the blast. The fuel used is bituminous coal. The reasons for the hot blast will be given in § 51.

The tools, a long handled, round-pointed shovel, a round bar, a paddle, and a square bar are shown in Figures 60 to 64. With

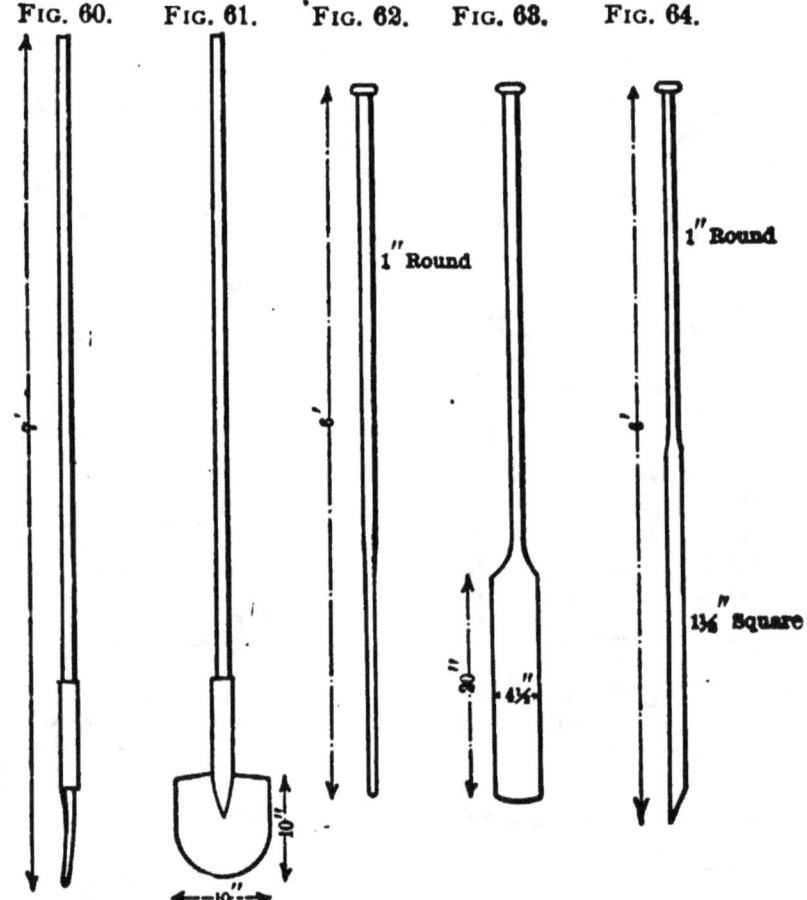

FIG. 60.　　FIG. 61.　　FIG. 62.　　FIG. 63.　　FIG. 64.

the Scotch[1] ore-hearth the round-pointed shovel is used at some works for ore only, and a square-pointed one for working in the furnace; then a short scraper is introduced through the feed-door to remove slag adhering to the tuyere nozzle.

§ 49. Mode of Working in the Ore-Hearth.—A fire is first kindled with wood; then coal is added, and the blast started. This will soon set the fuel in a blaze; more coal is then added, ashes and clinkers are removed, and in a short time a body of glowing fuel is obtained, filling the entire hearth. Some residue from a previous run is then spread over the back part of the fire

[1] Percy, "Lead," p. 282.

and the first charge, from ten to twenty pounds of ore, given. This soon becomes red-hot, and the lead that is set free trickles through the body of the fuel and collects on the bottom. More ore is added. The contents of the hearth are then lifted up with the bar and kept open in order that the heat may be distributed through the entire mass. Parts that have become melted and form lumps are drawn out with the shovel on to the work-stone, and the slag is separated from rich residue, which is returned to the furnace. Ore and fuel are again added, and the operations continued until the hearth is full of lead.

The hearth is now in normal working order; it is filled with lead, and glowing fuel, mixed with partly fused and partly reduced ore, floats on top of it. From twelve to thirty pounds of ore are mixed with $1\frac{1}{2}$ or 2 per cent. of lime, spread over the glowing floating mass, and then covered with a little fuel. It is now exposed for from three to five minutes to the action of heat and blast. After this one of the furnace-men inserts the bar at different places into the lead, loosens and stirs the charge, and raises it slowly, for which purpose the paddle is sometimes used. The other man draws the semi-fused mass out from below with the shovel upon the work-stone; this allows the ore on the surface to sink down. What has been drawn upon the work-stone is broken up and the slag separated from the half-decomposed ore. The former is thrown aside (sometimes into a water-box); the latter goes back to the furnace, slacked lime being spread over it, if necessary. Any slag adhering to the tuyeres is then removed; some fuel is distributed in front of them and over the charge; fresh ore is spread on the fuel, and this again covered with fine fuel; then all is ready for a second operation. As the smelting proceeds the bulk of the lead that is set free trickles through the charge into the hearth-box and overflows through the groove in the work-stone into the kettle. Here it is sometimes poled before being ladled, syphoned, or drawn off through a spout into moulds. Some lead passes off with the fumes and the rest goes into the slag.

Pattinson[1] calls attention to the following points in managing the ore-hearth. The amount of blast and its distribution through the entire charge should be carefully regulated, the half-reduced ore should be exposed on the work-stone to the oxidizing action of the air, and the additions of lime and fuel judiciously made.

[1] Percy, "Lead," p. 288.

Two men work as partners in eight-hour shifts. The results obtained with different furnaces are given in the following table:

EXAMPLES OF SMELTING IN ORE-HEARTHS.

NAME.	FURNACE. Front to back.	Width.	Depth.	Sides.	No. of tuyeres.	Ore in 24 hours, lbs.	PRODUCTS. Lead in 24 hours, lbs.	Gray slag in 24 hours, lbs.	Per cent. of lead direct from ore.	Men in 24 hours.	MATERIAL CONSUMED. Bitum. coal, lbs.	Peat, bushels.	Charcoal, bushels.	Wood, cords.	Lime, bushels.	Reference.
Keld Head Min'g Co., England	23″	21″	12″	Cast-iron	1	10,752	8,064	602	74.44	6	374	12	....	....	....	1
Rossie Works, N.Y.	24″	24″	12″	Air-jacket	1	9,988	7,389	....	73.88	4	....	....	....	¾	....	2
Granby, Mo.	22½″	24″	11″	Water-jacket	3	9,000	7,500	....	83.90?	6	....	....	27.6	....	....	3
Hopewell, Mo.	....	20″	12″	Water-jacket	1	6,000	1,464	....	73.20	6	....	....	20.0?	....	2.5	4
Lone Elm, Mo.	22″	48″	10″	Water and air jacket	7	27,000	12,032	5,247	45.00	6	2,160	....	....	....	5.40 lbs.	5

1. Percy, "Lead," pp. 278-283.   2. Kerl, "Metallhüttenkunde," p. 30.   3. Williams, "Industrial Report," p. 68.
4. Williams, Op. cit., p. 65.   5. Dewey, Trans. A. I. M. E., xviii, p. 674.

To the above statements of labor and fuel have to be added the engine-men and the fuel consumed under the boiler, which is one cord of wood for the three-tuyere Granby ore-hearth ("Broadhead's Report," p. 494).

The composition of some Missouri leads smelted in the ore-hearth is as follows:

	GRANBY.[1]	HOPEWELL.[2]	LONE ELM.[3]
As......	0.00124	0.00583	0.00011
Sb......	0.01085	0.00808	0.00146
Ag......	0.00057	0.00219	0.00056
Cu......	0.00780	0.00585	0.01782
Ni......	0.00087	trace	0.00077
Co......	—	—	0.00005
Fe......	0.00367	0.00145	0.00686
Zn......	trace	0.00156	0.00033
S.......	trace	—	—
Pb......	99.96905	99.97509	99.97204
by difference.	————	————	————
	100.00000	100.00000	100.00000

1. William's "Report," p. 63; 2. *Trans. A. I. M. E.*, v., p. 326; 3. *Trans. A. I. M. E.*, xviii. p. 687.

§ 50. **Treatment of Slags.**[1]—The gray slags obtained in smelting lead ores in the ore-hearth consist of lumps of a more or less scorified mixture of gangue and various lead (zinc) compounds, which contains mechanically enclosed lead and often particles of fuel. The mechanical analysis of Lone Elm slag[2] gave 21.45 per cent. metallic lead and the remaining 78.55 per cent. of pulp showed the following composition:

Residue.	Per cent.	Per cent.
SiO_2...................................	1.97	
$PbSO_4$...................................	0.24	
Fe_2O_3...................................	1.67	
Al_2O_3...................................	0.21	
ZnO...................................	0.57	4.66
$PbSO_4$...................................		4.94
Acetic Acid Solution.		
SiO_2...................................	10.73	
PbO...................................	33.55	
Fe_2O_3...................................	1.23	
Al_2O_3...................................	0.57	
ZnO...................................	18.96	
CaO...................................	11.49	
MgO...................................	0.12	71.65
Nitric Acid Solution.		
PbS...................................	14.73	
FeS_2...................................	0.67	
ZnS...................................	3.64	19.04
		100.29

[1] Bergen, in Commissioner Raymond's "Report," 1875, p. 424.
[2] *Trans. A. I. M. E.*, xviii., p. 685.

The gray slag is smelted in a low rectangular blast-furnace, usually 4 feet in height, having an external crucible (the cast-iron lead-pot) and one tuyere at the back. The bed-plate, sloping from back to front, carries the fire-brick lining of the furnace, and over-laps the lead-pot. A brasque bottom, consisting of equal volumes of clay and coke, is tamped on it, 5 inches at the back narrow-ing down to 1 inch at the front. The front of the furnace has a cast-iron plate with an opening near the bottom. This is closed by ramming in a breast of clay over a wooden plug. The lead-pot is divided into two unequal parts by a partition descending nearly to the bottom. The melted charge, "black slag," flowing through the opening made by the wooden plug, passes over the larger divis-ion, filled with charcoal, into a tank, through which a slow stream of water flows. The lead filters through the charcoal, collects at the bottom, and is removed at intervals from the smaller division. A hood serves to carry off the fumes. The dimensions of the furnace are:

	Inches.
Width of hearth	26
Depth, front to back	36
Height	46
Diameter of tuyere	3
Height of tuyere above bed-plate	10
Height of pozt	9
Width of pozt	12
Prolonged axis of tuyere, strikes front lining of fur-nace above pozt	2
Slope of bed-plate to 1 foot	1½
Size of bed-plate	40 x 48
Size of front plate	30 x 36
Long diameter of lead-pot	36
Front to back of lead-pot	12
Depth of lead-pot	12
Thickness of lining, back and sides	9
Thickness of lining, front	5

The starting of the furnace is very simple. A wood fire is kindled on the bottom, the furnace filled with clean charcoal, and some blast let on. When the charcoal is thoroughly ignited, a 6-inch bed of coke is placed on top and full blast (9 ounces) given. Clean slag of moderate fineness is then charged with the necessary coke, four shovels of slag to one of coke. To save the walls of the furnace from being eaten out too quickly, the slag is placed rather around the back and the sides of the furnace, and the coke in the front and in the centre. The wooden plug in the clay breast is

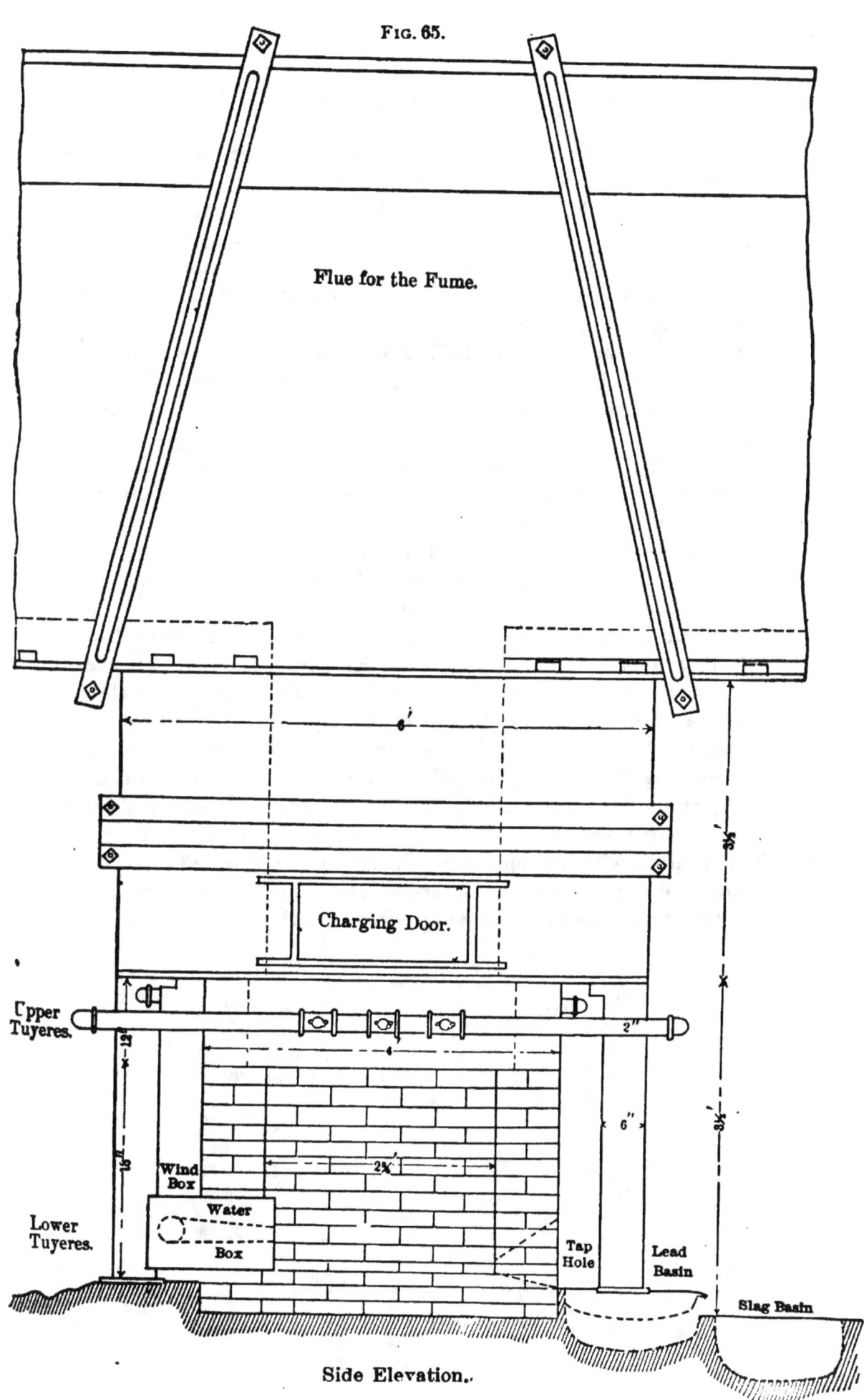

FIG. 65.

Flue for the Fume.

Charging Door.

Upper
Tuyeres.

Lower
Tuyeres.

Wind
Box

Water

Box

Tap
Hole

Lead
Basin

Slag Basin

Side Elevation.

withdrawn half or three-quarters of an hour after the full blast is given, when the "black slag" that had collected will begin to flow. During the run the charcoal in the lead-pot is removed every two or three hours. After smelting about sixteen hours the lining has become corroded, and infusible masses have accumulated on the bottom. The charging is then stopped and the furnace blown out by running off all the slag, tearing out the breast, drawing the rest of the charge, and chilling with water. After allowing it to cool for eight hours, adhering crusts are removed with steel bars, the furnace is patched, and it is then ready for another run.

From fifteen to Eighteen and a half tons of slag, assaying from 35 to 40 per cent. of lead are smelted in sixteen hours, giving 7,500 pounds of hard lead.

Two furnace-men, four helpers, and one roustabout form the crew; the fuel consumed is two tons of coke and twenty-two bushels of charcoal, besides that for the boiler.

A more modern slag-eye furnace, shown in Figures 65 to 67, is the one used at the Lone Elm works. It is in many respects similar to the one just described. The furnace has eleven tuyeres of $1\frac{1}{4}$ inches in diameter, passing through water-boxes; seven of these are below the charging-door and four above the bottom, the blast being heated. For the reason of this see next paragraph. The tools used are three 6-foot bars, three shorter ones, two shovels, and a ladle. A furnace runs from fifteen to twenty days without repairing. In the summer of 1891 the brick walls were replaced by water-jackets, which greatly prolongs the life of the furnace. The composition [1] of the final "black slag" is:

SiO_2	42.10
PbO	25.37
FeO	7.91
Al_2O_3	9.58
$(Ni,Co)O$	traces
MnO	0.27
ZnO	4.48
CaO	7.97
MgO	1.66
S	0.22
	99.56

[1] *Trans. A. I. M. E.*, xviii., p. 694.

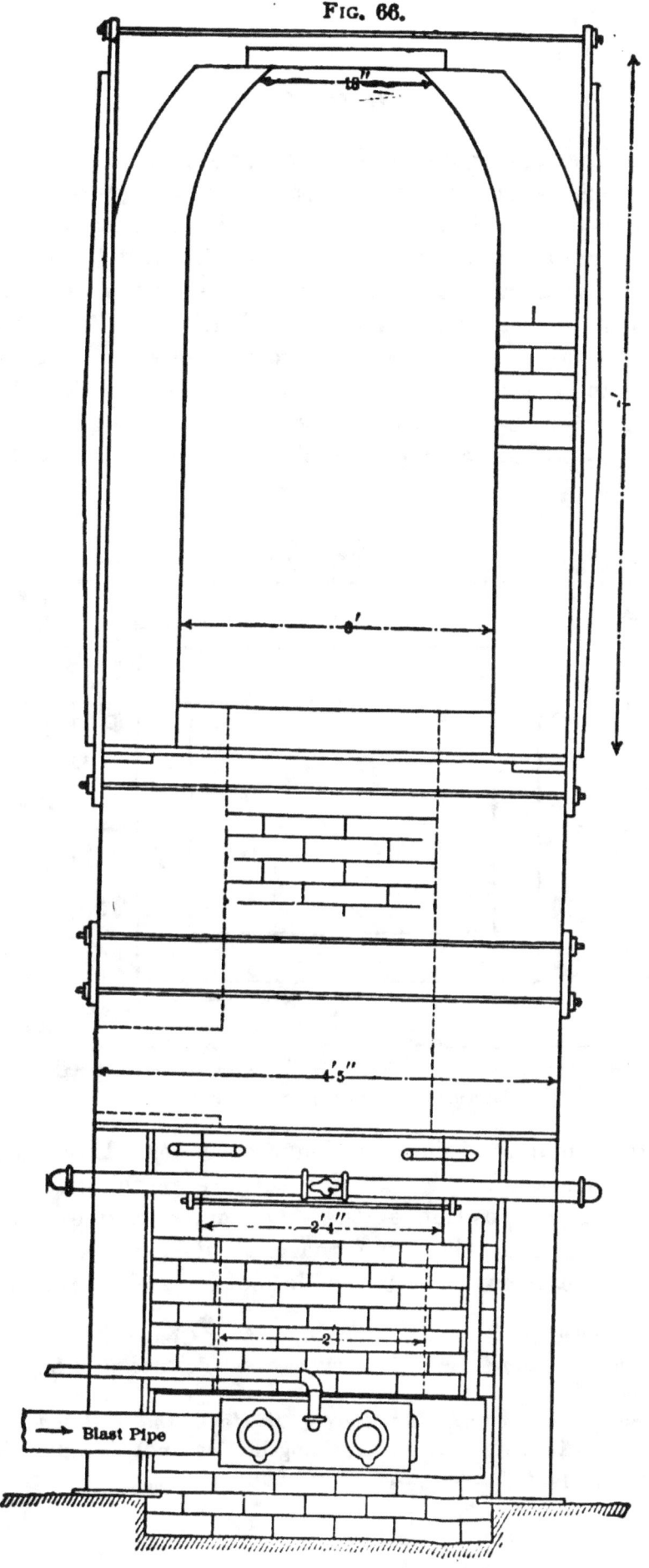

FIG. 66.

18"

6'

4' 5"

2' 4"

2'

→ Blast Pipe

Elevation of the back.

§ 51. **Recovery of Fluedust by the Lewis and Bartlett Bag Process.**[1]—One of the principal disadvantages of the hearth treatment is the loss in lead by volatilization. In some instances[2] condensing apparatus are used to save the flue dust. This subject will be discussed in connection with the smelting of argentiferous lead ores in the blast-furnace (§ 85). With most American ore-hearths the fumes are allowed to go to waste. The Lone Elm works began as early as 1876 to collect and utilize them for pigments. This has been carried so far that in smelting lead ores the aim is not to produce as much metallic lead as possible, but to ob-

FIG. 67.

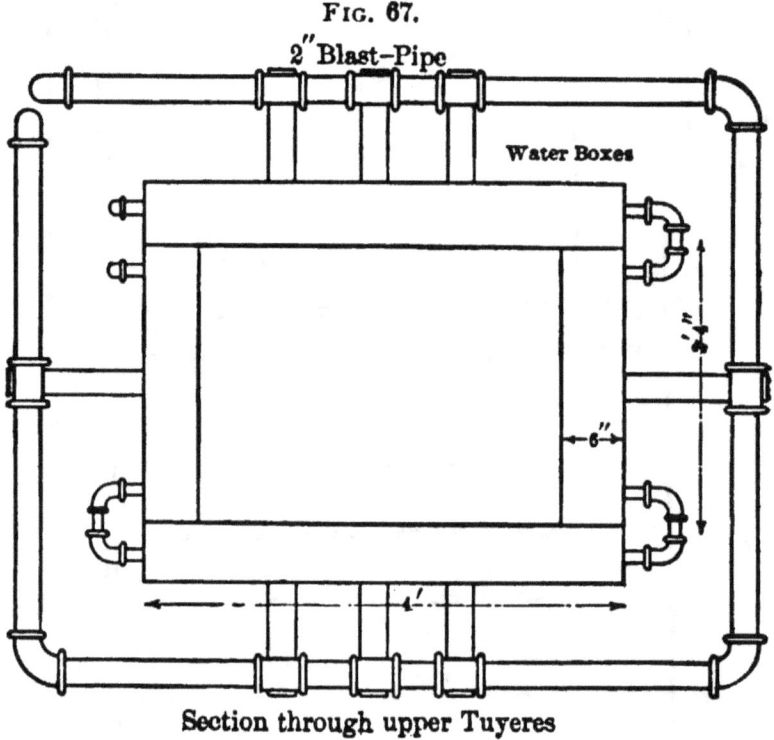

Section through upper Tuyeres

tain large amounts of fume, which are condensed, purified, and sold in the market. This explains the use of hot blast in both furnaces and the position of the seven upper tuyeres below the charging-door in the slag-eye furnace. This method of working permits the treatment of large amounts of ore and shows also why

[1] Dewey: *Trans. A. I. M. E.,* xviii., p. 674 ; Clerc : *Engineering and Mining Journal,* July 4, 1885 ; Ramsay : *Scientific American Supplement,* May 14, 1887, No. 598.

[2] Roesing : *Zeitschrift für Berg-, Hütten- und Salinen-Wesen in Preussen,* xxxvi., p. 103 ; (Lead Smelting in England) "English Government Report," quoted in § 48, a.

only 50 per cent. of the ore charged is converted into metallic lead in the ore-hearth.

The fumes are recovered by filtering them through woollen bags. This has been tried at the Grant smelter at Leadville,[1] but was given up again, as the flue dust saved, while running high in lead, contained only from 6 to 10 ounces silver per ton, and this did not pay at that time for the condensation. The process is, however, introduced at the Globe Smelting and Refining Works at Denver,

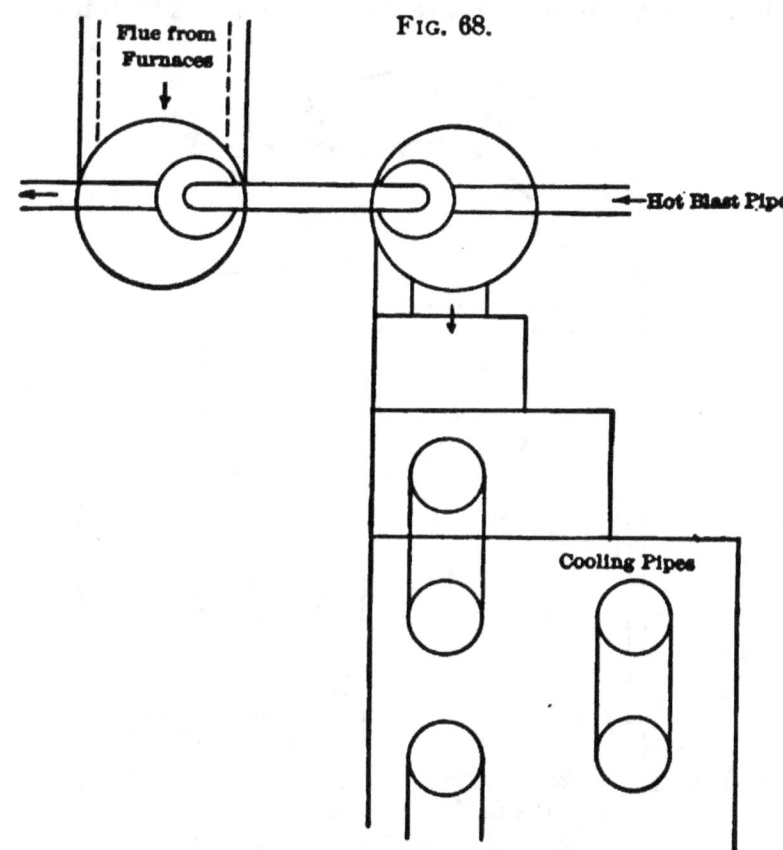

FIG. 68.

Plan of Cooling Cylinders etc.

and promises well. The conversion of the lead fumes into a marketable pigment comprises two operations:

1. Cooling and collecting the dark ore-hearth fume, "blue powder," in the first bag-house, the "blue room."

2. Refining the blue powder in the slag-eye furnace, which is followed by cooling and collecting the resulting "white paint" in the second-bag-room, the "paint-house."

[1] Guyard, in Emmons' "Geology and Mining Industry of Leadville," monograph xii. in "United States Geological Survey," pp. 673 and 717.

The fumes from the ore-hearth are drawn off by a suction-fan, 6 feet in diameter and 3 feet wide, which makes 290 revolutions per minute. They pass first through a brick dust-chamber (40 feet long, 19 feet high, and 6½ feet wide, with a door on one side), where any coarse-grained particles of more or less changed ore and fuel are collected. They thus pass out of the top of the chamber through a horizontal sheet-iron pipe, 5 feet in diameter, resting on 20-foot iron pillars, to the fan, and thence through a 4-foot pipe, resting on 12-foot pillars, to the blue room. The pipe is suf-

FIG. 69.

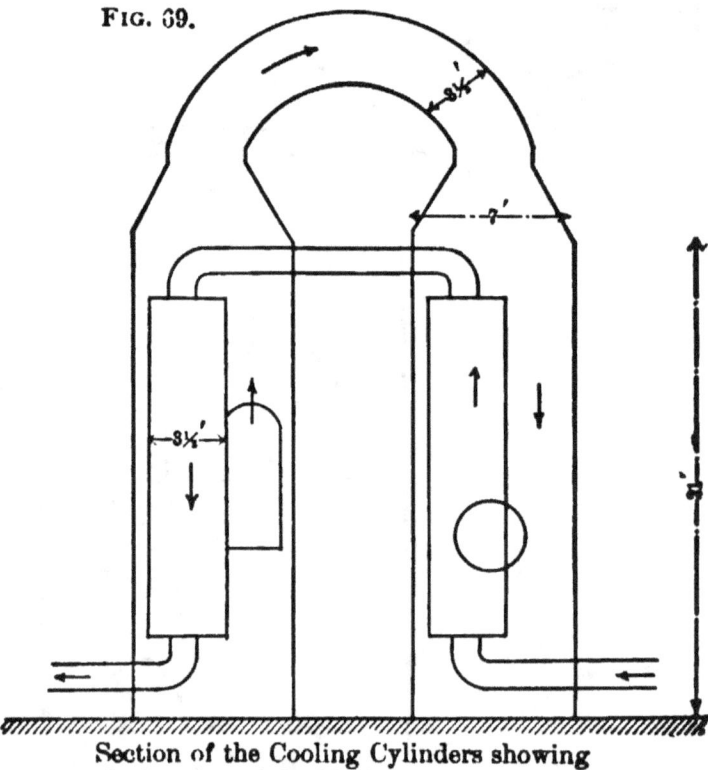

Section of the Cooling Cylinders showing the heating of the blast for the Slag-Eye Furnace.

ficiently long (350 feet) for the gases to cool in their passage through it.

The first bag-house is similar in construction to the second one, shown in cross-section in Figure 71. It is a brick building (95 feet long, 50 feet wide, and 45 feet high) divided into two compartments by a longitudinal wall, so that one may be shut off when it is necessary to gain access to the bags. Each compartment is divided into two stories, the lower being 12 feet high. The divisions (columns, beams, etc.) are all made of iron pipe. In fact everything in the building is either of brick or iron, except the filtering bags. The lower story contains four rows of sheet-iron hoppers, extending the length of the building, which serve to collect the fume that has accumulated. They have the form of a truncated pyramid and are closed at their lower face by a sliding damper. They stand on four iron pipes, 3½ feet long, encased in refractory clay pipes. The upper face of a hopper is covered with sheet iron $\frac{1}{16}$ inch thick. This has 16 holes, 18 inches in diameter, from which thimbles, 12 inches high, project upwards. Over these the lower ends of the bags, made of unwashed wool, 60 inches in circumference and 33 feet long (changing to 50 inches and 35 feet when in use), are slipped and tied fast. The upper ends are tied with strong cord, with which they are suspended from beams near the roof. There are 800 bags in the bag-house, each costing $9. Between every two rows of bags is an iron scaffolding with iron footways placed at convenient heights, so as to make all parts of the building accessible.

The cooled gases, being pressed through the main pipe, enter four branch pipes, each of which passes through and connects a set of hoppers. The gases, laden with the dust, ascend into the hanging bags, where they are filtered, the fumes falling into the hoppers below. These are emptied once in two days, when the bags are also shaken to detach adhering fume. For this purpose the current of the gas is shut off, and men with aspirators pass quickly through the building, giving each bag a quick shake.

The collected fume is a very fine bluish-gray powder (blue powder), consisting mainly of lead oxide and sulphate, with some lead sulphide.

BLUE POWDER.

	1st Hopper, 1st Row.	6th Hopper, 2d Row.	1st Hopper, 4th Row.	10th Hopper, 4th Row.
Pbs.	6.18	10.37	5 19	8.61
$PbSO_4$	45.34	43.55	46.88	43.57
PbO	44.44	44.48	45.08	44.18
Zn	.61	.34	.43	.61
$(FeAl)_2O_3$	.10	.05	.07	.07
CaO	.21	.01	.03	.02
SiO_2	.17	.11	.12	.14
CO_2	.23	.19	.26	.08
SO_2	.96	.12	.68	.44
On combustion yielded :				
H_2O	2.12	1.57	2.33	1.71
CO_2	3.62	2.98	3.22	2.92

The fume let down from the hoppers is spread over the floor in piles and set on fire with oil-waste. It burns for about ten hours and does not flame, but liberates a good deal of heat and some sulphurous acid. The fine, loose blue powder is hereby converted into a porous, pinkish-white crust that is friable but sufficiently coherent to stand handling and charging. The roasted blue powder is free from carbonaceous matter and lead sulphide.

ROASTED BLUE POWDER.

$PbSO_4$	48.76
PbO	46.82
Fe_2O_3	0.32
Al_2O_3	0.05
ZnO	0.27
CaO	0.48
SiO_2	0.10
CO_2	0.90
SO_1	1.65
H_2O	0.37
	99.72

In refining the roasted blue powder in the slag-eye furnace, the object is to oxidize all the components of the charge as much as possible; hence little metallic lead is produced. To prevent any carbonaceous compounds from injuring the color, Connelsville coke is used as fuel.

The average daily charge for a furnace is made up of 2,800 pounds slag, 1,000 blue powder, about 600 dry-bone (carbonate ore), 450 fume from the cooling pipes, and the necessary coke. For every pound of metallic lead about 1.6 pounds of paint are obtained, and the product of a furnace is 4,250 pounds of paint.

In running the furnace, it is important to have a hot top, a liquid slag, and no stoppages or irregularities, otherwise the paint obtained will be of inferior grade. Thus, when a furnace is started, the product of the first four or five hours is classed as blue powder, and that of the next twenty hours is still below the standard.

A No. 5 Baker blower furnishes the blast for three slag-eye furnaces. From these the gases are drawn by a fan (6 feet in diameter, 3 feet wide, making 290 revolutions per minute), first through a set of cooling pipes which surround the blast-pipe, and then through a second set built over brick dust-chambers, where any heavy particles are collected to go back to the furnace, while the gases sufficiently cooled pass off into the second bag-room. The first set—shown in vertical section in Figure 69 and connected

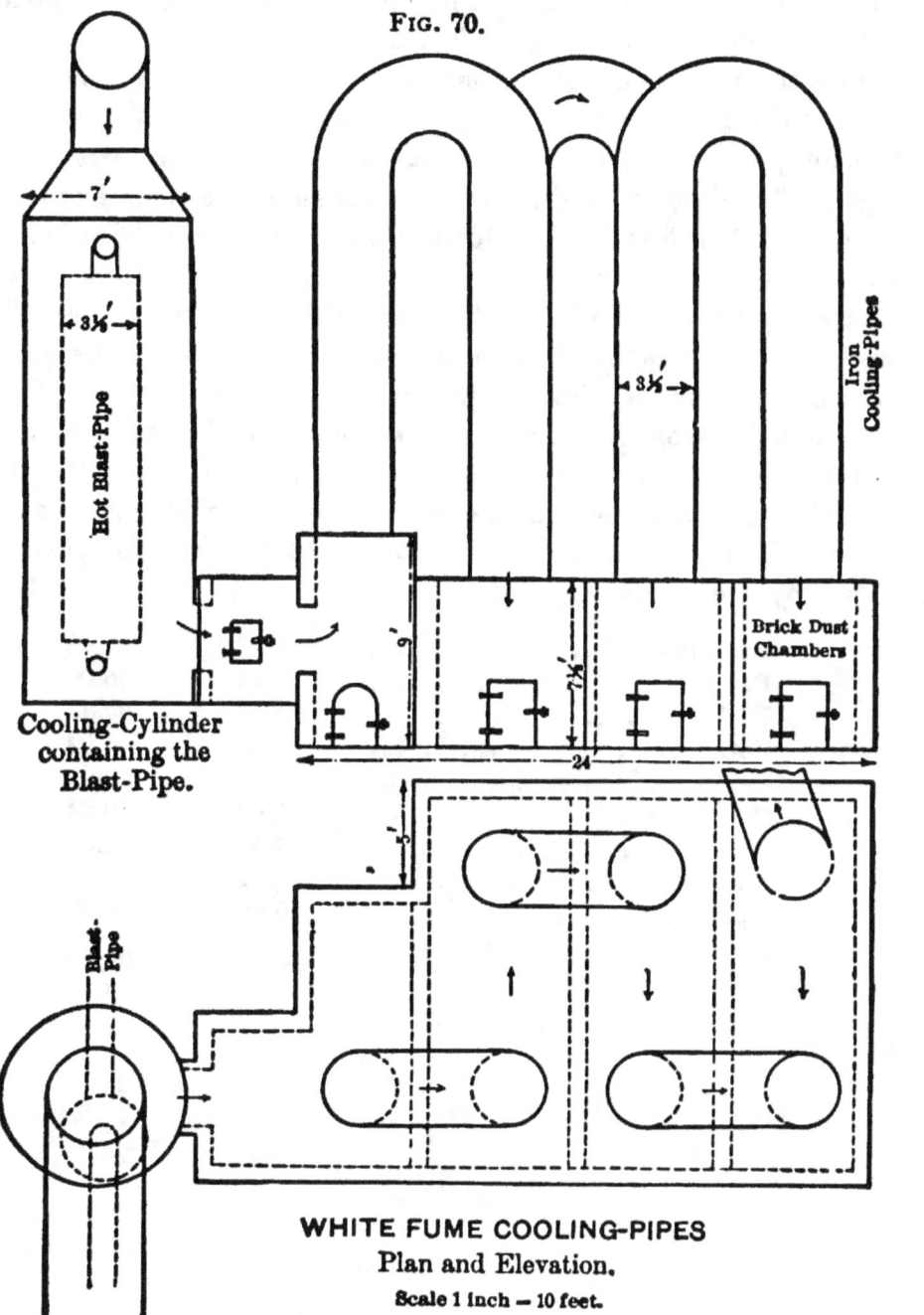

FIG. 70.

Hot Blast-Pipe

Cooling-Cylinder containing the Blast-Pipe.

Iron Cooling-Pipes

Brick Dust Chambers

Blast-Pipe

WHITE FUME COOLING-PIPES
Plan and Elevation.
Scale 1 inch = 10 feet.

with the second set in Figure 68 (plan) and Figure 70 (plan and section)—consists of two iron cylinders, 7 feet in diameter and 20 feet high, lined with fire-brick and connected by a pipe 3½ feet in diameter. The blast-pipe, 18 inches in diameter at its entrance near the bottom, widens to 3 feet, to take up more heat from the surrounding gases, and before passing out is again contracted to its normal size. The second set, four U-shaped iron cooling pipes, are 3 feet in diameter and 20 feet high.

The second bag-room or paint-house, shown in Figure 71, is a brick building without any partition wall. It is 40 feet wide, 90 feet long, and 45 feet high. It has two stories; the lower is 9 feet high. The divisions and general arrangements are similar to those in the first bag-house. The lower floor, however, is boarded. The hoppers are made of wood and lined with iron. They are suspended by iron straps (12 inches apart) from 1½-inch pipes. These are laid across the 2-inch pipes at intervals of 2 feet. Beneath the hoppers are wooden bins closed with canvas.

The paint from the bins is packed into barrels, each holding 500 pounds. It has a good body, a good color, and mixes well with oil. It is also used as weighting material. Its price in 1884 was $3.50 per hundred pounds, with pig-lead at $3.35. Its composition is shown by the following analyses :

Insoluble	0.08	0.08
$PbSO_4$	65.46	65.00
PbO.	25.85	25.89
ZnO	5.95	6.02
Fe_2O_3	0.03	0.03
CaO	0.02	0.02
CO_2	1.53	2.00
SO_3	0.04	none
H_2O	0.69	0.85
	99.65	99.89

FIG. 71.

9 Hoppers in each row.

SECTION OF THE BAG-HOUSE
FOR WHITE FUME.
Scale 1 inch = 10 feet.

Am. Bk. Note Co. N. Y.

CHAPTER VIII.

SMELTING IN THE BLAST-FURNACE.[1]

§ 52. **Introductory Remarks.**—The treatment of lead ores in the blast-furnace is generally discussed under three heads :

Precipitation,
Roasting and Reduction, } for Sulphide Ores.
General Reduction, for Oxidized Ores.

These processes are quite distinct in European practice, where large mines furnish certain smelting works with uniform ores. For instance, in the Hartz Mountains, in Prussia, concentrated galena ores have been for years, and still are, smelted raw in the blast-furnace. In other districts of Germany as well as in France the sulphide ores are always roasted before they come to the blast-furnace, and oxidized ores are of such rare occurrence that they are hardly treated separately.

In the United States this regularity of treatment can seldom be pursued, as the ores a smelter has to treat show the greatest variety, as can be seen from § 35, Purchasing of Silver-Lead Ores. For this reason the chemical classification suited to European practice will be set aside and smelting in the blast-furnace treated as one process, in which lead and silver are extracted from the ore in the form of base bullion by means of carbon, iron and lime.

All smelting ores can be worked in the blast-furnace ; any ore containing over 4 per cent. of silica must be so treated. If it is a sulphide, it is generally first roasted in the reverberatory furnace ; if a carbonate or a mixture of carbonate and sulphide, the carbonate prevailing, it is smelted at once. If the preceding analyses of argentiferous lead ores of the Rocky Mountains (p. 33) and the Pacific (p. 40) be referred to, it will be seen that they usually contain much over 4 per cent. silica. This explains the universal use of the blast-furnace in the West.

---

[1] Hahn, "Mineral Resources of the United States," 1882, p. 325 ; also *Engineering and Mining Journal*, August 25, September 1 and 8, 1883. Henrich, *Engineering and Mining Journal*, October 8, 1883 ; July 17, 1886. Guyard, in Emmons' "Geology and Mining Industry of Leadville, Col.," monograph xii., United States Geol. Survey, 1886, p. 618.

§ 53. **Lead Slags.**[1]—As the object of smelting is to separate by fusion the lead as metal from its ore, the other constitutents, the acid silica and the bases, iron, lime, etc., have to be combined in certain proportions to form a slag. If this does not occur naturally, either silica or the various bases, *i.e.*, fluxes, will have to be added in the requisite quantities.

Composition of Lead Slags.—The main slag for the lead smelter is the singulo-silicate slag,

$$2 \overset{\text{II}}{R}O + SiO_2 \text{ or } 2\overset{\text{III}}{R_2}O_3 + 3SiO_2,$$

with the oxygen ratio of bases to silica as 1 : 1.

In practice the slags are made slightly more acid than the formula calls for. Another slag that is sometimes made is the sesqui-silicate

$$4\overset{\text{II}}{R}O + 3SiO_2 \text{ or } 4\overset{\text{III}}{R_2}O_3 + 9SiO_2,$$

with the oxygen ratio of bases to silica as 2 : 3, written sometimes

$$(2\overset{\text{II}}{R}O + SiO_2) + 2(\overset{\text{II}}{R}O + SiO_2), \text{ or } (2\overset{\text{III}}{R_2}O_3 + 3SiO_2) + 2(\overset{\text{III}}{R_2}O_3 + 3SiO_2),$$

as if consisting of a singulo- and a bi-silicate. This slag, not being so readily fusible as the singulo-silicate, is made, if a high temperature is desirable, *e.g.*, if much lead sulphide is to be decomposed by metallic iron resulting from ferric compounds. With a readily fusible slag the ferric compound would be reduced only to a ferrous compound and then slagged. Henrich[2] recommends a slag more acid than the singulo-silicate for arsenical ores in order to keep the speise liquid.

The following tables by Balling[3] give the necessary proportions of silica and bases required to form singulo-silicate, bi-silicate and sesqui-silicate slag, one table having silica for a unit, the other the bases. The base baryta has been added.

---

[1] Hahn, *Op. cit.*, p. 325 ; Guyard, *Op. cit.*, p. 701 ; Iles, " Mineral Resources of the United States," 1883–84, p. 440 ; Schneider, *Trans. A. I. M. E.*, xi., p. 56.

[2] *Engineering and Mining Journal*, October 6, 1883.

[3] "Compendium der Metallurgischen Chemie," Bonn, 1882, p. 98.

One Part by Weight of Silica Requires	Parts by Weight of Bases.	One Part by Weight of Bases Requires	Parts by Weight of Silica.
For Singulo-Silicates :		For Singulo-Silicates :	
Lime	1.86	Lime	0.535
Baryta......	5.10	Baryta................	0.196
Magnesia	1.33	Magnesia	0.750
Alumina...............	1.14	Alumina...............	0.873
Ferrous oxide..........	2 40	Ferrous oxide.........	0.416
Manganous oxide.......	2.36	Manganous oxide.......	0 422
For Bi-Silicates :		For Bi-Silicates :	
Lime	0.98	Lime........	1 070
Baryta	2.55	Baryta	0 392
Magnesia	0.66	Magnesia	1.500
Alumina............. ...	0.57	Alumina...............	1.747
Ferrous oxide..........	1.20	Ferrous oxide.........	0 883
Manganous oxide.......	1.18	Manganous oxide.......	0.845
For Sesqui-Silicates :		For Sesqui-Silicates :	
Lime....................	1.24	Lime	0.803
Baryta.................	8.40	Baryta................	0.294
Magnesia	0.88	Magnesia	1.125
Alumina..........	0.76	Alumina.....	1.310
Ferrous oxide..........	1.60	Ferrous oxide..........	0.625
Manganous oxide ..:.....	1 57	Manganous oxide........	0.633

This table is to facilitate the study of slags that do not correspond to any of the typical slags but nevertheless give satisfactory results in the furnace.

Typical slags are definite combinations of silica, iron, lime and sometimes alumina, the success of which has been proved and which therefore become representative. They fulfil the requirements of a good slag, which are, according to Eilers,[1] that it should not contain over $\frac{3}{4}$ per cent. of lead or $\frac{1}{2}$ ounce of silver to the ton, provided that the base bullion does not run higher than 300 ounces, nor have a density over 3.6, nor permit accretions in the hearth, thus keeping the lead red-hot, nor any creeping up of over-fire.

For instance, a typical slag discovered by Eilers is formulated thus:

$$6FeO, 3SiO_2 + 2CaO.SiO_2.$$

In percentage it gives:

$$30.61SiO_2 + 55.10FeO + 14.29CaO = 100.$$

After deducting 10 per cent. for other ingredients of ore and fluxes, such as alumina, zinc oxide, etc., that cannot be brought

---

[1] *Engineering and Mining Journal,* April 9, 1881.

under the head of ferrous oxide, like manganous oxide, or of lime, like baryta or magnesia, there remains:

$$27.55SiO_2 + 49.59FeO + 12.86CaO = 90,$$

which gives, in round figures,

$$28SiO_2 + 50FeO + 12CaO + 9Al_2O_3, \text{ etc.} = 100.$$

Another slag, first brought into prominence by Eilers, has the formula,

$$6FeO.3SiO_2 + 4CaO.2SiO_2$$

and therefore the following percentage of composition:

$$31.38SiO_2 + 45.18FeO + 23.44CaO = 100.$$

Deducting again 10 per cent., there remains

$$28.24SiO_2 + 40.66FeO + 21.10CaO = 90,$$

which practical experience has altered to

$$30SiO_2 + 40FeO + 20CaO + 10Al_2O_3, \text{ etc.} = 100.$$

The following slags have been thoroughly tested and are in successful use. The first seven (A to G) are those so designated by Iles in his paper already quoted.

Authority.	Type.	SiO_2.	Fe(Mn)O	Ca(Ba, Mg)O.	Total.
Iles......................	A*	32	52	6	90
Iles	B*	35	45	10	90
Eilers..........	C	28	50	12	90
Iles	D*	34	34	16	84
Eilers.....................	E	30	40	20	90
Schneider.................	F	33	33	24	90
Raht	G	35	27	28	90
Hahn	H	34	50	12	96
Page	I	33	36	16	85
Livingstone	J	30	36	20	86
Hahn......................	K	36	40	20	96
Iles........	L	32	33	23	88
Murray....................	M	40	34	26	100

* No longer recommended by Iles.

These typical slags are the general guides for the lead smelter. He chooses the one best suited to the character of his ore. He need not, however, adhere strictly to it. Slight variations are always liable to occur, and need give no uneasiness, if the furnace runs well and the slag analysis shows a low amount of lead and silver.

The old rule of making a slag with from 30 to 34 per cent. silica, an equal amount of metallic iron, and from 8 to 12 per cent. lime is practically obsolete.

In reviewing the table, it will be seen that the composition of the slags ranges between

SiO_2.................. 28 and 36 per cent.
FeO.................. 24 " 52 " "
CaO.................. 6 " 30 " "

Slags have been made that run as high as 40 per cent. silica, for instance, $40SiO_2$, 40FeO, 10–12CaO, but they require much fuel, and are slightly viscid even then. The lowest practical limit in silica is probably reached with 28 per cent. The percentages of ferrous oxide represent also extremes. As regards the lime, 30 per cent. is the highest Schneider[1] gives in his experiments, but 6 per cent. is too low A modern American lead slag does not contain less than 10 per cent. of lime; European slags, as quoted by Kerl and Percy, run as low as 5 per cent., although the tendency there is towards a higher percentage.

PHYSICAL PROPERTIES OF LEAD SLAGS.

The fusibility of a slag depends on the percentage of silica and the character of the bases; the effect of the latter will be discussed in the next paragraph. The more fusible a slag the larger will be the amount of charge smelted per unit of fuel. With the correspondingly low temperature in the furnace the reducing power will be diminished; therefore a very readily fusible slag is not always desirable.

The liquidity of a slag depends largely on its fusibility. A slag ought to be sufficiently liquid to allow a perfect separation of lead in the furnace and of matte in the slag-pot. A correctly composed slag, which would otherwise be liquid, becomes viscous if the weather suddenly becomes wet or cold; in such cases more fuel is required. It is difficult, even for the practised eye, to say always whether the viscidity comes from lack of fusibility or from a slight insufficiency in fuel. As a rule, singulo-silicate slags containing earthy and metallic bases solidify quickly without first becoming pasty. When the tap-hole has been closed, the slag, if good, will drop slowly into the pot, drawing a small thread as it leaves the spout; a slag with an excess of base will fall off quickly in little round drops. Most slags have certain characteristics in

[1] *Trans. A. I. M. E.*, xi., p. 57.

their manner of running, which have to be studied by observation. In close connection with the running of the slag from the furnace is the manner of its rising in the slag-pot and the appearance of the surface when it has just solidified and is still red. Thus many slags show very characteristic surfaces.

Well-composed slags have a decided tendency to crystallize. The centre of a cone of slag is generally more crystalline than any other part, because the cooling is slower. Slags that crystallize usually become amorphous if chilled suddenly, and crypto-crystalline if not given sufficient time to develop crystals. A small percentage of zinc oxide in the slag interferes greatly with the crystallization. Iles thinks that the form in which a slag crystallizes stands in some relation to the percentage of lime it contains. He says (see Figure 72):

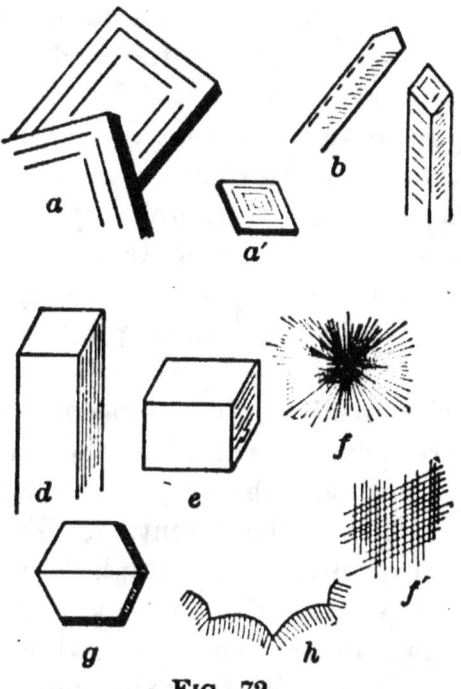

FIG. 72.

Slags with from   3 to   5 per cent. of lime crystallize like a.
"   "   8 " 12   "   "   "   " a'.
"   "   15 " 18   "   "   "   " b.
"   "   19 " 22   "   "   "   " c.
"   "   23 " 25   "   "   "   " d.
"   "   25 " 27   "   "   "   " e.
"   "   30 " 35   "   "   "   " g.

The forms f f' h represent crystals formed in attempting to slag very siliceous ores with lime alone. It is to be noted that the same slag will melt at a lower temperature when glassy than when crypto-crystalline.

There is some variety in the tenacity of slags. Siliceous slags are generally tougher than those where the base prevails. The more crystalline a slag the more brittle. A pot of slag may show brittleness in one part and toughness in another. Slags are not always pure silicates as expressed in the chemical formulæ, they are often mixtures of different silicates or even chemical silicates holding indefinite compounds of silica and base in solution. If, in these, crystals form, they represent the true silicate, being surrounded by a mass of an entirely different composition and different physical properties. This is important to remember in taking a sample for slag analysis (see § 81).

The specific gravity of a slag is an important factor in its separation from the matte. Iles gives, as extreme figures, 3.3 and 4.16; as a common range for good slags, 3.4 and 3.6. The average specific gravity of one hundred determinations of good slags gave him 3.691, and 3.6 is accepted as the highest specific gravity a good slag ought to have.

Slags do not possess to any extent the property of transmitting light. Single crystals are in exceptional cases transparent; sometimes they are translucent, but generally opaque.

Lead slags are usually black from their high percentage of iron. Thin pieces sometimes appear greenish-black; a large amount of iron will give a brownish hue. Lime produces a bluish or grayish tone.

The lustre of slags varies. It is occasionally metallic, but generally vitreous to resinous. Very often slags are dull.

Last may be mentioned the magnetic property of some lead slags, to which Iles first called attention. This is caused by the presence of magnetite[1] or magnetic sulphide of iron (Fe_8S_9). How the magnetic oxide gets into the slag is a matter for further investigation. Hahn suggests an incomplete reduction of ferric oxide; Guyard thinks it results from the oxidation[2] of metallic iron by lead oxide:

$$4PbO + Fe_3 = Fe_3O_4 + 4Pb.$$

§ 54. Action of Fluxes and Influence of Foreign Matter.—The fluxes usually required are iron ore, manganese ore, limestone, dolomite, rarely fluorspar, and slag.

a. *Iron.*—An iron flux acts in three different ways. It gives a base for the silica in the ore,

[1] Guyard, *Op. cit.*, p. 702.
[2] Berthier, "Traité des essais par la voie sèche," Liége, 1847, i., p. 854.

$$FeO.SiO_2 + FeO = 2FeO.SiO_2,$$
$$4FeO.SiO_2 + 2FeO.2SiO_2 = 6FeO.3SiO_2.$$

Being reduced to metallic iron by means of carbon, it acts as a precipitating agent (§ 7),

$$4PbS + 2Fe_2O_3 + 3C = 4Pb + 4FeS + 3CO_2,$$
$$2PbS + 4FeO.SiO_2 + 2C = 2Pb + 2FeS + 2FeO.SiO_2 + 2CO.$$

It liberates lead oxide from its combination with silica, after which it is then reduced by means of carbon,

$$4FeO.SiO_2 + 2PbO.SiO_2 + C = 2Pb + 4FeO. 2SiO_2 + CO_2.$$

The larger the amount of iron, within reasonable limits, a slag contains the greater will be its fusibility and liquidity and the higher its specific gravity. Slags rich in iron are dangerous, as they are liable to cause the formation of crusts in the furnace. A slag high in iron is, however, a necessity if much zinc is contained in the ore, as iron favors the slagging of zinc oxide. The purer an iron ore the greater will be its fluxing power, as only that part of the iron is available which is not required by its own silica, sulphur or arsenic. The silica not only limits the fluxing power, but also consumes limestone to form the slag. With the increase of slag in the blast-furnace the amount of lead ore of course diminishes.

These remarks are equally applicable *mutatis mutandis* for the other fluxes.

ANALYSES OF IRON FLUXES.

	Bruce Iron Ore, Col.	Madonna Mine, Col.		Bruce Iron Ore, Col.	Madonna Mine, Col.
Fe	66.44	42.05	SiO_2	2.39	5.10
Mn	0.01	3 12	CaO	0.12	..
Zn	0.02	1.86	MgO	0.62	..
Cu	0.02	..	Al_2O_3.........	0.04	..
Pb	..	4.0–9.0	Ag, oz. p. t....	0.13	2.0–5.0
As..........	0.01	..	Au, oz. p. t....	0.06	..
S..........	..	0.02			
Reference.....	1.	2.	Reference.....	1.	2.

[1] Guyard, *Op. cit.*, p. 647.
[2] Dewey, Bulletin No. 42, " United States National Museum," p. 47.

b. *Manganese.*[1]—The fluxing properties of manganese are similar to those of iron, and it may be substituted for the latter if de-

[1] Iles: " *School of Mines Quarterly,*" v., p. 217, and *Engineering and Mining Journal*, March 12, 1881 ; Eilers: *Engineering and Mining Journal*, April 9, 1881.

sired. It makes a slag more fusible and liquid, so that with large amounts of manganese it is advisable to choose a slag that does not require much metallic base. The commonest manganese mineral is pyrolusite, MnO_2. Its oxidizing power has been said to be the cause that certain slags rich in manganese have so high a tenor in silver. This seems to be contradicted by the fact that slags rich in manganese have been made by Church[1] that ran low in silver and lead, the base bullion averaging 314 ounces to the ton.

SiO₂	FeO	MnO	CaO	MgO	Al₂O₃	Ag. ozs.	Pb
29.60	11.56	43.25	7.50	trace	6.84	trace	1.4
33.00	14.22	25.78	13.10	1.00	4.20	0.5	1.0

The oxidizing power of oxides of manganese on blende in the reverberatory furnace, when sulphide copper ores are smelted, is mentioned by Pearce,[2] who obtained a slag of SiO_2, 48; MnO, 30; and ZnO, 12.5 per cent.; some manganese (3 per cent.) also entering the matte (50 to 60 per cent. copper) as sulphide. Iles believes that in the blast-furnace the amount of matte and speise formed diminishes with the increase of manganese in the charge.

Harbordt[3] gives it as his experience that the percentage of matte formed is not affected unless the manganese is present in considerable quantities. Another peculiarity of manganese is that it reduces the dissolving power of a slag for zinc oxide, magnesia, and barium sulphide.

According to Chisolm,[4] 60,000 tons of manganiferous iron ores were smelted in Colorado in 1887.

ANALYSES OF MANGANESE ORES AND MANGANIFEROUS IRON ORES.

	Montana.[5]	Colorado.[6]
SiO₂................ ...	6.60	8–15
Fe₂O₃...............	3.20	35–50
Mn₂O₃.............	88.40	8–15
Ag. oz............. ...	9.00	7–12

c. *Lime and Magnesia.*—The manner in which lime replaces

---

[1] *Trans. A. I. M. E.*, xv., p. 612; *School of Mines Quarterly*, v., p. 322.
[2] *Trans. A. I. M. E.*, xi., p. 59.
[3] Private communication, July, 1891.
[4] "Mineral Resources of the United States," 1887, p. 153.
[5] Peters : "Mineral Resources of the United States," 1883–84, p. 380.
[6] Chisolm : *Ibid.* 1885, p. 348.

iron in a slag can be expressed by changing the formula given above to

$$4FeO.2SiO_2 + 2PbS + 2CaO + 2C = 2Pb + 2FeS + (2CaO.SiO_2 + 2FeO.SiO_2) + 2CO.$$

Burnt lime is rarely, if ever, used in the blast-furnace. The effect of lime is to decrease the fusibility and the specific gravity of slags. Slags rich in lime require more heat, *i.e.*, more fuel, a stronger blast, and consequently a higher furnace than if iron predominates. They generally give a good separation of slag and matte, and limestone being in nearly all cases cheaper than iron ore, the tendency in most large smelting works is to make slags rich in lime.

Schneider[1] found that with slags containing much lime less matte is formed than when they are rich in iron, and also that the matte is lower in lead and higher in silver. He explains it by saying that calcium sulphide is formed and then dissolved by the slag. Its presence in Leadville slags has been proved by Guyard.[2] The use of lime in slags is limited by foreign matter in the ore, and especially by the presence of zinc. In a general way it may be said that the less lime and the more iron a slag contains the better will the furnace work. It does not seem advisable to go beyond 16 per cent. of lime in a slag, if (say) from 9 to 12 per cent. of zinc is present in the charge. With 28 per cent. of lime the zinc simply refuses to enter the slag; it is volatilized and incrusts the furnace.

Magnesia is undesirable in a lead furnace, as it makes a slag pasty and streaky, but in many cases the only available limestone is dolomitic, and it must be made the best of. This undesirable property of magnesia is especially observable if the slag contains zinc. Magnesia and zinc oxide appear to intensify each other's property of being difficult to slag. In a slag containing 8 per cent. zinc and from 2 to 3 per cent. baryta, very common just now in Colorado, from 2 to 3 per cent. magnesia shows a decidedly bad effect, and 5 per cent. causes a great deal of trouble.

Magnesia is generally figured in a slag as replacing lime; it may be doubted whether this is justifiable (Harbordt[3]).

[1] *Trans. A. I. M. E.*, xi., p. 58.
[2] *Op. cit.*, p. 78.
[3] Private communication, July, 1891.

ANALYSES OF LIMESTONE.

	Cañon City, Col.	Glass Mine, Col.	Carbonate Mine, Col.	Iron Co., Mo.	Saint Jo-seph, Mo.	Mine La Motte, Mo.
$CaCO_3$	88.90	49.57	57.95	47.50	60.34	55.2
$MgCO_3$......	6.30	37.08	39.65	42.19	32.17	40.9
SiO_2.........	3.10	4.22	0.76	5.11	3.77	1.27
$(FeAl)_2O_3$ ...	1.50	..	..	F_2O_3 4.67	6.88	..
$FeCO_3$	..	6.23	..	..	..	..
Reference....	1.	1.	1.	2.	3.	4.

1, Guyard : Emmons, *Op. cit.*, p. 646 ; 2, Gage, *Trans. A. I. M. E.*, iii., p. 117 ; 3, Monell, *School of Mines Quarterly*, ix., p. 214 ; 4, Neill, *School of Mines Quarterly*, ix., p. 214.

d. *Fluorspar* is of slight importance in smelting lead-silver ores in the blast-furnace. It forms with barium and calcium sulphate readily fusible compounds, and assists in fluxing zinc either as sulphide or oxide. This is mainly due to the fact that fluorspar, when once melted, is very liquid, and thus assists other less fusible compounds to enter the slag. Its chemical action in volatilizing silicon as silicon fluoride need not be taken into consideration. Foehr[1] makes a great many claims for fluorspar: that the use of from 1 to 5 per cent. in roasting ores in a reverberatory furnace saves fuel, that adding it to the charge in refining lead prevents shots of lead from being retained by the litharge, etc.

e. *Slag.*—There are four reasons for the use of slag in a blast-furnace charge. (1) It may contain too much lead or silver to be thrown away. (2) It makes the charge less dense. (3) It helps the actual smelting process because the slag having been already melted, it will re-melt easily and promote the smelting of the ore itself; and (4), if it be more acid or basic than the slag that is being formed by the smelting mixture, it will act as an acid or basic flux.

With a furnace running in a normal way, some rich slag is always produced. This is especially the case when the last slag in the furnace is being tapped and the blast passing through the tap-hole blows out valuable parts, which enrich the slag in the pot (blow-pot). Then again, when much matte comes out with the slag, it is liable not to settle out perfectly.

With a coarse easy charge the addition of slag is not absolutely necessary, although some is always given—say from 150 to 200 pounds for 1,000 pounds of ore. With fine ore the addition of slag is necessary, as otherwise the blast would not penetrate evenly

[1] *Engineering and Mining Journal*, June 21, 28, 1890.

through the smelting mixture, but form blow-holes. As much as 25 per cent. of the charge may have to consist of slag. With ores rich in zinc, slag is added to increase the fusibility.

Slag more basic than the normal slag comes into play where, for example, matte is being concentrated in a reverberatory furnace, and the resulting slag contains much iron that is available for the blast-furnace.

In smelting the by-products of desilverizing works in the blast-furnace without ore, the amount of slag added goes up as high as 50 per cent.

f. *Alumina.*—From the composition of the typical lead slags it will be seen that the place alumina occupies in lead smelting is generally a subordinate one. When it is present in large quantities, it becomes a question whether it acts as an acid or a base. It is known in a general way that with a high percentage of silica, alumina acts as a base ; with a low percentage it acts as an acid. Iles [1] gives this as his experience in lead smelting. Hahn,[2] however, thinks that alumina always acts as a base, and says that an increase of alumina requires also an increase of silica, or, what would be the same, a decrease in the bases.

Schneider [3] found that as a general rule an increase of alumina called for an increase in the proportion of lime, which means a decrease in silica, the alumina acting as an acid. Howe,[4] summing up the statements of Hahn and Schneider, suggests that the part played by alumina may depend upon the proportion of the other two fluxes, lime and iron, and that in calcareous singulo-silicates, low in iron, alumina may act as an acid and in ferruginous slags, low in lime, as a base. The idea seems to be confirmed by the experience of Peters [5] in smelting Mount Lincoln [6] ores in Colorado. The writer's own impression is that alumina acts as an acid in ferruginous slags, as in smelting an ore resembling an argentiferous clay he could not get his furnace to work with a slag that ran higher than 28 per cent. silica.

Henrich,[7] in a very interesting paper on smelting sulphide copper

[1] "Mineral Resources of the U. S.," 1883-84, p. 433.
[2] "Mineral Resources of the U. S.," 1882, p. 328.
[3] *Trans. A. I. M. E.*, xi., p. 57.
[4] *Engineering and Mining Journal*, November 17, 1888.
[5] *Engineering and Mining Journal*, November 24, 1888.
[6] *Trans. A. I. M. E.*, ii., p. 310.
[7] *Engineering and Mining Journal*, July 17, August 26, October 2, 1886 ; December 27, 1890.

ores rich in silica, alumina, magnesia, and low in iron, came to the conclusion that alumina always acts as an acid, and recommends two general types of silicate-aluminates for ores rich in alumina, on the hypothesis that $2Al_2O_3$ are equivalent to $3SiO_2$.

Type I. $\left\{ \begin{array}{l} \text{8 metallic oxide} \\ \text{4 earthy oxide} \end{array} \right\} : \left\{ \begin{array}{l} 2Al_2O_3 \\ 3SiO_2 \end{array} \right\}$ Mono-Silicate-Aluminate.

12 oxygen of base : 12 oxygen of acid $= 1:1$.

Type II. $\left\{ \begin{array}{l} \text{8 metallic oxide} \\ \text{4 earthy oxide} \end{array} \right\} : \left\{ \begin{array}{l} 2Al_2O_3 \\ 9SiO_2 \end{array} \right\}$ Bi-Silicate-Aluminate.

12 oxygen of base : 24 oxygen of acid $= 1:2$.

The theoretical composition of these two types is :

	I.	II.
SiO_2	16.1	35.0
Al_2O_3	18.2	13.2
FeO	25.7	37.3
CaO	40.0	14.5

By a combination of them he thinks it possible to meet all cases where the percentage of alumina is high. Stone[1] and Elbers[2] have discussed the matter from the iron-metallurgist's point of view.

To sum up: the question, under what conditions alumina ceases to be a base and becomes an acid, is not settled. It has to be determined for each slag by synthetical experiments in the laboratory.

g. *Barite.*—Barium sulphate in the blast-furnace is partly converted into barium sulphide and partly into barium silicate. In both forms it causes an imperfect separation of matte and slag. The reaction is usually expressed by the following formula:[3]

$$2BaSO_4 + 8Fe + 4SiO_2 = (BaS.FeS) + 7FeO.BaO.4SiO_2.$$

This would account for the formation of barium sulphide, even if no barium sulphate were reduced by means of carbon. It would also lead one to suppose that all barium sulphide entered the matte, but the fact is that, while a very little can be discovered in the matte, most of it is dissolved by the slag. Further, the proportions given do not correspond to the results obtained in a lead-furnace. That barium sulphide can enter the matte under suitable conditions is proved by the mattes made in the Altai mountains, which contain, according to Jossa and Kurnakoff,[4] from 4 to 22 per cent. barium.

[1] *Engineering and Mining Journal*, November 24, 1883.
[2] *Engineering and Mining Journal*, March 27, July 31, 1886.
[3] Balling, "Metallurgische Chemie," p. 89.
[4] *Berg- und Hüttenmännische Zeitung*, 1886, p. 547.

According to Schweder[1] there is another and more important reaction:

$$3BaSO_4 + 4SiO_2 + FeS = 3BaSiO_3 + FeSiO_3 + 4SO_2$$
$$BaSO_4 + 2SiO_2 + Fe = BaSiO_3 + FeSiO_3 + SO_2.$$

This means that in the presence of a metallic sulphide or a metal, barium sulphate is readily decomposed by silica; the liberated sulphur trioxide is decomposed at the elevated temperature into sulphur dioxide and oxygen, and this has an oxidizing effect on the iron sulphide or the metallic iron.

Schweder found in his experiments that calcium sulphate and sodium sulphate act in the same way.

Silicates of iron and baryta form very liquid slags.

In computing a charge the baryta is figured as replacing part of the lime.

From what has been said, it will be clear that all the sulphur contained in the barite cannot be figured into the charge. In fact, if 10 per cent. of the sulphur is figured as combining with iron to form iron sulphide, the amount will be covered.

h. *Pyrite.*—If an ore containing pyrite be smelted directly in the blast-furnace, it will consume iron as shown by the formula:

$$FeS_2 + Fe = 2FeS.$$

In calculating a charge this iron has to be added. There are, however, cases in smelting oxidized ores when part of the sulphur passes off as sulphurous acid and comparatively little matte is made. Henrich[2] gives his successful experience in smelting carbonate ores at Benson, Ar., consisting of galena (15 per cent.) and anglesite (75 per cent.), with silver-bearing pyrite. He obtained very little matte (20 pounds from 13 or 14 tons of ore), but considerable sulphurous acid. This he explains as having been caused by the following reactions:

$$2FeS_2 + 5PbSO_4 + SiO_2 = 5Pb + Fe_2SiO_4 + 9SO_2.$$
$$PbSO_4 + PbS = 2Pb + 2SO_2.$$

He says that the furnace ran rapidly and became very hot, so that the fuel (coke) had to be cut down from 12.5 to 11 per cent., the pressure of the blast being $1\frac{3}{8}$ inches mercury.

i. *Chalcopyrite and Copper Ores.*—In the blast-furnace the aim is always to carry the copper into the matte, which it enters as

---

[1] *Berg- u. Hüttenm. Ztg.*, 1879, p. 38.   *Iron*, 1879, xiii., p. 387.
[2] *Engineering and Mining Journal*, September 22, 1883.

cuprous sulphide. Copper, having a greater affinity than any other metal for sulphur, will generally take up all the sulphur in the charge to form Cu_2S, and what is left is then available for iron, lead, etc. If a charge does contain copper and not enough sulphur to form cuprous sulphide, some copper will be reduced to metal and be alloyed with the lead. The alloy hardens, sinks to the bottom, and closes up the passage of the lead-well. There is one case where even with sufficient sulphur to form cuprous sulphide the copper combines with the lead. This is when matte is concentrated in the blast-furnace with a highly ferruginous slag. The affinities of copper and iron for sulphur and silica seem to become disturbed.

If the slag be too basic the iron takes up some sulphur and goes into the matte instead of separating out and forming a crust; thus some sulphur belonging to the copper may be taken away, and this alloys with lead. Another way of explaining the fact would be that reactions between sulphides and oxides of copper take place similar to those of the reverberatory furnace, and the resulting metallic copper becomes alloyed with reduced lead. Whatever may be the theory, the fact remains that any excess of iron has to be avoided in the slag if the copper is to be concentrated in the matte and not partly driven into the lead. This is liable to occur when the matte contains about 12 per cent. of copper, and increasingly so with the increase of copper.

j. *Blende and Zinc Oxide.*—Zinc is the metal that causes the greatest difficulty in the blast-furnace in whatever form it may occur. Blende[1] is decomposed by iron oxides and silicates, the resulting zinc oxide entering the slag; metallic iron liberates metallic zinc, but most of the zinc sulphide entering the blast-furnace remains undecomposed and enters the matte as well as the slag. Generally the percentage of zinc found by analysis in matte and slag will be about equal. Blende makes matte less fusible, obstructs the separation, and carries other metallic sulphides into the slag. With ores containing little zinc this imperfect separation can be remedied by an addition of chalcopyrite[2] to the charge. If blende is present to any considerable extent, the ore must be roasted before smelting, to form zinc oxide. In order to carry this off, the slag has to be very fusible, as zinc silicate proper is infusible, and thus lowers the fusibility of any other silicate. If zinc

[1] Plattner, *Berg- u. Hüttenm. Ztg.*, 1854, p. 81.

[2] Hahn, " Mineral Resources of the United States," 1882, p. 343.

oxide is to be slagged care must be taken that it is not reduced to metal, hence the smelting has to be done quickly and at a low temperature. This requires a slag not high in silica and with a preponderance of iron. If, however, a slag high in lime has to be used, it has become the practice to figure one-half of the zinc (ZnO) as replacing an equivalent amount of lime (CaO) in the slag. The amount of lime that the slag may retain to work well in the furnace then depends mainly on the percentage of matte that is formed; the higher this is the lower must be the lime. In any case, in ore smelting, a slag ought not to contain more than from 6 to 8 per cent. of zinc, as otherwise there is too great a loss in metal. Slags are made that contain more than 8 per cent. zinc, but that this is commercially good practice nobody will maintain. If zinc oxide be reduced to metal in the lower parts of the furnace by carbon, or to a slight extent by metallic iron, it becomes volatilized. The vapor ascending carries with it lead and silver, and, being oxidized higher up by carbonic acid or water-vapor or by the oxygen taken from any readily reducible oxide, carries off metal as fluedust, and forms accretions on the sides of the furnace, which grow in thickness up to the charging-door. If zinc oxide, not reduced to metal on its downward course in the charge, comes, with the presence of carbon, in contact with lead sulphide or sulphate, it is converted into sulphide. If sufficient iron is present it will decompose the lead sulphide and the zinc oxide will remain unchanged.

The aim must therefore be to remove as much zinc as possible from the ore before it is smelted. The various attempts to do this have not been very successful, with the exception of mechanical concentration, and this with very fine-grained ores is out of place on account of the intimate contact of blende with galena, and with oxidized ores the losses in precious metal would be too great.

A number of plans for removing zinc from ores not suited for concentration have been tried or suggested. Thum[1] gives a way of distilling the roasted ore in inclined cylindrical retorts, the zinc vapors being carried off at the upper end and the lead compounds and residue removed at the lower. Lumaghi[2] proposes to work on a similar plan. Binon and Grandfils[3] describe a furnace with vertical retorts, which are charged from the top and discharged at the bottom,

[1] *Berg- und Hüttenmännische Zeitung*, 1875, p. 1.
[2] *Engineering and Mining Journal*, August 23, 1890.
[3] *Revue universelle des mines*, 1879, vol. v., p. 228.

while the zinc-vapors are drawn off by a horizontal condenser near the charging opening. Chenhall[1] suggests a similar method. Another dry method is that of Simmonet,[2] who roasts finely pulverized lead and silver-bearing blende with coarse-crushed limestone. Most of the lead and silver pass into the limestone, which has been in part changed to caustic lime, calcium sulphate, and sulphide, and can be separated from the fine zinc ore by screening. The writer once experimented with this method on several kinds of ore, but too little silver passed into the limestone to make it a success. The method is, however, worth trying, as it succeeded with Simmonet, and may again with suitable ore.

One of the latest suggestions is by Bartlett,[3] who wants to apply the ore-hearth method of Joplin, Mo. (§ 48, c.), in a modified form, and expects not to lose any silver, which is to be concentrated in a more or less melted mixture of matte and slag.

One class of wet methods aims to convert the zinc sulphide into zinc sulphite and sulphate, and to leach these with or without the addition of sulphuric acid. This is represented by Parnell,[4] Chenhall,[5] Siemens and Halske,[6] Croselmire,[7] Lowe,[8] West,[9] and Létrange.[10] In the Lower Hartz[11] (Prussia) the old process of heap-roasting the mixed sulphide ores and leaching out the zinc sulphate has lately received a fresh impetus from the manufacture of a white paint, called "Lithopon," which is a mixture of barium sulphate and zinc sulphide resulting from the double decomposition of barium sulphide and zinc sulphate.

The other method of making zinc soluble is to convert it into chloride. Maxwell-Lyte[12] and Wilson[13] have based their processes

---

[1] *Oesterreichische Zeitschrift für Berg- und Hütten-Wesen,* 1880, p. 462.

[2] *Annales des mines,* 1870, vol. xvii., p. 27.

[3] *Engineering and Mining Journal,* August 3, 1889.

[4] *Berg- u. Hüttenmännische Zeitung,* 1881, p. 254 ; 1883, p. 242.

[5] *Ibid.,* 1884, p. 465; *Iron,* xxii., p. 465; *Chemical News,* xlii. p. 201.

[6] *Wagner's Jahresberichte,* 1888, p. 368.

[7] *Engineering and Mining Journal,* September 29, 1888 ; February 9, 1889.

[8] *Ibid.,* January 5, 1889, p. 17.

[9] *Ibid.,* March 14, 1891.

[10] *Berg- u. Hüttenmännische Zeitung,* 1882, p. 489 ; 1883, p. 287.

[11] *Zeitschrift für Berg-, Hütten-, und Salinen-Wesen in Preussen,* xxv., p. 144; *Berg- u. Hüttenm. Zeitung,* 1890, p. 131 ; 1891, p. 184.

[12] *Engineering and Mining Journal,* March 10, 1883.

[13] *Chemiker Zeitung,* 1884, p. 1670.

on hydrochloric acid as a solvent. Electrochlorination has been tried by Slater.[1]

In smelting zinc-bearing by-products from refining works in a blast-furnace without ore the same slag is run over and over again until it becomes saturated with zinc. When a ferruginous slag contains 10 per cent. of zinc oxide it begins to be very mushy, and with 16 per cent. a gritty mass forms on top of the lead in the furnace, and the crucible cannot be kept open for any length of time. As far as the writer's experience goes, 16 per cent. is the maximum amount of zinc oxide that a singulo-silicate slag rich in iron can bear in a furnace having an Arents syphon-tap.

k. *Antimony.*— Antimony occurs either as a sulphide or an oxide. The antimony sulphide behaves on the whole like lead sulphide, but is much more volatile. If decomposed by metallic iron, the resulting metal is more likely to combine with the lead than it is to form a speise with any excess of iron that may be present. It may also be volatilized. The oxide is generally present as an antimoniate of lead or iron, and this, being reduced to an antimonide, combines with the lead or the speise, if any is made, or with the matte. The two main injurious effects of antimony, therefore, are that it causes loss by volatilization and impairs the character of the lead. Antimonial speise is rare, and in making up an ore-charge no account need be taken of the small quantity of iron likely to be consumed by the antimony.

Two difficulties have to be contended with in treating the antimonial by-products of refining works in the blast-furnace. If the slag contains but little iron, thus requiring a high temperature, much antimony and lead are volatilized ; if rich in iron, some speise is liable to form, which either separates out, causing the loss of the antimony, or becomes mixed with the slag, making it rich. A ferruginous slag is generally preferred as the lesser evil of the two (see § 119).

l. *Arsenic.*—Arsenic occurs very frequently in argentiferous lead ores and must not be neglected in computing a charge. It causes loss by volatilization and combines with the lead, but not to such an extent as antimony, as it has a great affinity for iron and forms a speise. Arsenic combines with iron in various proportions, from Fe_2As on. Speise having a composition that lies between Fe_2As and Fe_4As is fine grained; it is not readily fusible, and not liquid when melted; it is likely to form hearth-accretions (" speise-

---

[1] *Engineering and Mining Journal*, June 2, July 14, 1888.

sows ") and to prevent a good separation of lead. Speise of the composition Fe_5As has large cleavage planes, similar to spiegeleisen; it is moderately fusible and pretty liquid when melted, does not readily form accretions, and retains only very few shots of lead. A speise of the composition Fe_4As and upward seems to be just as harmful as one that contains less iron than Fe_5As. In computing a charge, some metallurgists add enough iron to form Fe_4As, keeping probably in mind that some arsenic is volatilized or combines with the lead; others assume Fe_5As. In making up a charge with an ore that contains arsenic, this has to be considered in choosing the slag. By taking a readily fusible slag the speise, floating on the lead beneath the matte and far removed from the hottest part of the furnace, is liable to chill. This can only be prevented by selecting a slag that requires a great heat for its formation. Thus the fusibility of the slag must stand in an inverse ratio to the amount of speise formed. With large amounts, even a sesqui-silicate slag may be necessary.

Sometimes old iron castings are added to loosen up any speise-sows that have been formed in the hearth. This is useful if the speise does not contain enough iron, as it supplies a deficiency, but does no good if the accretion is caused by too low a temperature.

§ 55. **Fuels Used in the Blast-Furnace.**—The fuels used in lead smelting are coke, charcoal, and a mixture of the two. Experiments have been made to replace part of the coke by bituminous coal and anthracite; gaseous fuel has also been tried.

a. *Coke.*—The coke to be suited for smelting purposes ought to be hard enough to bear the burden, slightly porous and low in ash. As regards the strength of the cell walls, Connelsville coke is the best, gas coke the worst. But Connelsville coke is very dense, and consequently does not oxidize readily. In order to overcome this, a high pressure of the blast is needed, which is on the whole not desirable. Gas coke might form a suitable fuel as regards its porosity, if the pores were not too large; they become partially clogged and the lumps of coke become glazed so as to be not only useless but harmful. A coke, therefore, that combines strength with a moderate degree of porosity is what is wanted. The disadvantages of Connelsville coke are, however, more than made up for by the fact that it is not much broken up in transportation and by its low percentage in ash and its uniform quality. Thus, if coke has to be brought from a distance, Connelsville coke is the best.

The amount of ash in coke varies from 10 to 20 per cent.; one cubic foot of dry coke weighs about 50 pounds; it has about 50 per cent. of cell-space, and can bear about a 90-foot charge without crushing.

COMPOSITION OF SOME COKES USED IN LEAD DISTRICTS.

	Connels-ville, Pa.	Cardiff, Wales.	Grand River, Col.	El Moro, Col.	Crested Butte, Col.
Fixed carbon.........	87.46	95.00	98.75	87.47	92.03
Moisture..............	0.49	0.01	..	..	..
Ash..................	11.32	4.26	5.49	10.68*	6.62*
Sulphur..............	0.69	0.68	0.76	0.85	..
Phosphorus..........	0.029	0.02	0.10	..	..
Volatile matter	0.011	..	..	1.85°	1.35°
Reference............	1.	1.	1.	2.	2.

° Including moisture.   * Generally higher.

ANALYSES OF COKE ASH.

	Connellsville, Pa.	El Moro, Col.	South Park, Col.
SiO_2....................	44 64	84.50	29.10
Al_2O_3..................	25.12	8.40	23 10
Fe_2O_3..................	22.73	7.10	47.80
CaO....................	6.95	....	....
MgO...................	1.91	....	....
Reference............	2.	3.	3.

1, "Mineral Resources of the United States," 1887, p. 396 ; 2, "Tenth Census of the United States," 1880, xv., p. 72 ; 3, Emmons, "Geology and Mining Industry of Leadville, Col.," p. 642.

In computing a charge the ash of the coke has to be taken into account.

Coke, used alone as fuel, gives clean slags and little fluedust.

Before the coke is fed into the furnace all the fine parts have to be removed. This is done with a coke-fork having prongs ½ inch apart, while removing the coke from the sheds. Sometimes a scoop is used and the fines screened out by dumping the coke over a grizzlie, which discharges the coarse coke on the feed-floor and the fines into a bin, whence they are removed to be burned under the boiler.

b. *Charcoal.*—As regards porosity, charcoal is the best fuel for a lead furnace, as it consists[1] of a large number of small cells joined to each other by porous walls. Hence, being readily oxidized, it is a good reducing agent for oxidized ores, and requires only a low

[1] Thörner, "Stahl und Eisen," vi., p. 71.

blast, which is an advantage. Its greater porosity is one of the reasons why weight for weight charcoal smelts more ore than coke. It also causes greater bulk (3 : 1), thus making the charge looser. This is favorable for quick smelting. The other reason is the low percentage of ash. The great disadvantage of charcoal is that very few kinds can bear any heavy burden; it breaks up and crumbles. Fine charcoal is not only worthless as a fuel, but it is a bad conductor of heat. It makes unclean slags and also causes loss in metal by increasing the amount of fluedust.

Nut-pine (piñon) charcoal is the best, but it has to be well burned. Charcoal from lighter woods, such as yellow and white pine, quaking aspen, and cottonwood cannot be used alone in the furnace, and even with coke only a small percentage is allowable ; some metallurgists condemn it entirely. Mesquite makes a good charcoal, but it is obtained with difficulty in large pieces. Charcoal from hard woods, such as mahogany, cedar and oak, decrepitates in the furnace. When charcoal is exposed for any length of time to the open air it breaks up and the amount of fines becomes large. On the other hand, its quality is said to be improved by storing, through the oxygen that it absorbs. Lead smelters do not like to have large amounts of charcoal on hand. It should be stored where it is not exposed to the sun.

Lightwood charcoal contains about 2 per cent. of ash, consisting principally of alkali and alkali-earth carbonates, and does not affect a charge to any appreciable extent. One bushel weighs about 14 pounds. The height of charge it can bear varies too much with the different kinds of charcoal to give a general figure; it is less than with coke.

c. *Coke and Charcoal.*—From what has been said it is obvious that the ideal fuel for the lead smelter must combine the strength of coke and the porosity of charcoal; thus a mixture of coke and charcoal will put through more charges in a given time than either alone. The coke bears up the charge and prevents the charcoal from being crushed. This burns quickly, helps to ignite the coke, and, having hardly any ash, leaves hollow spaces for the blast to penetrate.

d. *Coke and Bituminous Coal.*—Neill[1] has succeeded in replacing part of the coke by a non-caking or only slightly-caking bituminous coal, using separately lump, nut, and pea coal. He gives as results of his experiments that, besides the direct saving

---

[1] *Trans. A. I. M. E.*, xx.

in substituting the cheaper bituminous coal for the coke, jackets, slag and lead appeared hotter, the tuyeres brighter, and the crucible kept open better. The slag assays ran lower than with the usual coke and charcoal mixture and the separation of slag and matte was good. On the feed-floor the charges settled more evenly, as fewer zinc accretions were formed, and the top was cooler than usual; while the volume of smoke was larger, there was no greater loss in the fluedust on account of the charge being cooler. In his furnace, 36 by 78 inches at the tuyeres, and 12 feet from the tuyeres to the charging-door, he uses coke and charcoal in the proportion of 3 : 1, and is able to have 27 per cent. of bituminous coal in his fuel charge. He expects to reach 33 per cent., and thinks that with a higher furnace one-half of his fuel might be bituminous coal.

e. *Coke and Anthracite.*—Dwight[1] publishes some experiments made by Rapp in substituting anthracite of goose-egg size for part of the coke, no charcoal being used. The furnace was 36 by 80 at the tuyere level and 9 feet from there to the feed-floor. The result was that the smelting power of the furnace was reduced as the proportion of anthracite was increased, *e.g.*, one-third with anthracite as 60 per cent. of the fuel; otherwise the furnace remained in good condition. The top kept cool and the crucible open; there was a good reduction shown by a clean slag (0.4–0.8 % lead), a matte low in lead (8 % lead, 4 % copper), and a speise having a coarsely crystalline structure. Finally, less zinc accretions were formed than when coke alone was used, the charge containing 7.5 per cent. zinc.

f. *Gaseous Fuel.*—This has been used in a single instance. At the works of the Pennsylvania Lead Company[2] Blake introduced natural gas with the blast by inserting a gas-pipe through the tuyere-pipe. The amount of natural gas was regulated by stopcocks and the blast-pressure increased so as to supply sufficient air for the combustion of the gas. Thirty per cent. of coke was successfully saved in this way. By replacing sixty per cent., the top of the furnace became too hot. That solid fuel cannot be entirely replaced by gaseous is clear from the reactions going on in a blast-furnace that require solid carbon. For pecuniary reasons it is improbable that any artificial gaseous fuel will ever be used in the blast-furnace.

[1] *Trans. A. I. M. E.,* xx.
[2] *Trans. A. I. M. E.,* xv., p. 661.

The weight of the fuel required in a lead furnace is generally expressed in terms of percentage of the total weight of the charge (ore, flux, and slag), sometimes by saying that one part of fuel bears so and so many times its weight of charge; the latter is more common in copper smelting. Misunderstanding occasionally arises from deducting the pounds of lead contained in the charge and referring the percentage of fuel only to the slag and matte material. Often the percentage of fuel used refers to the weight of the charge excluding the slag that is added. The amount of fuel required varies with its character, with the fusibility of the charge, with the time of year, and the altitude at which the smelting is carried on.

As regards the character of the fuel : coke that is rich in ash is not only an inferior fuel in proportion to the smaller amount of carbon it contains, but a considerable quantity of this carbon is consumed to melt the ash and the fluxes necessary to slag it. For this reason a smaller amount of charcoal than of coke would seem to be required. The exact opposite is, however, the case. This is probably because the charcoal crumbles and is crushed in its descent in the furnace.

The richer an ore is in lead and the more fusible the rest of the charge, the less fuel will be needed. For instance, an ore containing zinc requires more fuel than one that is free from it ; a calcareous slag requires more fuel than one that is ferruginous ; a coarse and open charge requires less fuel than one that is fine and dense.

In summer less fuel is generally required than in winter, not so much owing to the higher temperature as to the more rapid evaporation of the moisture contained in ore, flux, and fuel. The difference may be as much as 5 per cent.

The altitude at which an ore is smelted makes a great difference in the amount of fuel required. Hahn,[1] for instance, gives the figures of 14 and 17 per cent. in Salt Lake City (4,000 feet above the level of the sea) as against 22 and 24 per cent. in Leadville (10,000 feet) ; the lower figure refers to the summer, the higher to the winter season. The only explanation that is at least in part satisfactory is the one given by Headden at a meeting of the Colorado Scientific Society, where the matter was informally discussed. A cubic foot of air entering the blast-furnace under a certain pressure will expand more at a high elevation, where the air

---

[1] "Mineral Resources of the United States," 1882, p. 339.

is rarefied, than at sea-level; consequently more heat will be consumed, and this has to be made up by an extra amount of fuel. In the same way more power and consequently more fuel is required at a high elevation to obtain this cubic foot of compressed air.

Thus it will be seen that from various causes the percentage of fuel required in a blast-furnace must show great differences. For coke and a mixture of coke and charcoal (3 : 1 or 4 : 1) from 12 to 16 per cent. is the common figure, rising sometimes to 18, rarely to 22. With charcoal alone it goes as high as 26 and 28 per cent.[1]

§ 56. The Roasting of Ores.

a. *Introductory Remarks.*—The object of roasting ores previous to smelting is to drive off as much sulphur and arsenic as possible, thus reducing the amount of iron otherwise necessary to decompose sulphides and arsenides and with it the amount of matte and speise, thus also permitting an increase in the amount of ore smelted.

To decide whether it is necessary to roast an ore or whether it can be smelted raw, the character and amount of sulphides, the richness of the ore, and the cost of roasting have to be considered. As a general rule, any ore containing 15 per cent. sulphur is best roasted before it is smelted. To some extent the richness of the ore in silver may modify this rule. On account of the loss in silver endured in roasting, it may in some cases be better to smelt an ore raw which contains more than 15 per cent. sulphur. An ore running 100 ounces silver to the ton is rarely roasted; some metallurgists[2] draw the line at 50 ounces silver, which, however, seems rather low.

The deleterious influence of blende in the blast-furnace has been previously emphasized (§ 54, j). If it is present to any extent, the ore will have to be roasted, for if it is smelted raw a large percentage of slag would necessarily have to be added to the charge to diminish the relative amount of blende and thus reduce its bad effect. In a general way it may be said that the higher the percentage of lead in a charge the more blende is permissible. For instance, if with mixed sulphide ores containing little or no pyrite the amount of lead present is twice that of zinc, the ore is smelted raw; if zinc and lead are present in equal amounts, or if there is more zinc than lead, it is best to roast the ore before smelting.

[1] Habn, *Op. cit.*, p. 889.
[2] Newhouse, *Engineering and Mining Journal*, February 28, 1891.

With pyritic ores there is more margin. In smelting the ore raw in the blast-furnace pyrite consumes iron and thus reduces the furnace's capacity for ore; it forms much matte, consequently quite a percentage of lead and silver is not directly recovered in the form of base bullion. The matte has to be roasted before resmelting. The iron previously consumed in the blast-furnace becomes again available, so that the actual consumption of iron in smelting ores raw is not so great as is generally assumed. Whether a pyritic ore shall be roasted or not is decided by the percentage of sulphur it contains above that which is required for the amount of copper present. A large quantity of matte is, however, to be avoided, as the silver entering the slag increases with the percentage of matte formed. This bad effect begins to show itself with 10 per cent. of matte in the charge.

Pure argentiferous galena ores rarely come in such quantities to a smelter as to make their separate treatment necessary. They are generally added to the charge in the raw state in such quantities as to make up for the common deficiency of lead. Impure galena ores are usually mixed with sulphuretted ores that are free from lead before roasting. It has already been shown (§ 9) that in roasting lead sulphide this is converted into oxide and sulphate, and that the presence of other sulphides increases the proportion of sulphate. Of the principal metallic sulphates[1] silver, iron, copper, zinc, nickel, cobalt, manganese, and lead sulphate, silver sulphate loses its sulphuric acid at a very low temperature, and lead sulphate gives it up only to a very small extent at a high temperature. If it is to be decomposed, the sulphuric acid has to be expelled from its combination by the stronger acid silica, which combines with the lead oxide and forms a silicate (§ 6 and 8). Zinc sulphate requires a bright-red heat and a considerable time to be converted into oxide. For this reason ores rich in blende are best roasted separately from those containing little of it. If several sulphides are roasted together, the order in which the sulphates give up their sulphuric acid differs from the one just stated, and will be iron, copper, silver, zinc, manganese, the silver being sulphatized again by the sulphuric acid set free through the decomposition of the iron and copper sulphate.

The roasting of lead ores can be carried on in heaps, stalls, kilns, and reverberatory furnaces. So-called mixed ores consisting mainly of galena, pyrite, chalcopyrite, blende, and containing com-

[1] Kerl, "Grundriss der Metallhüttenkunde," Leipsic, 1881, p. 70.

paratively little gangue, are sometimes roasted in heaps and stalls, the sulphurous acid being allowed to go to waste, or in kilns, when the sulphurous acid is to be converted into sulphuric acid. As this roasting is comparatively rare with lead ores, but very common with copper ores, and as the apparatus is practically the same, the method is best omitted here and the reader referred to the works on the metallurgy of copper [1] and the manufacture of sulphuric acid. [2] Here only the roasting in the reverberatory furnace will be considered.

To roast an ore in the reverberatory it is charged at the cool flue end and gradually moved towards the hottest part, next to the fire-bridge, whence it is withdrawn as a pulverulent, agglomerated, or slagged mass, according to the prevailing temperature and the fusibility of the charge. As regards the subsequent smelting, it is best to slag the ore, as by obtaining the roasted ore in lump-form the disadvantages of treating fine ores in the blast-furnace would be overcome; but other considerations often prevent this. The principal ones are the loss in lead and silver and the increase of cost. Newhouse [3] gives as a result of a series of experiments in roasting ores containing from 12 to 18 per cent. lead a loss of from 15 to 18 per cent. of lead and of 2 per cent. of silver with subsequent fusion, and from 2 to 5 per cent. of lead and none in silver without it. By agglomerating the ore the loss will be only slightly higher than when it remains pulverulent. The sulphur will not be so effectually removed as when the ore is slagged, but more so than when it remains a powder. For instance, slag-roasted ore contains about 1 per cent. sulphur, while roasted pulverulent ore retains from 3 to 7 per cent. sulphur. The loss increases on the whole with percentage of lead in the charge. As a general rule, it may be said that an ore with 10 per cent. lead or less can be safely slagged; with from 10 to 20 per cent. it is advisable only to agglomerate it; should the lead run over 20 per cent. the temperature is best kept so low that the roasted product remains pulverulent, or is only slightly adhesive when drawn from the furnace.

Making up the sulphide ore-beds will therefore be regulated to

---

[1] Howe, "Copper Smelting," Bulletin No. 26, U. S. Geol. Survey, Washington, 1885 ; Peters, "Modern American Methods of Copper Smelting," New York, 1891.

[2] Lunge, "Sulphuric Acid and Alkali," London, 1891, vol. 1.

[3] *Engineering and Mining Journal*, February 28, 1891.

some extent by the amount of coarse ore necessary to give the blast-furnace the required smelting power. Newhouse[1] advises the mixing of galena concentrates with pyrite, in order that the charge may contain 9 per cent. of lead, which is fused to obtain coarse lumps for the blast-furnace, and reducing the lead contents of the remainder down to 2.5 per cent., and roasting this without fusion, whereby sufficient lead for the blast-furnace would be obtained. At some works ores are fused that are entirely free from lead. In this case matte is liable to form in the fuse-box ; this, however, does not assist the volatilization of silver.

The greater cost of fusing over roasting is caused by the extra labor, the larger quantity and better quality of fuel that is required, and the greater wear and tear of the furnace, owing to the high temperature and the corrosive action of slagged ore.

All the statements made about the loss in lead and silver refer to the mixed sulphide ores treated by Western smelters, which as a rule run low in lead and high in silver. They are not intended for the pure galena concentrates free from silver (as in the Mississippi Valley) or low in silver (as in most European silver-lead works), because in both instances the ores are always slagged to a greater or less degree. The charges running 50 and 60 per cent. lead and free from impurities require when roasted a very slight increase of temperature to be slagged, care being taken to keep it as low as possible. Therefore the loss in lead and silver is slight, although the percentage of lead is high. Cramer von Clausbruch[2] states that at the Altenau smelting and refining works (Hartz Mountains) he obtains the best results in treating his galena ores if the charge contains 15 per cent. silica and from 55 to 60 per cent. lead. If there is more silica a base has to be added to effect a complete slagging at a reasonably low temperature; if there is less, some lead sulphate remains undecomposed. He notes the interesting fact that, if the roasted ore is not completely slagged but retains parts of sulphides and sulphates that have been only agglomerated, the silver and copper will be concentrated in the agglomerated part. One hundred parts of his charge give 85 per cent. thoroughly slagged ore, 10 per cent. of a mixture of slagged and agglomerated ore, and from 2 to 3 per cent. of unroasted agglomerated galena, the loss in roasting varying from 2 to 3 per cent. The slagged

[1] *Loc. cit.*

[2] *Zeitschrift für Berg-, Hütten- und Salinen-Wesen in Preussen,* xxxi., p. 26 ; *Engineering and Mining Journal,* May 24, 1883.

part of the charge contains one-half of the silver and only a trace of copper, while the other half of the silver and all the copper are concentrated in the rest of the charge.

In making up charges for fusing-furnaces that contain 10 per cent. and less lead, quite different standards have to be followed. The principal base to combine with the silica of the sulphide ore will be iron, and the next important lead. Charges are made up so that they may be readily fusible and sufficiently acid not to corrode the bottom and side-walls of the fuse-box. The silica in them varies from 25 to 32 per cent., and the iron calculated as metallic iron is made to equal the silica. It is not common to add lime to a charge containing lead. With charges free from lead, proportions like

SiO_2	FeO	CaO
32	32	24
36	36	24

are sometimes made up, resembling closely blast-furnace slags.

The results obtained in roasting an ore depend not only on its chemical composition but also upon the size to which it has been crushed, the thickness of its bed in the furnace, the amount of rabbling it receives, the time it remains in the furnace, and the temperature to which it is exposed.

As galena oxidizes but slowly when heated with access of air, a large number of surfaces (fine-crushing) are necessary, if the roast is to have the desired result. Then, as the roasting proceeds in each particle from the surface to the centre, it is probable that, if the galena is too coarse, a reaction may take place where the oxide and sulphate found at the surface come in contact with undecomposed sulphide at the centre, and the resulting metallic lead causes a considerable loss in lead and silver—another reason for fine-crushing. Ores that do not roast readily, *i.e.*, ores rich in galena and blende, are crushed through an 8-mesh sieve. Ores that roast easily, *e.g.*, pyritic ores and iron matte with 10 per cent. lead, are crushed through a 4-mesh sieve. The oxidation with these is rapid, and the roasted product less fusible and more porous than it would be if richer in lead.

The thickness of the charge on the hearth and the amount of necessary working depend also upon the character of the ore. The richer it is in lead the thinner must be the charge and the more work does it require. The thickness of the bed varies from 3 to 6 inches, the rabbling being repeated from every $\frac{3}{4}$ to every $1\frac{1}{2}$ hour.

The time required to roast the ore depends upon the readiness with which it is oxidized and its fusibility when roasted. Ores rich in blende require a considerable time and a high temperature before they are fused, if the zinc sulphide is to be completely converted into sulphate and this fully decomposed. Pyritic ores can be roasted quickly, and there is no danger of the half-roasted ore becoming sticky and adhering to the hearth of the furnace. Ores in which galena prevails require a very slow roast and a low temperature throughout, as even with the most careful roasting it is impossible to prevent the roasted ore from retaining undecomposed lead sulphide.

b. *Roasting Furnaces.*—The furnace that has survived all other modifications of open reverberatories constructed for roasting is the "Fortschaufelungsofen" or open, long-bedded, calcining furnace. The characteristic of the American type is a roasting hearth from 14 to 16 feet wide and from 40 to 60 feet long, with working-doors on either side. The hearth terminates at one end in a flue leading to the dust-chambers, at the other in a small vertical flue leading to the fuse-box or slagging hearth, which is about four-fifths as wide as the roasting hearth and one-sixth as long. The slagging-hearth with discharging doors on either side abuts on the fire-bridge.

The old furnaces, about 6 feet wide, with working-doors only on one side, are probably not to be found now. Furnaces having two hearths, one on top of the other, can still occasionally be met with—for instance, at the large silver-lead works of Mechernich,[1] Prussia. Having a double hearth has the advantage that the longitudinal extension is only one-half as great as with one hearth, and therefore, if cramped for space, one may be justified in putting it up. But a double hearth requires a much more solid construction, and is therefore more expensive; then if any repairing has to be done on the lower hearth, which is often the case, that part of the upper hearth situated above the place to be repaired has to be torn out to permit work. Finally, with a double hearth the workman, to turn over and move the ore on the upper hearth, has to stand on a high truck that runs on rails parallel with the furnace. Standing on a shaky platform the man cannot do as good work as when he is on solid ground, and it is difficult to inspect his work.

[1] *Berg- u. Hüttenmännische Zeitung*, 1875, p. 129; 1886, p. 434; *Engineering and Mining Journal*, March 3, 1877.

The consequence will be that all the ore is not moved towards the fire-bridge; particles will remain behind, and if the charge be rich in lead it will adhere to the hearth and gradually form a crust, which will have to be cut out. This requires shutting down the furnace in order to build a wood-fire on the hearth near the crust, in order that the flame may pass over it and soften it. It is claimed that fuel is saved by a double hearth, as much less heat is lost through the roof than is the case with a single hearth, but the same advantage can be gained by placing a layer of sand on the roof of a single-hearth reverberatory.

It is a great improvement on the ordinary furnace to separate the roasting hearth from the slagging hearth (or fuse-box) ; this was first done in Colorado. It effects a sudden transition from powdery to really pasty ore, which is desirable. In the ordinary furnaces the hearth near the fire-bridge has a slight depression or sump, in which the roasted ore is melted down. The consequence of this is that the ore in front of the third and fourth doors from the fire-bridge is pasty, and if the fire has not been carefully watched, the heat may be excessive up to the fifth door. Not only does this interfere with a good roast, but it also renders the moving of the ore with the paddle a very arduous piece of work. To counteract this, it has been and often still is the custom with such furnaces to collect the ore from the third and fourth doors in a heap in front of the second door, and to melt it down into the sump, whence it is removed through the first door.

This gradual passage from slagged ore through a sticky stage to pulverulent ore is the reason why revolving roasting cylinders discharging directly into a stationary slagging hearth have, as a rule, been a failure. For instance, the "revolving cylindrical roasting furnace with slag hearth" of the old Swansea Silver Smelting and Refining Works,[1] described in every text-book, was a failure from the start. Brückner[2] wanted to use at the Germania Works, Salt Lake City, Utah, two of his cylinders placed at right angles to each other. He proposed to set the upper one 4 feet higher than the lower one and to let it discharge into the latter by means of a chute. The lower one would discharge into a slag hearth, while the flame passed from the slag hearth through the lower and then through the upper cylinder. This plan was not carried out.[3]

[1] *Trans. A. I. M. E.*, iv., p. 42.
[2] *Engineering and Mining Journal*, January 15, 1887.
[3] Terhune, *Trans. A. I. M. E.*, xvi., p. 20.

Another combination of a Brückner roasting cylinder and a stationary slagging hearth is said to be at work at Spezia,[1] Italy. The cylinder, 15 feet long and 5 feet in diameter, is fired with producer gas and heated air. The gas-currents can be reversed as in a Siemens' regenerative furnace. By this means the ore is said to be uniformly roasted and sintering avoided. The roasted ore is discharged into a wagon and transported to the slagging hearth. The waste heat from this is used to warm the air for the roaster.

c. *Roasting Furnace with Fuse-Box.*—Figures 73 to 78 represent the construction of the calcining furnace which is used at most Western silver-lead smelting works. It is generally spoken of as a 14 by 60 foot furnace, which refers to the dimensions of the roasting hearth. A detailed description is not necessary, as the drawings can be understood without it. A few remarks, however, may be in place. The roasting hearth is in four separate planes, divided by 3-inch offsets, which serve to keep the charges apart. The distance between roof and hearth is thus diminished by stages, leaving the former horizontal. This can also be done with a single inclined hearth, which is preferred by Hodges,[2] who gives it as his experience that the offsets, furnishing points of attack, lead to the injury of the hearth and are not required to separate one charge from another. Only that part of the roof above the lowest roasting hearth is built of fire-brick, the rest is of red brick. In the end view (Figure 76) are seen four openings for admitting air into the roasting hearth. The additional air required enters through the two doors next to the flue, which leads into the fuse-box, the door-lids being left slightly ajar. The working bottom of the fuse-box used to be (and is still sometimes) made of quartz sand, to which small amounts of slag are added after the sand has been put into the furnace and heated until it becomes slightly sintered on the surface. This bottom is represented in the drawing. It has not proved as satisfactory as was expected, and has been generally replaced by a 9-inch fire-brick bottom, built slightly concave. The bottom (Figure 78) rests on two arched roofs, and is thus cooled by air circulating below it. In the fire-bridge (Figure 74) there is on one side of the air-space, a heavy cast-iron bridge-plate to bear the longitudinal stress of the hearth. The parts of the furnace that wear out fastest are the flue leading from the fuse-box to the roasting hearth and the fuse-box itself; the for-

---

[1] *Engineering and Mining Journal*, December 20, 1884.
[2] *Engineering and Mining Journal*, October 24, 1885.

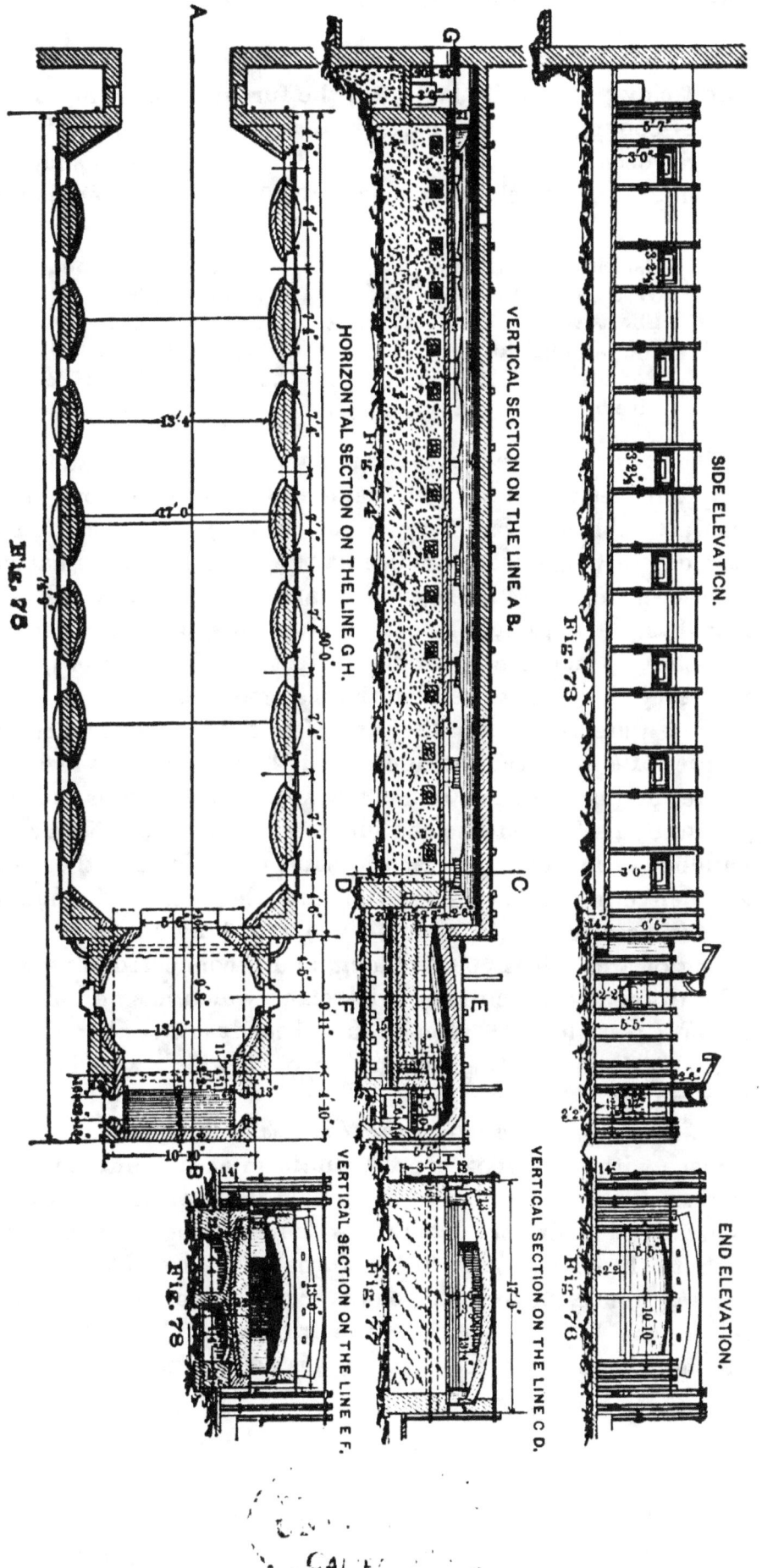

SIDE ELEVATION.

Fig. 73

END ELEVATION.

Fig. 76

VERTICAL SECTION ON THE LINE C D.

Fig. 75

VERTICAL SECTION ON THE LINE A B.

Fig. 74

VERTICAL SECTION ON THE LINE E F.

Fig. 77

HORIZONTAL SECTION ON THE LINE G H.

Fig. 72

Fig. 78

mer is patched during the run with raw clay mixed with some burnt clay ; to repair the latter, the furnace has to be shut down. Water-cooling of these parts has, as far as the writer is aware, not yet been tried. The cost of building a furnace, as shown in the drawings, is in Pueblo or Denver $3,000. The materials required are:

Cast iron	12,000 pounds.
Wrought iron	4,000 "
Sheet iron	400 "
Old rail buckstays	10,000 "
Red brick	86,000 bricks.
Fire-brick	15,000 "

The tools required by each roaster-man are: 2 paddles (blade 5 by 8 inches of ⅛-inch iron, handle 10 or 12 feet long of 1-inch iron); 2 rabbles (head 3 by 9 inches of ⅛-inch iron, handle 12 feet long of ⅜-inch iron); 1 slice-bar (1½-inch iron stem flattened to a chisel-point 3 or 4 inches wide), and 2 door hooks. The front-man has 2 scoops for the coal, 2 slice-bars, 3 rabbles (head 4 by 9 inches, ½-inch iron; handle 10 feet, of ⅞-inch iron), and the necessary slag-pots to receive the slagged ore.

In Figure 79 is given the plan of two furnaces forming part of a series of six. Near the flue end of the furnaces, 10 feet 8 inches above the furnace floor, runs a track, over which pass the ore-buggies to be discharged into the furnaces. At the opposite end of the building is another track at the same elevation for the coal-buggies delivering the necessary fuel into the bins. The flues of the six furnaces lead into a dust-chamber, 10 by 10 feet, up to the spring of arch. It is 500 feet long, and is connected through a flue 4 by 10 feet with a circular brick stack which has an inner diameter of 7 feet and is 85 feet high. The dust chamber has sliding doors to discharge the fluedust, two of which are shown in the drawing.

d. *Mode of Conducting the Process.*—The mode of working a furnace with a fuse-box is pretty uniform at the different smelting works. The ore is sometimes dried on the roof of the furnace, but generally it is directly dropped, without drying, through the hopper into the coolest part of the furnace, the weight of the charge varying from 2,400 to 3,300 pounds according to the thickness of bed the ore can bear. Sometimes the ore is shovelled onto the hearth through the last two doors, but this is only permissible when dropping through a hopper is impracticable. The charge is spread uniformly with paddle and rabble over the hearth at the highest point. If this is not separated by steps, but merely inclined, the charge is so spread that it shall lie in front of two doors. There it remains until the slagged ore is drawn from the fuse-box, when it is moved down the furnace to its second place on the next hearth, or in front of the next two doors. During its journey to the fuse-box it is not only turned over with the paddle while being moved, but is raked with the rabble once, twice, or three times, according to the interval of time between the movings. Before the charge is transferred to the fuse-box the latter receives some siliceous ore to protect the bottom. After dropping the charge the fire is urged. The liquefying begins at the surface, and much rabbling is required to bring the unfused parts from the bottom to the top. This is done at intervals of half an hour or fifteen minutes at a time at the beginning, later on for ten minutes, and towards the end for five. When the charge is fused it is drawn into slag-pots. This may be done in two ways, either all at once or in three instalments. The former method is used with charges containing 10 per cent. of lead, the latter with those running high, say 50 and 60 per cent. lead, for if they remain any length of time in contact with the acid hearth material they will take up silica and corrode the hearth unnecessarily. As soon as the front-man has drawn the slagged ore from the fuse-box the roaster-men drop in the next charge and begin to transfer the other charges downward, until the hearth near the flue is emptied and ready for a new charge. When the fused ore has become cold it is dumped from the slag-pots, broken up, and transferred to the feed floor of the blast-furnace.

e. *Results.*—A furnace like the one shown in the drawings roasts and fuses in twenty-fours from to 6 to 8 charges (varying in weight from 2,400 to 3,300 pounds), consumes from 3 to 4 tons of bituminous coal, which is half lump and half pea, and requires 3 men in a 12-hour shift—one front-man, who attends to the fuse-box and firing, and 2 roaster-men, who do the work on the roaster hearth. If the ore is not fused two men are sufficient to do the work. The cost of roasting and fusing with coal at $1.75 a ton, and labor, $2.25 for 12 hours, is $2 a ton.

f. *Roasting Furnace without Fuse-Box.*—The following is a description of a roasting furnace without a fuse-box where a galena concentrate is slag roasted. It is at Mine La Motte, Mo.[1] The hearth, 55 x 11½ feet on the inside, is slightly inclined from the flue to the bridge, making the respective distances from the hearth to the horizontal roof 15 and 22 inches. The top of the bridge-wall 22½ inches wide, is 9 nches above the hearth and 13 inches below the roof. The grate, 10 feet by 21 inches, is 3 feet 6 inches below the top of the bridge, this depth being necessary on account of the fuel used, which is wood. Of special interest is the construction of that part of the hearth where the ore is fused and of the bridge. The former is built into a wrought-iron pan resting on brick pillars between which the air circulates freely. It is formed by a full course of fire-brick, and is slightly concave. The air-cooling has proved very effective in preventing the corrosion of the brick. A similar result is produced by the air-flue passing through the bridge. When this was at first constructed in the usual way it was found that the middle part of the bridge was apt to be eaten through by the slagged ore. As a central wall divides the fireplace into two parts, the idea was conceived of closing the air-flue in the middle and erecting a small chimney on the roof, communicating with both parts of the air-flue. By this means a strong current of air could be passed through the flue. The experiment was a success, and the improvement has now been used for ten years.

The writer, has used a water jacket in fusing antimoniate of lead in a reverberatory furnace, and has found that it stopped all leakage at the bridge.

The furnace is charged every 6 hours with 2 tons of galena concentrates, to which some sand is added as acid flux. The thickness of bed is 6 inches. The galena runs from 40 to 70 per cent. lead, and from nothing to 25 per cent. iron, and is crushed to pass a 12-mesh sieve. The slagged ore retains from 4 to 6 per cent. sulphur; 4 men work on a 12-hour shift, and 0.42 cords of wood are burned per ton of ore; a second roasting furnace, without fuse-box, is described and illustrated in § 76.4.

g. *Products.*—The two products obtained by roasting are roasted ore and fluedust. The composition of some European ores running high in lead that have been agglomerated or completely slagged is given below. Complete analyses of slag-roasted ores running about 10 per cent. lead are not made at the smelting works, and therefore cannot be quoted.

---

[1] Private communication from J. T. Monell, May, 1891.

	Rodna, Transylvania.		Mechernich, Prussia.		Freiberg, Saxony.	Hall Valley, Col.	Mine La Motte, Mo.
	Raw Ore.	Agglomerated Ore.	Raw Ore.	Slag-roasted Ore.	Slag-roasted Ore.	Roasted Ore.	Agglomerated Ore.
Pb	47.29	54.27	60.40	62.08	....	....	75.95
PbO	....	....	....	....	22.0	42.04	....
Ag......	0.059	0.061	0.0105	....	0.13	Ag_2O,0.21	....
Au	0.0001	0.0001	....	....	....	....	Ni-Co1.11
Cu	trace	0.02	0.17	0.14	....	....	....
Cu_2O....	....	....	....	....	0.3	....	....
CuO	....	....	....	....	....	1.71	....
As......	0.34	0.030	NiO,10	n. d.	....	....	....
As_2O_5 ..	....	....	....	....	1.1	....	....
Sb	0.02	0.027	0.07	0.08	....	....	....
Fe......	20.36	24.06	0.80	0.56	....	....	....
FeO	....	....	....	....	....	3.59	4.5
Fe_2O_3 ..	....	....	....	....	33.3	....	....
Zn	0.67	0.87	0.15	n. d.	16.0	....	....
ZnO	....	....	....	....	....	....	....
Al_2O_3 ..	0.11	0.23	3.60	4.24	1.8	8.11	....
CaO	trace	trace	0.88	1.28	2.0	0.42	0.76
MgO ...	trace	trace	....	....	0.5	....	....
BaO	....	....	....	....	....	12.05	....
SiO_2	0.49	0.80	22.05	22.77	17.4	22.71	7.21
CO_2	....	....	....	....	....	....	....
SO_3	....	2.25	....	....	trace	6.31	....
S	29.86	2.72	9.72	0.60	8.6	2.94	7.41
O.......	....	13.41	....	....	....	....	....
Reference:	1.	1.	2.	2.	3.	4.	5.

REFERENCES : 1, *Oesterreichisches Jahrbuch*, xxix., p. 27 ; 2, *Berg- u. Hüttenmännische Zeitung*, 1875, p. 129 ; 3, *Oesterreichisches Jahrbuch*, xvi., p. 397 ; 4, *Trans. A. I. M. E.*, v., p. 568 ; 5, *School of Mines Quarterly*, ix., p. 216.

The amount of fluedust carried off with the gases is about two per cent., and is practically all collected in the dust chambers; of the metal volatilized in fusing very little if any is recovered. Fluedust from furnaces when the ore is simply roasted has a brownish color ; if slagging is carried on, it is gray from volatilized lead and zinc. The composition of fluedust from some European works is subjoined.

	Friedrichs-[1] hütte, Silesia.	Freiberg,[1] Saxony.		
Pb.........................	58.0	26.27	21.37	16.27
Ag, oz. per. ton..............	23.33	....	....	....
Cu.........................	....	....	....	0.02
Cd.........................	....	....	....	....
As	....	7.56	37.5	46.41
Zn..	2.56	19.10	2.11	0.45
Fe_2O_3	....	....	4.36	1.63
CaO........................	1.96	0.80	0.61	0.43
MgO	....	0.50	0.25	0.15
SiO_2	....	8.90	4.81	2.53
SO_3........................	25.80	28.14	8.23	7.95
S	....	....	1.34	0.60
C	....	....	3.40	3.75
H_2O........................	1.50	....	1.46	3.27

[1] Hering, " Die Verdichtung des Hüttenrauches," Stuttgart, 1888, p. 34.

Iles [1] gives as the average assay of fluedust obtained from roasting and from fusing ores the following figures:

	ROASTING.	FUSING.
Ag...................	23–28	20–37
Au...................	0.5–1.5	0.10–0.75
Pb...................	5–10	19–40
Zn...................	none	7–8

The large quantity of gold contained in it is probably due to gold-bearing pyrite concentrates from Colorado gold mill, used as iron flux. The manner of working fluedust is discussed under the head of fluedust from blast-furnaces, § 85.

§ 57. **The Selection of a Furnace Site.**[2]—In selecting a furnace site, economical as well as technical considerations must come into play. Some of these are:

a. Are the conditions such as to justify the erection of a permanent plant, or is the structure to be only a temporary one? If the former, the plant is built of stone, brick, and iron; all arrangements are so planned as to allow enlargement by simply adding to the old plant. If the plant be a temporary one, as little money as possible will be spent on the building, which is usually a light wooden one. Often the necessary capital is not available to build properly from the start, and a temporary structure has to grow by its own profits into a large, permanent plant. In such a case the general plan will commonly be the same as that of the permanent plant, the temporary structure being replaced as opportunity permits, although in some instances it may prove more advantageous to erect the permanent plant independently of the temporary one, this continuing to run till the other is ready to be started up.

b. What is the precise nature of the work to be done? Do the ores arrive in large or small quantities, at regular or irregular intervals, and through the whole year or only during a part of it? Upon these points it will depend whether the ore, as it comes, can be directly conveyed to ore-beds, whether bins are sufficient to store it, or whether a storage yard has to be provided. If the last, the ore may generally lie uncovered, but sometimes it has to be protected from snow. For fuel, protection is always needed. The size of the plant and the richness of the ore will determine whether

[1] *Engineering and Mining Journal*, January 30, 1886.
[2] Locke, J. M., " Smelting Plants," Cincinnati, 1883.

hand sampling or mechanical sampling or a combination of the two is most desirable. The character of the ore must decide whether it will be necessary to provide a place for ore-roasters and their crushing plant.

c. What is the character of the ground on which the works are to be erected? A hillside is always chosen, if possible. The questions to be settled are: how much fall is desirable and how much can be had? With an insufficient fall the main stress will be laid on having at least two terraces for the ore floor and the furnace floor with a convenient slag dump. In some instances works have to be built where there is no fall whatever, which requires additional elevators.

If excavating has to be done, the nature of the soil must be considered. If it be rocky, it requires blasting ; if earthy, grading with a scraper will be sufficient, but will necessitate the erection of retaining walls.

d. What water supply can be had for boiler and jackets? If there is no natural flow from an elevated point, a well will have to be sunk or the water obtained from a creek below, which requires pumps and additional boiler power to raise the water to the tank. The character of the water, if hard or soft, clear or muddy, may also be an important consideration.

§ 58. **General Arrangement of a Smelting Plant.**[1]—In laying out a smelting plant, the principal aim must be to simplify as much as possible the handling of materials. This is done in two ways: first, by utilizing every possible opportunity to discharge the materials by gravity into and from the truck, when transported from one place to another ; second, by making the run-ways as short as possible without too much crowding of apparatus, which must be avoided on the all-important score of necessary ventilation, and also to give room for moving freely about.

If there is a good natural fall, the arrangement will be as follows: taking the furnace floor as base-line, there will be on one side the slag dump, with a fall of 20 or more feet; on the other, the furnace, reaching to the feed floor, 16 or more feet above. The roasting furnaces, will be on the third level, if the ore is all to be roasted; otherwise at the height of the feed floor. Next comes the track from which the crushed ore is discharged into the hoppers of the roasters, 8 or more feet above the roaster floor and below the

---

[1] Locke, J. M., "Smelting Plants," Cincinnati, 1883.

discharge of the sampling mill, through which most of the ores that come to the smelter pass. It will be the level of the ore yard and the storage place for fluxes and fuels.

The precise way in which the floors are placed will vary with the configuration of the ground, as it is necessary not only to have the right fall, but also convenient grades for bringing in the materials and carrying away the products.

The situation of the machinery for driving the blowers, elevators, and dynamos, and that of the pumps and of the machine and blacksmith shops, also varies, although it is usually on the furnace floor.

The machinery for crushing is all found in the sampling de-

Fig. 80

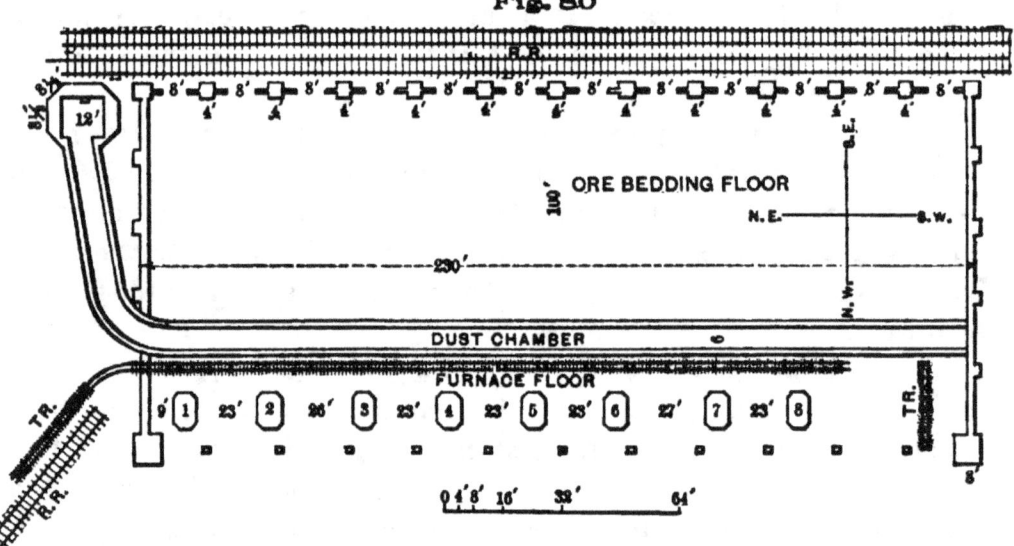

partment. The steam is usually all furnished from one set of boilers.

The office and laboratory are ordinarily near the ore yard.

The general arrangement of the works of the Globe Smelting and Refining Company, at Denver, Col., as planned by the superintendent, Dr. M. W. Iles, is based on the following scheme: if two intersecting lines are taken, running east and west and north and south, the ground-level will be represented by the southeast and southwest fields, containing respectively the furnaces facing south in a row, and the boilers and machinery for blowing, lighting, etc. The upper fields will show in the northeast field the calcining furnaces, and in the northwest field the sampling department, on the level of the feed floor, the only other level. Two sets of tracks on the upper level, running east and west, bring in ore, flux, and fuel on either side of the sampling and calcining departments; another track on the furnace floor takes away the base bullion produced. An inclined elevator running north and south brings the foul slag, matte, and fluedust from the furnace floor to an elevated track between the calcining and sampling departments and dumps the three products in the places where they are to be further treated, *i.e.*, the slag near the ore beds, the matte near the sampling mill to be crushed before roasting, and the fluedust near the fusing furnaces, where it is to be slagged.

Figures [1] 80 and 81 represent the general arrangement of the smelting department of the Grant and Omaha Smelting and Refining

Fig. 81

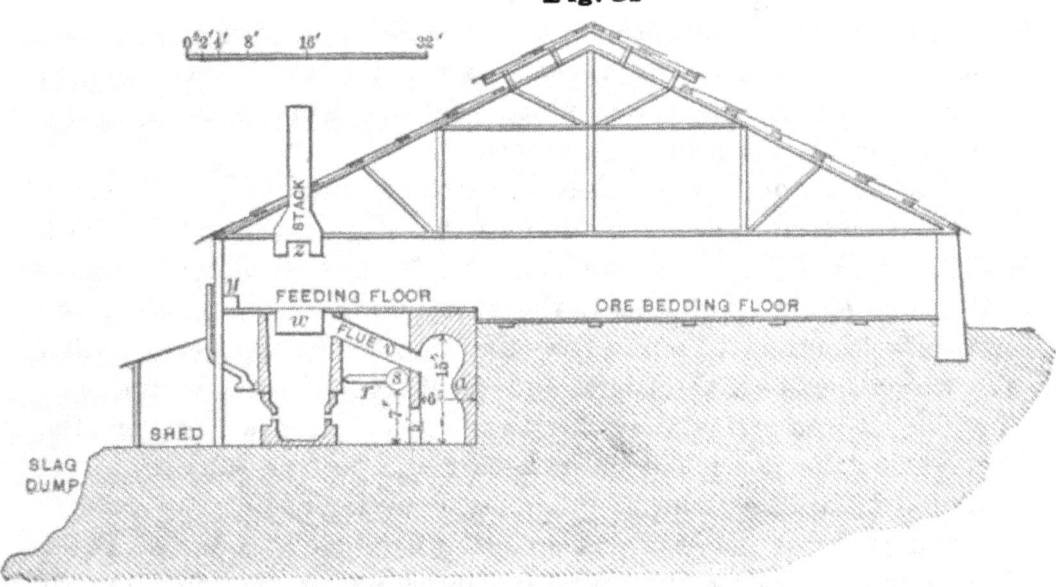

Company's works at Denver, Col. A few changes in detail have been made, but the general outline is correct. This department corresponds to that one of the Globe Works represented in the southeast field just referred to, and is similar in construction.

LEGEND.

Nos. 1–8—.Blast-furnaces.

a = Projection in the dust chamber (abolished).

r = Bustle-pipe.

s = Induction-pipe.

v = Flue carrying **fluedust from the blast-furnace into the dust chamber.**

w = **Sheet-iron curtain through which the charges are fed into the blast-furnace (abolished).**

[1] Frost, *Engineering and Mining Journal*, March 24, 1883.

$y =$ Elevated tramway for the fuel trucks that the contents may be dumped on the feed floor.

$z =$ Telescope stack, used in blowing in or out, to carry explosive or hot gases out into the open air (abolished).

$RR =$ Broad-gauge track on feed floor on which ore and fluxes arrive.

$TR =$ Narrow-gauge track on furnace floor delivering the base bullion to the broad-gauge track R. R.

$TR' =$ Tramway for raising slag, etc. (altered).

To be noted especially is the large ore-bedding floor. It is the practice to make two large ore beds, each occupying nearly one-half of the floor, about 8 feet high and holding about 3,000 tons of ore. All the furnaces receive their ore from one bed, and while this is being consumed, the other bed is made.

Another feature not to be overlooked is the position of the fronts of the furnaces in regard to the points of the compass. Facing northwest they are as much as possible in the shade, an important consideration in hot weather.

The sheet-iron hood z', placed in front of each furnace to carry off the fumes that arise on tapping the slag, never did its work satisfactorily. The hood now ends in a horizontal pipe (x', Figure 96), which either terminates in the dust-chamber or in a galvanized iron pipe, common to a number of furnaces, and connected with a fan which sucks off the fumes and discharges them into the open air. Sheet-iron plates hung on either side of the hood prevent the draught on the furnace floor from carrying off the fumes into the building before they can be taken away by the hood.

The works of the Montana Smelting Company, at Great Falls, Montana, planned by Mr. A. Eilers, are shown in plan and vertical section in Figures 82 and 83. Their general arrangement differs very much from that of the two works previously discussed. In the latter the longest extension is parallel with the row of furnaces; at the Montana works this is reversed. A striking feature is the extensive roasting plant with its separate ore yard. The study of the plans will show that great attention has been given to the handling and storing of large amounts of ore. The loaded cars, arriving on the main track, are weighed on track-scales and then switched off on side tracks, which lead to the "crushing house and sampling works," to the "bins for sulphide ores" (tracks 9 and 10), and to the "bins for carbonate ores" (tracks 2–6). At the sampling works all ores requiring crushing are received, and the sulphide bins are filled with concentrates, which are sampled by fractional

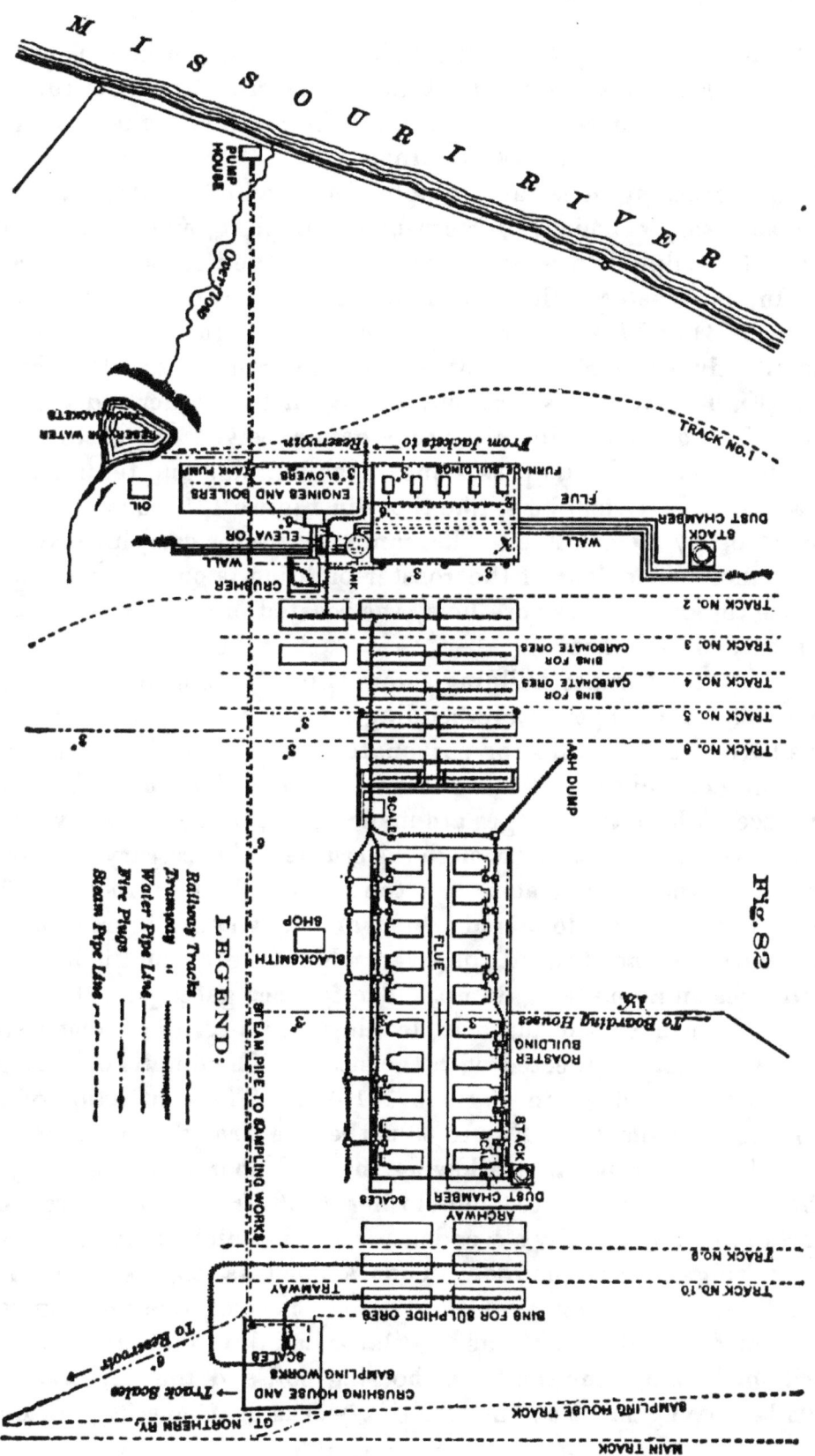

FIG. 82

selection while they are being unloaded. The lower bins are for ores, sampled in the same way, that go straight to the blast-furnace. Track No. 1 brings the coal for the boilers and takes the base bullion produced in the blast-furnaces.

. The crushing house and sampling mill, which is fitted up with the necessary machinery for crushing, sampling, and grinding, delivers the sulphide ores over a tramway to their respective bins.

In the roaster building is room for twenty calcining furnaces (see § 76.4) and fusing furnaces. They deliver their gases into two parallel flues (see Figure 163) running along the centre of the building (Figure 82). These are built one against the other and are combined into one main flue after they have received the gases of the single furnaces. This passes out of the building and terminates in a series of dust chambers, like those in Figure 171, connected with the stack by a small flue. The tramways are for carrying ore, fuel, and ashes. The floor of the roaster building is on the same level as the top of the lower ore bins; the roasted ore can thus be easily discharged into them.

The blast-furnace building shows three floors instead of two, as is usual. The upper or ore-bedding floor is on the same level as the bottom of the bins for carbonate ores. From the beds made on it the ore mixture is dumped onto the feed floor near the single furnaces, where the charges are made up. The fluxes, arriving on track No. 2, are passed through the crusher, if necessary; the fuels arrive on the same track. On the furnace floor is room for five blast-furnaces, the details of which are shown in Figures 99–120. The furnaces face toward the river, where the ground is high and furnishes an ample slag dump. The furnace gases pass through a wide flue back of the furnaces, leading into a series of dust chambers, which are connected with the stack. The details of flue and chambers are given in Figures 171–177. The ventilation of the furnace floor shown in Figure 83 is the one strongly recommended by Eilers. It consists in allowing the feed floor to extend only to the front of the furnaces and closing it off from the entire front part of the building by a wooden partition, which slants backward to the ridge-pole of the roof. Thus a hood as long as the furnace building, and having a width of about 12 feet, reaches from the level of the feed floor to the ventilator and draws off all the vapors and smoke from the tap-hole, the slag-pot, and the lead-well. A similar arrangement at the smelting-works in Clausthal,[1] Prussia,

[1] Private notes, 1890.

was very efficacious in removing the danger of lead poisoning, which had previously been a common thing.

Next to the furnace room are placed the boilers, engines, blowers, etc. The drawing, Figure 82, shows the blacksmith-shop, the pump-house, the water pipes and fire-plugs, the water-tank and lower reservoir, which receives the overflow of the jackets, but not the upper reservoir.

§ 59. The Blast-Furnace and Its Accessory Apparatus.[1]

a. *Introductory Remarks.* The blast-furnaces used in lead smelting are of the most various description. Taken in cross section they are square, polygonal, circular, oblong, and elliptical. In vertical section they are prismatic or with sides tapering towards the bottom, and in addition may or may not have a bosh. Then, the smelting zone may be enclosed by water jackets and the crucible may be internal, or partly internal and partly external. Furnaces with a detached crucible have been constructed and patented,[2] but

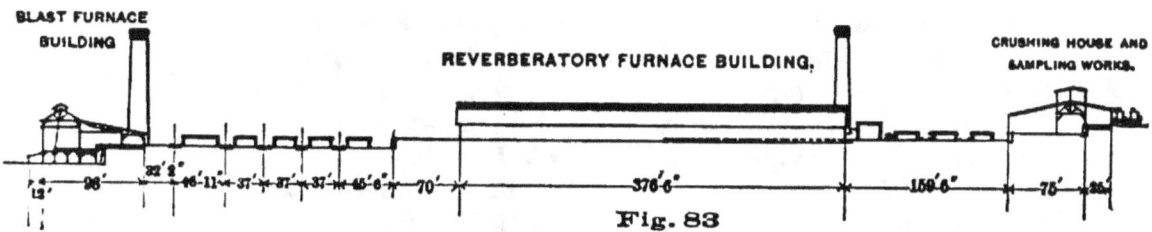

Fig. 83

are not, as far as the writer is aware, in use. Finally, the lead may either be tapped from the bottom of the crucible or removed by means of Arents' automatic tap.

Only two kinds of furnaces are now in use in this country, both having an internal crucible, water-jackets and Arents' automatic tap. The one is circular, has the form of an inverted truncated cone, and is used for smelting small quantities of ore or by-products of refining works ; the other (Figures 84–120) is oblong, its sides are either vertical or only slightly inclined, it always has a bosh, and is the common ore-smelting furnace. Oblong furnaces have almost entirely replaced circular furnaces.

In the subjoined legend the letters used refer to the similar

[1] Working tools are discussed in § 63–67.

[2] Devereux : 1887, December 6, No. 374,239 ; 1888, April 17, Nos. 381,118 and 381,119 ; June 12, No. 384,349 ; 1889, July 23, Nos. 407,335, 407,336, and 407,337 ; December 17, Nos. 417,314 and 417,315 ; 1890, May 6, No. 427,058. Konemann : 1888, October 9, No. 390,785. Wilson : 1889, May 21, No. 403,815, and others.

parts of the four blast-furnaces (Figures 84–120) chosen as characteristic types.

LEGEND.

A. Cast-iron breast jacket, right.
B. Cast-iron breast jacket, left.
c. Shaft, red brick shell.
c'. Shaft, fire-brick lining.
E. Cast-iron side jacket.
F. Wrought-iron side jacket.
G. Wrought-iron back jacket.
H. Wrought-iron front jacket.
a. Slag-spout.
b. Crucible.
c. Lead-well or basin of Arents' automatic tap.
d. Syphon, the inclined channel of Arents' automatic tap.
e. Water jackets.
f. Cast-iron water feeder.
g. Lateral water-supply pipe.
h. Water-feed pipes.
i. Water-overflow pipes.
j. Galvanized-iron water trough.
k. Cast-iron drain pipe.
l. Breast of furnace.
m. Tap-hole.
n. Tapping jacket.
o. Tuyere.
p. Blast-pipe.
q. Wind-bag.
r. Bustle-pipe.
s. Induction-pipe.
t. Cast-iron collar, supporting—or carrier-plate.
u. Cast-iron pillar.
v. Down-comer.
w. Sheet-iron curtain.
x. Main water-supply pipe.
y. Lugs fastened by bolts.
z. Sheet-iron hood.
a'. Lead-spout.
b'. Wrought iron bolts.
c'. Corner-irons for tie-rods.
d'. Tie-rods.

e'. Interior of water jackets.
f'. Channel of water feeder.
g'. Brick arch.
h'. Crucible-castings.
i'. Strengthening-ribs of crucible-castings.
j'. Telescope stack.
k'. Chains.
l'. Counter-weights.
m'. Feed-door.
n'. I-beams.
o'. Capital of pillar.
p'. Brass nozzle.
q'. Hand-hole.
r'. Wrought-iron pipe connecting the single jackets.
s'. Eye or peep-hole.
t'. Cast-iron flange.
u'. Hog-chain.
v'. Iron band.
w'. Top-plate.
x'. Pipe leading to fan.
y'. Bed-plate.
z'. Hood leading into pipe x'.
a''. Tuyere-box.
b''. Steel rails.
c''. Wrought-iron rods.
d''. Expansion-space.
e''. Sliding sheet-iron door.
f''. Angle-iron ring.
g''. Angle-iron damper.
h''. Crank with nut.
i''. Circular guide with slot.
j''. Groove for damper.
k''. Water feeder.
l''. Peep and poking-hole.
m''. Outlet for furnace gases.
n''. Swinging valve.
o''. Cross-pin.
p''. Movable weight.

Figures 84–88 represent the blast-furnace of the Omaha and Grant Smelting and Refining Company's Works at Denver, Col., of

1883; Figures 89–95, the furnace of the Globe Smelting and Refining Co., at Denver, Col., 1891, designed by Iles; Figures 96–98,

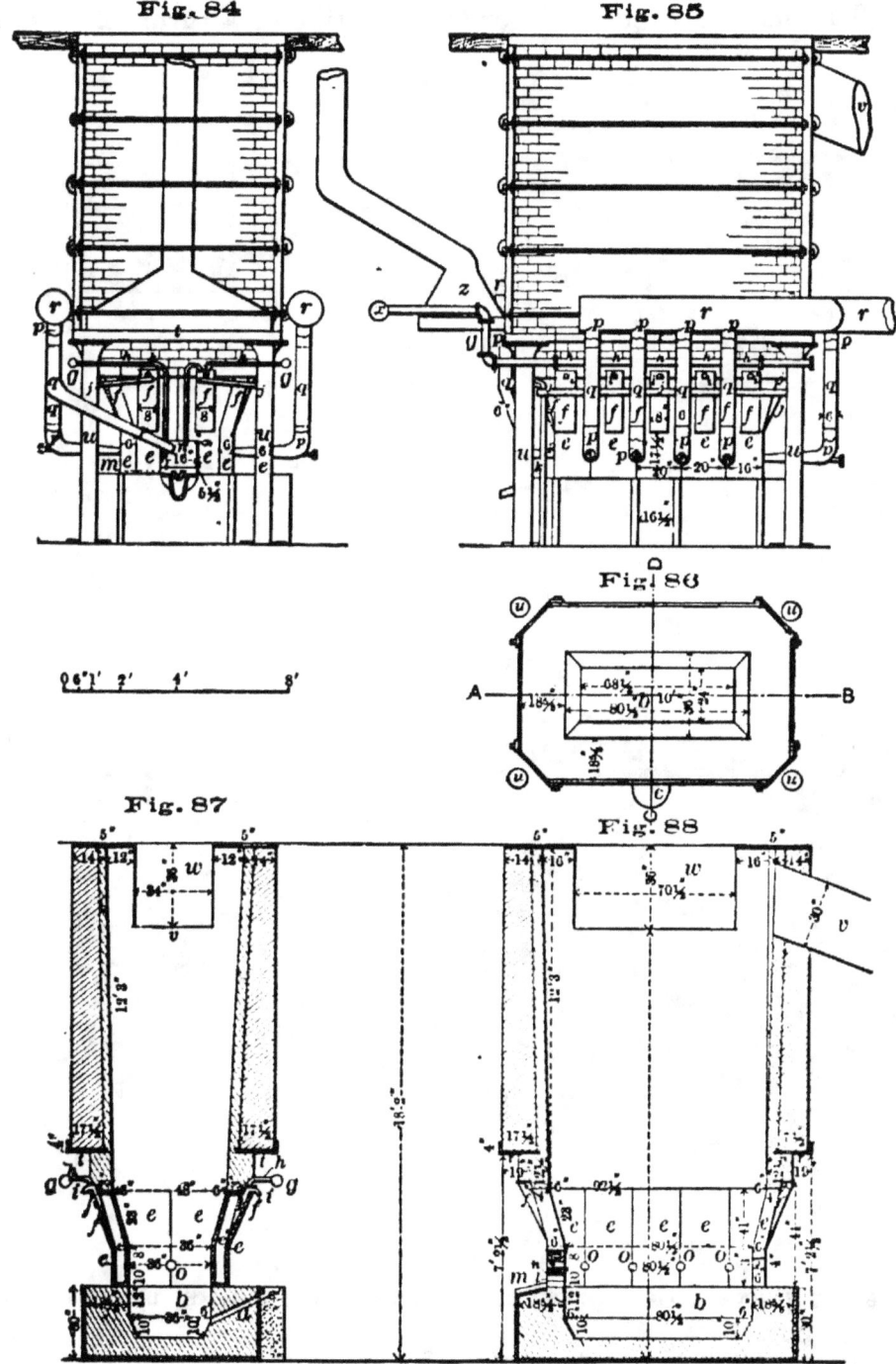

Fig. 84   Fig. 85

Fig. 86

Fig. 87   Fig. 88

a furnace designed by the Colorado Iron Works at Denver, 1891; Figures 99–120, the furnace of the Montana Smelting Co., Great Falls, Mont., 1891, designed by Eilers.

No drawing of the circular furnace is given, as it is little used now; its advantages and disadvantages will, however, be discussed in connection with the oblong furnaces.

The materials required for the erection of a furnace, as shown

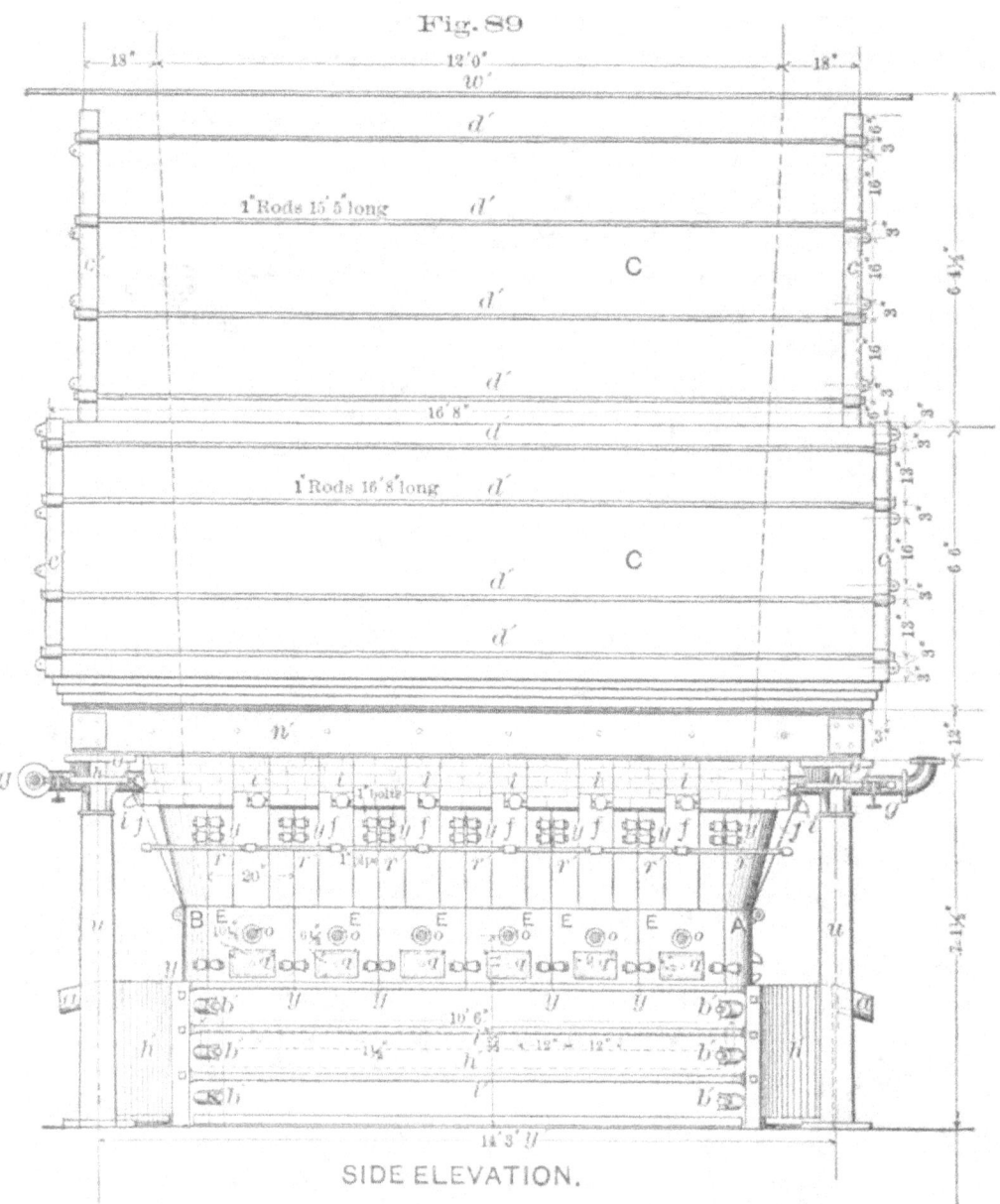

Fig. 89

SIDE ELEVATION.

BLAST-FURNACE OF THE GLOBE SMELTING CO.'S WORKS, DENVER, COL.

in Figures 89–95, are : cast iron, 27,300 pounds ; wrought iron, 3,200 pounds ; steel beams, 4,250 pounds ; fire-brick, 9,500 bricks ; red brick, 17,000 bricks. Where a "telescope" stack is used, 1,600 pounds of wrought iron must be added to the above figures. The cost of erecting the furnace, excluding all the fittings for blast and water, at Denver, was $1,200, one-quarter of which went for labor.

b. *Foundation.*—The first thing in erecting a furnace is to have a solid foundation. Its depth will depend on the character

Fig. 90

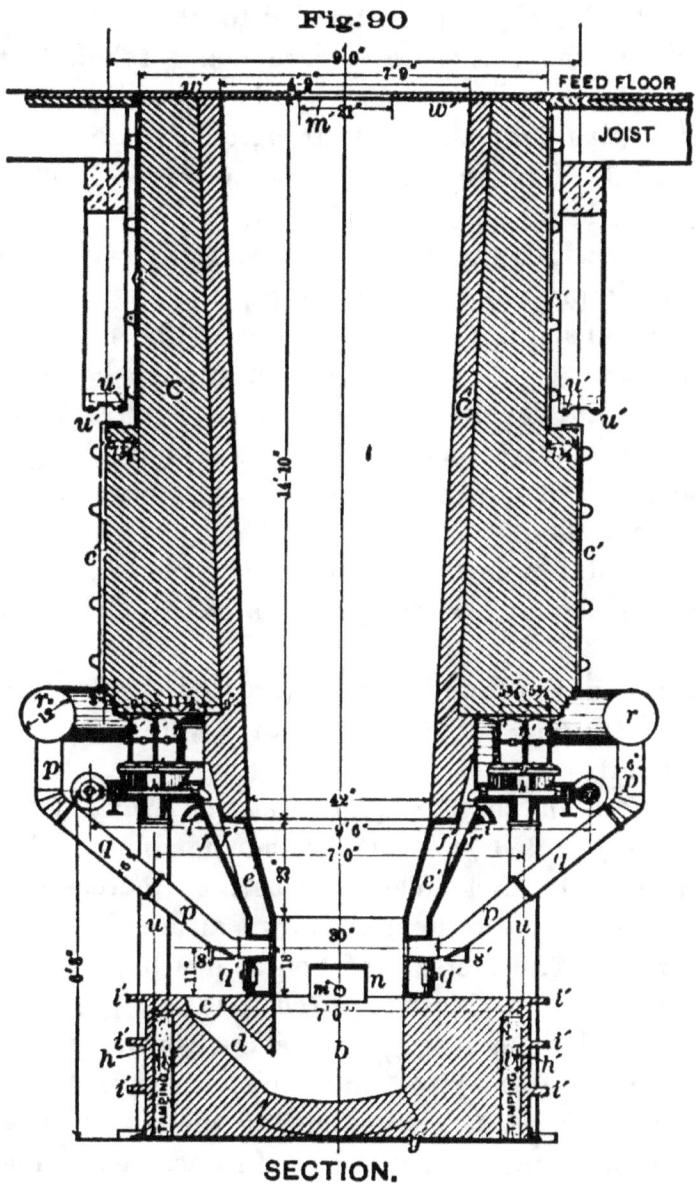

SECTION.

BLAST-FURNACE OF THE GLOBE SMELTING CO.'S WORKS, DENVER, COL.

of the subjacent ground. If there is exposed bed rock this will furnish as good a foundation as can be wished for. If there is loose soil or gravel covering bed rock for not more than 10 feet, it is best to excavate until this is reached; otherwise a depth of 5 feet will usually be sufficient to start the masonry below frost line, and to give the foundation the requisite strength. With very loose soil it is sometimes advisable to place in the bottom of the pit two layers of 3- or 4-inch planks spiked crosswise to each other, and upon that to build the foundation, which should extend from 2 to 3 feet beyond the bed plate and the four pillars. It is built up of undressed rock, well rammed into place, the largest pieces being used for the corners, and care being taken to fill up the crevices and joints with as many spalls as possible; the whole is well grouted with a mixture of four parts of lime mortar and one part of cement. The topmost course must be absolutely smooth and horizontal, being generally of brick.

If one furnace is already in operation and a second one is to be erected, the simplest way of obtaining a good foundation is to empty the liquid slag into the place that has been excavated, or to throw in broken up slag and to cement it with liquid slag. The top is evened and levelled by making shallow rectangular areas surrounded by sand or pieces of iron rails, and filling them level with liquid slag. Any ridges or other rough parts that remain are removed by chipping.

On the foundation is spread a thin clay mortar, upon which the wrought-iron bed plate y' (Figures 89, 90, 96, 98, 99, 100) is placed.

c. *Shaft.*—On the foundation are erected the four hollow cast-iron pillars u (Figures 84–86, 89, 90, 92–94, 96–101), which are to support the shaft.

The *height* of the shaft, *i.e.*, the distance from the centre of the tuyeres to the feed floor, has been somewhat increased of late. It used to be from 10 to 12 feet; it was then increased to 14 feet, which is the common dimension now, although occasionally it reaches 18 feet. The increase of height has been necessitated by the greater pressure of blast required for the highly siliceous and calcareous slags. The ferruginous slags, formerly made, needed only a pressure of from $\frac{3}{4}$ to 1 inch quicksilver, instead of from 1 to 2 inches as at present. By enlarging the distance between the tuyeres, to increase the capacity of the furnace, it became necessary also to increase the pressure from 2–2$\frac{1}{2}$ inches quicksilver. This is a disadvantage, as will be shown further on.

The *horizontal section* of the shaft is either a circle or an oblong. When a square or polygonal furnace is blown out, the inside will have a similar appearance to that of the circular, as the corners very soon fill up. The same holds good with the oblong furnace.

The circular furnace gives, as regards the quality of work, sastisfactory results; there is an even distribution of blast and heat, and as it offers the largest surface for the smallest circumference, the loss of heat by radiation is the least possible. The drawback lies in the quantity of the work, which is limited, since the diameter at the tuyere-section must not exceed 42 inches (36 inches being the common dimension), as too great a pressure of blast would be required. For large quantities of ore the oblong form is therefore the proper one, as the area can be enlarged by making the furnace longer without increasing the distance between the tuyeres. Thus the length of some oblong furnaces has been doubled in the last ten years (from 60 to 120 inches). In increasing the width the pressure of the blast has to grow, and with it the height of the furnace. With high-pressure blast the heat creeps up and smelting begins at the top of the jackets, in which case the walls are liable to be eaten out, instead of just above the region of the tuyeres, which of course becomes cool and causes rich slags and rich mattes. Another result is a greater loss of metal, as the retarding effect of the boshes on the hot gases becomes weakened and the working height of the furnace diminished. Furnaces have been built 48 inches wide at the tuyeres, the water-cooled nozzles protruding 6 inches into the furnace, thus making the distance between the jackets 60 inches. In order to prevent a wedge of unmelted charge from forming along the centre, the bulk of the fuel has to be fed there in order that the powerful blast, which has more than $2\frac{1}{2}$ inches quicksilver pressure, may penetrate the charge. If a furnace 60 inches wide at the tuyere-section smelted twice as much charge as do two 30-inch furnaces, its size might be justifiable, as in the end it might be cheaper work. But it does nothing of the kind. It puts more charges through than one set of furnace hands can handle in a twelve-hour shift, and not enough for two sets, and is more expensive than smaller-sized furnaces. It is improbable that 60-inch lead furnaces will again be built.

For the usual run of ores the width at the tuyeres may range from 30 to 36 inches, the height above the tuyeres from 11 to 14 feet, and the blast pressure from 1 to $1\frac{1}{2}$ inches quicksilver. A furnace 33 by 100 inches at the tuyeres, with five $3\frac{1}{4}$-inch tuyeres on either side, 12 feet active height, will smelt with $1\frac{1}{8}$ inches quicksilver pressure as much medium-coarse charge as one regular crew of men will be able to handle in a twelve-hour shift, *i.e.*, 60 tons of charge, **or about 45 tons of ore.**

The *vertical section* of all oblong furnaces shows the bosh, and with it the contracted tuyere-section. This last secures a more perfect and rapid combustion and thus a concentrated, intensified heat, with the result of a quicker fusion and a more complete decomposition of sulphide and arsenide of lead. If somewhat higher up the width of the furnace is suddenly enlarged by the bosh, the zone of fusion will be narrowed; further, the gases generated at the tuyeres will be evenly diffused, thus checking the velocity of their upward motion; by gradually giving up their heat they prepare the charges for the subsequent smelting process and decrease the amount of fluedust formed.

That a circular furnace having the form of an inverted cone cannot fully possess the same advantages is evident, but whether the enlargement between the tuyeres and the throat of the furnace be sudden or gradual, the relative areas of the hearth and of the throat remain nearly the same—1 : 2 or $2\frac{1}{2}$.

The shaft (c.c. Figures 90, 98–100, 102) is made of common brick c, and lined with fire-brick c′ up to the feed floor. It rests on four supporting plates t, which used to be (Figures 84, 85, 87, 88, 98) fastened to the capitals o′, of the pillars u, but the effect was to loosen the pillars by the unequal expansion of brickwork and cast-iron plates, and thus endanger the safety of the shaft. To relieve the pressure upon the plates, brick arches (g′, Figure 96) were introduced, supporting the walls of the shaft and throwing the weight upon the pillars. To counteract the lateral thrust the lower part of the shaft is enclosed in cast-iron plates (Figures 84, 85, 96, 98) firmly bolted together, in addition to which the plates t (Figure 98) sometimes have flanges t′. In the latest furnaces a set of three I-beams (n′, Figures 89, 90, 93, 99, 100) on each side of the furnace form the support of the cast-iron plates t (Figures 99, 100), being firmly bolted to each other and screwed tightly to the capitals o′ of the pillars. The cast-iron plates t are in no way fastened to the beams, but rest freely upon them. By this arrangement supporting plates and shaft can expand independently of one another without endangering the stability of the shaft. The plates, being supported by the I-beams, no longer need to be reinforced by the arches, but can bear the weight of the shaft safely.

The walls of the modern furnaces are made very thick at the bottom in comparison with older ones ($32\frac{1}{2}$–39 inches against $17\frac{1}{2}$ inches, Figures, 89, 90, 99, 100, against 87, 88), decreasing toward the feed floor, which causes a considerable saving of fuel.

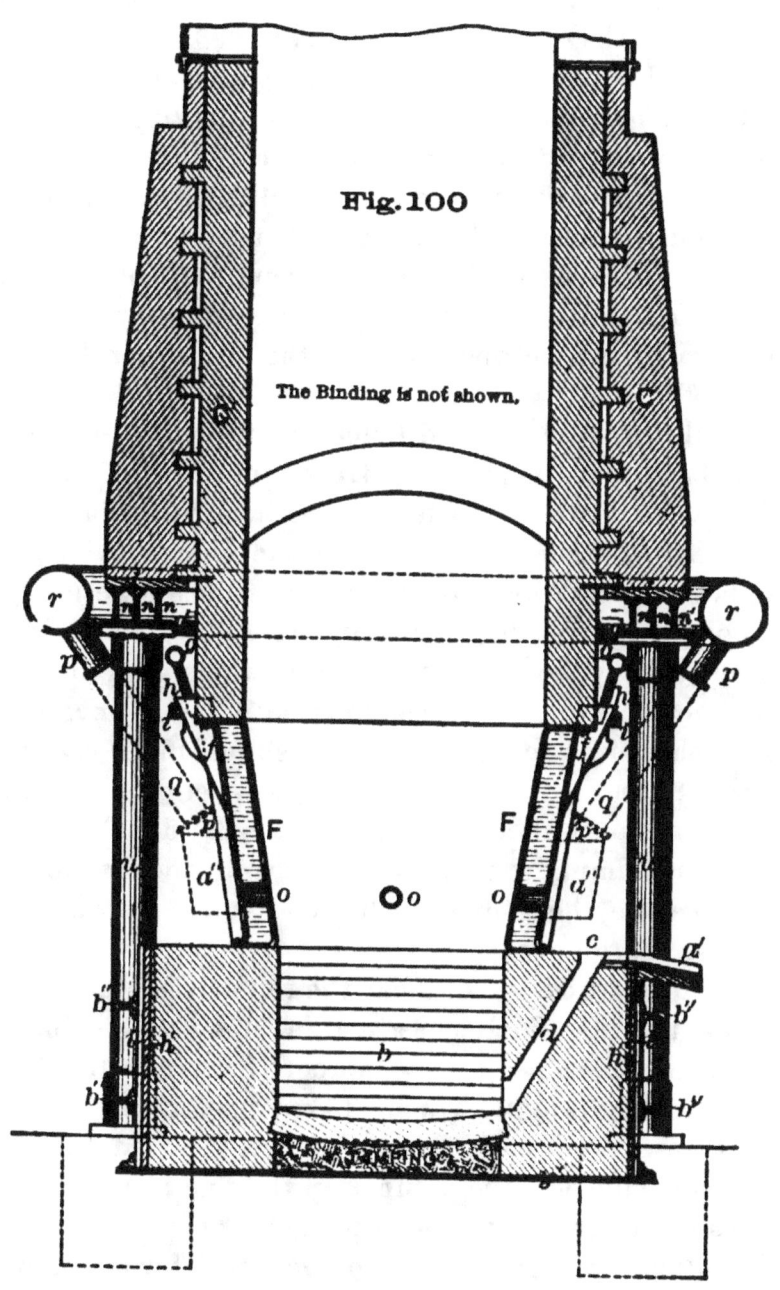

Fig. 100

The Binding is not shown.

BLAST FURNACE,

MONTANA SMELTING CO.

The entire shaft is well braced with tie-rods d', secured in corner-irons c' (Figures 84, 85, 89, 90, 95, 96).

d. *Feed Holes and the Collecting of Fumes.*—There are two general arrangements for feeding and carrying off fumes. The first is to have a feed hole on either side of the furnace m' (Figure 102). The stack or chimney c (Figures 102, 103) is of brick, forming the continuation of the shaft; it is contracted at the upper end to about $3\frac{1}{2}$ feet square (inside measurement), and then passes through the roof. The top is closed by a swinging damper resting in the groove j'', and can be opened from the feed floor by means of a damper rod.

In small furnaces the feed holes are placed in the middle of the sides, in large ones nearer the front and back of the furnace respectively, *i.e.*, not opposite each other, but diagonally. This makes it much easier to distribute the charge evenly and to bar down the wall accretions, dividing the furnace practically into two. The doors (Figures 108, 109, 110) are 5 or 6 feet high, in order that a man may be able to stand in them and direct the bar. During the run they are closed to from 18 to 24 inches by letting down a sheet-iron curtain e' that slides in a cast-iron frame (Figure 108) and is balanced by counter-weights. The sill of the feed floor is placed about one foot above the floor level, so that the feeder cannot simply shovel the charges into the furnace, but, being obliged to raise the shovel each time, will be more likely to distribute the charge evenly.

The fumes are drawn off near the top of the stack by a circular sheet-iron flue v, which passes at a steep angle into the dust chamber. In the flue is the damper (Figures 102, 104, 105, 106) to regulate the draught, which is kept just strong enough to prevent the gases from passing out through the feed holes. In blowing in or blowing out, the damper in the flue is closed and the one on the top of the stack thrown open.

The other arrangement is to feed the furnace from the top (Figures 87–91, 96–98). This is covered in part by cast-iron plates w' (Figures 91, 98), leaving an opening through which the charge is introduced. Formerly the Pfort method (Figures 87, 88) of carrying off the gases was in general use. It consists in suspending from the cast-iron top-plates w' an iron curtain w, so as to leave room between it and the walls of the furnace for the gases, whence they pass off through a flue v into the dust chamber (Figure 81), the charge filling the inside of the curtain up to the feed floor. While this arrangement proved very satisfactory in a good many ways, it had one great disadvantage, that it lengthened the time required for barring down wall accretions, because it was necessary to remove the curtain before and to put it back again afterward. For this reason the curtain has been almost universally discarded and the gases are simply drawn off by a flue v (Figures 96, 98) at the back of the furnace, or better by two on the sides. To avoid sucking in air or letting out fumes, the feed opening m' in the cast-iron top plate w' (Figures 90, 91, 98) is made rather small; it is, however, large enough for the feeder to be able to spread his charge in any way that may be necessary, and to reach any part of the side walls with a bar when cutting out wall accretions. The flue leading to the dust chamber, which used to be sheet iron (Figure 88), is now commonly built of brick (Figure 96). It rests on heavy rails and is thoroughly bound with buckstays and tie-rods.

Formerly a " telescope stack " (z, Figure 81, and j', Figures 96 and 98) was suspended by chains k' over every furnace. It is a sheet-iron pipe reaching through the roof and balanced by counter-weights l'. Its lower part is enlarged to the oblong form of the feed opening and has a small feed door m' on either side. The stack is lowered when the furnace smoke is not sufficiently drawn off by the flue, or when the furnace is blown in or out, to carry off the gases into the open air.

At present, large smelting plants have (say) two of these sheet-iron stacks suspended from a traveller, to be used in case of necessity, and in some instances this stack has been thrown off entirely (Figures 89–95). Instead, above each furnace is suspended a $\frac{3}{4}$-inch cast-iron plate sufficiently large to close the feed opening m'. It is lowered when the furnace is being blown out, to prevent the fumes from passing to the feed floor; the joint is made air-tight by spreading moistened fine ore over it.

Another manner of collecting the waste gases with open-mouthed furnaces may be mentioned here. This is the Darby tube, a wrought-iron pipe of small area, in comparison with that of the throat of the furnace (280 square inches as against 3,267). It is hung in the middle from the girders, some distance (5 feet $2\frac{1}{4}$ inches) down into the furnace. This tube is used in the Upper Hartz Mountains, where fine galena concentrates are smelted raw. The advantages claimed for it are that the charge is less liable to pack toward the centre, and that the gases, being drawn off there, are prevented from rushing up at the sides, and penetrate the charge more evenly.

The idea of closing the throat of the furnace with cup and cone and feeding automatically has been often suggested, *e.g.*, by Hahn [1] but not tried until lately; the results of the experiments have, however, not been made public.

e. *Hearth with Arents' Automatic Tap.*—The bottom of the hearth is formed by a bed plate of boiler iron y' (Figures 89, 90, 96, 98–100), which is to prevent any lead from percolating downward. It is placed on the foundation, as indicated in § 59, b. Care must be taken to have its centre coincide with that of the shaft by dropping a plumb-line from the feed floor. The bed plate sometimes has an angle-iron rim enclosing the bottom course of brick (Figures 89, 90, 99, 100); sometimes it reaches beyond the castings (Figures 96 and 98), enclosing the hearth, which rests upon it.

These castings h' (Figures 86, 89, 90, 92, 96–101), reaching to the top of the hearth, have been (and still are) a great trouble, as they are very liable to crack. At first they were made 1 inch thick; later the front and back plates were strengthened by ribs i (Figures 89, 90, 92, 96–101); then the side plates, castings, and ribs have been made thicker and the bevelled corners have been fastened together by special tie-rods b' (Figures 89, 92). Still there is danger of their cracking, so that in some furnaces a wrought-iron band v' (Figure 90) is screwed to the sides to hold the casting together. It would seem as if making the outer wall of the hearth oval and enclosing it with a ⅜-inch wrought-iron plate, as is done in the modern large iron blast-furnaces, would be the way to solve the difficulty.

To the casting on the front is fastened with bolts the slag spout a (Figures 89, 91, 96, 97, 99, 101), and to the side the lead spout a' (Figures 98, 100, 101), if the well is confined within the hearth plates, as is now usual. It is not often that the slag is tapped alternately from the front and the back of the furnace, requiring two slag spouts (Figures 89, 92, 96, 97), and that the lead is removed from the two sides (Figure 98). Two slag taps have been used at some furnaces to counteract the forming of a crust at the back of the furnace, where it usually begins growing toward the front, from which it cannot easily be reached. If the slag is tapped from both front and back, the danger of crusting at the back is in

---

[1] "Mineral Resources of the United States," 1882, p. 343.

part at least avoided, and if an obstruction forms there, it is easily removed. Having two lead-wells is simply a waste of heat, for when one becomes clogged up, the other will also.

The hearth walls and bottom are of fire-brick. They usually rest on the bed plate (Figures 87, 88, 90, 98). Sometimes on the bed plate and below the crucible a 6-inch layer of ground brick and raw clay (3 : 2, by volume) is beaten down firmly in the form of an inverted arch (Figures 99, 100) on which the bricks forming the bottom proper are placed. In building the side walls it is better not to place the bricks in direct contact with the castings (Figures 87, 88, 98–100,) but to leave a small space of (say) 2½ inches and tamp it out with brasque (equal volumes of ground coke and clay) while the bricks are being placed (Figure 90). In this way the crucible, when it expands, will simply pack the brasque tighter, and thus relieve the castings from at least part of the strain.

Arents' Automatic Tap, or siphon tap, which forms part of the side wall, consists of an inclined channel, the siphon *d* (Figures 87, 90, 98, 100), 3 or 4 inches square, running from the lowest part of the crucible wall inside to the top on the outside, where it is enlarged into a dish-shaped basin—the lead-well *c* (Figures 87, 90, 97, 98, 100, 101), the length, of course, depending on the depth of the crucible, which varies from 22 to 30 inches. The tap is usually in the middle of one of the sides (Figures 96, 97) although sometimes placed nearer the front (Figure 101); while the furnace is running, the crucible remains nearly full of lead, that in the automatic tap standing a little higher on account of the pressure of the blast. From the well the lead, as fast as it is made in the furnace, is either ladled into moulds or it overflows into the cooling-pot, or it is periodically tapped into it, and is thence ladled into moulds. When first used,[1] Arents' tap consisted of a 3-inch wrought-iron pipe which terminated a foot or more below the upper rim of a sheet-iron cylindrical shell rammed full of fire-clay and bolted to the casting, the well being afterward cut out. The wrought-iron pipe has been universally abandoned. The remains of the sheet-iron cylinder are still found in some instances in a half-cylinder that is bolted to the casting (Figures 86, 87). In most furnaces to-day the lead-well is enclosed in the crucible wall (Figures 90, 96, 98, 100, 101). The advantages of this improvement are that the siphon is shortened and the lead is kept hotter. In smelting

---

[1] Hahn, Eilers, Raymond, *Trans. A. I. M. E.*, i., p. 108.

charges that run high in lead, the lead in the crucible is frequently exchanged, and a slight loss in heat does not make itself felt; the siphon can be long, and the well, not being close to the tuyere-pipes, will leave these cool. With charges low in lead, all loss of heat must be avoided. The increased thickness of the side wall, necessitated by the enclosed lead-well, is altogether an advantage, because the loss of heat is diminished. In any case, the side walls ought never to be less than 22 inches thick.

The advantages of the automatic tap are many. Without it the lead and matte are tapped from the bottom of the crucible. To do this the blast is shut off, the blast-pipes are removed, and the tap-hole is opened, which is often done with difficulty. Lead and matte run out into a shallow tapping-kettle, and the moment the slag appears the opening must be closed with a stopper of brasque or clay. Then the crucible is cleared by inserting through the fore-hearth (if it has one) a curved iron bar, the tuyeres are cleaned with iron rods, the blast-pipes are put in place, the blast is then turned on again slowly, and smelting is resumed. This stoppage takes considerable time, and therefore cools the furnace. Then into the crucible (now free from lead and matte, although it retains some fuel) falls an equivalent amount of half-melted charge, which has gradually to be lifted up by fresh lead and matte, when tapping begins again. These half-melted masses thus have a chance to adhere to the bottom of the crucible, and are very apt to be the beginning of a bottom crust. If this has once started, it is nearly sure to grow and gradually freeze up the furnace.

With Arents' tap there is no stoppage when the lead is removed from the lead-well or when matte and speise are tapped with the slag at intervals into a slag-pot, where they settle out according to their specific gravities. The furnace therefore runs more regularly, and the first formation of slag or matte accretions at the bottom is prevented, as the crucible is always filled with lead.

The claim originally made that the lead from the automatic tap is purer than that from the tapping-kettle will hardly be maintained to-day. It was based on the theory that the metal, being taken continuously in small quantities from the bottom of the crucible, would be purer, as the heaviest, *i.e.*, the purest lead, would gather there, and the impurities would float to the surface, to be taken up by matte and slag. This presupposes that the lead in the crucible is sufficiently undisturbed to permit liquation. The facts are that there is a constant current in the lead which prevents this separation of dross and lead. The dross is disseminated through the lead; some reaches the surface and is taken up by matte and slag; but a large part of it rises in the siphon and collects in the well. Thus the bars from the lead-well are often less pure than those from the tapping-hearth, as with the latter the dross adheres to the cake of matte floating on the still liquid lead. However, by skimming the dross from the well or cooling-pot, clean bars can be obtained and the dross returned at once to the charge, thus involving little loss in metal. With the tapping-hearth, the dross adhering to the matte must undergo all the operations with it, and thus much lead and silver are lost. The automatic tap, therefore, insures a considerable saving in metal.

There is one case where tapping from the bottom is to be preferred to the automatic tap. It is in smelting coppery ores—as an alloy of lead and copper separates out from the lead, adheres to the bottom of the crucible, and grows upward, filling it (see § 54, i). The trouble is remedied by adding sufficient sulphur in some form or other to form matte. As soon as this runs 12 per cent. copper the difficulty begins again to make itself felt. In concentrating lead matte from 12 per cent. copper upward, the automatic tap is out of place, and the ordinary copper furnace with internal crucible is used.

f. *Water-Jackets.*—These (*E*, Figures 89, 90, 92, 96, 98) are water-cooled iron shells that enclose the smelting zone of the furnace to protect it from the corrosion of the slag. Since about 1873 they have come into more general use, and have now entirely replaced the brick walls at the region of the tuyeres, whenever there is sufficient water to warrant their use. Only where this is very difficult to obtain, sandstone, fire-brick, or other refractory material is reverted to.

Quite a discussion arose in 1885–86 [1] as to the date of the invention of the water-jacket furnace and the inventor. The writer has failed to find any reference to water-jackets in the treatises of Karsten [2] and Scheerer. [3] The earliest mention of their use is made

[1] *Engineering and Mining Journal,* 1885 : July 25 (Harnickel, F., Rolker); August 1 (Courtis) ; August 15 and 29 (Kleinschmidt); August 22 (Editor); September 12 and 26 (Williams) ; October 10 (Hahn) ; October 24 (Arents) ; October 31 (Douglas) ; November 7 (Courtis, Daggett) ; November 14 (Curtis) ; November 29 (Kleinschmidt) ; 1886, January 2 (Tew).

[2] "System der Metallurgie," Berlin, 1832 ; and "Handbuch der Eisenhüttenkunde," Berlin, 1841.

[3] "Lehrbuch der Metallurgie," Brunswick, 1846–53.

by Overman,[1] who describes and illustrates a refinery furnace, the sides of which consisted of water-cooled cast-iron shells, through which water-cooled tuyere nozzles protruded into the furnace. Douglas[2] says that J. Williams built near Drontheim, Norway, in 1852 "sectional water-jacket furnaces consisting of a circle of long, narrow water-backs, perforated by tuyere-holes." About the year 1865 the same J. Williams erected a number of water-jacket blast-furnaces at Houghton, Lake Superior. According to Arents,[3] N. Haskell built in 1865 a water-jacket furnace in California. Kerl,[4] in describing the improvements made in smelting in the Hartz Mountains, records the introduction in 1864 of "*water-blocks* to cool the hearth and to serve as a support for the water-cooled tuyere nozzles," but these had been used in refinery furnaces for a very long time,[5] and are not to be confounded with water-jackets. The latter never were and are not to-day in use in the Hartz Mountains. Courtis, who made a drawing of the Pilz furnace at Freiberg in 1866, says that the tuyeres and the front of the furnace were water-cooled. Water-jackets have been introduced at Freiberg and Pribram only since they became common in this country; they were used in the Saint-Louis Smelting Works near Marseilles before 1878.[6] Spray-jackets were used at La Pise as early as 1862.[7]

The water-cooled tuyere nozzles, which resist the action of heat and slag so well, while the brick walls are eaten out, appear to have suggested to different persons the idea of extending the water-cooled iron surface, and thus caused the construction of the water-jacket. This would give several men the credit of having invented it.

The water-jackets (*E*, Figures 90, 98 ; *F*, Figures 90, 100) are placed on top of the hearth-walls, forming their continuation on the inside. Their height has varied from 2 to 4 feet ; 3 feet 6 inches is an ordinary measure. They thus reach from the top of the hearth to within about 12 inches of the cast-iron carrier-plate (*t*) or the I-beams (*n′*) which support the shaft. The centre of the

---

[1] "Treatise on Metallurgy," New York, 1852, p. 556.
[2] "Mineral Resources of the United States," 1882, p. 268.
[3] *Berg- und Hüttenmännische Zeitung*, 1866, p. 316.
[4] *Ibid.*, 1867, pp. 6 and 47.
[5] Percy, "Metallurgy of Iron and Steel," London, 1864, pp. 584 and 625.
[6] Grüner, "Traité de métallurgie," Paris, 1873–78, vol. ii., p. 391.
[7] Grüner, *Annales des mines*, 1868, xiii., p. 364.

tuyeres is placed 10 inches higher than the bottom of the jackets, and from 8 to 10 inches above this, begins the bosh, the amount of which varies from 6 to 10 inches.

Jackets are made of cast iron and wrought iron.

Cast-iron jackets (*E*, Figures 84, 85, 87–89, 90, 92, 96–98) are generally 6 inches thick, the sides being of $\frac{3}{8}$- or $\frac{5}{8}$-inch iron. Each jacket has a special water-feeder *f*, which begins 8 or 10 inches above the centre of the tuyere and runs from 3 to 4 inches above the top of the jackets, extending outward about 4 inches. This insures the complete filling of the jacket with water. As the top of the feeder is closed only by a lid, tools can be introduced to scrape off scale. The feeder was formerly cast in one piece with the jacket, but now is a separate casting, which is fastened on with screws or bolts. At first there was no opening at the lower end of the jacket (Figures 84, 85, 96) to remove mud or scale that had collected; now there is usually a hand-hole (*q'*, Figures 89, 90) for this purpose, and thus the life of the jacket is much prolonged. The tuyeres (*o*, Figures 84, 85) always used to be at the junction of two jackets, each having a semicircular recess. This was for fear that an opening through the centre of the jacket might weaken it; but it has not proved to be the case, and the tuyeres are now (Figures 89, 92, 96, 97) generally made in that way. Tuyeres between jackets are sure to cause considerable leakage of air. When a furnace is new, and the two semicircular recesses are just opposite each other, and a brass nozzle (*p'*, Figure 98) is inserted, which receives the galvanized iron blast-pipe, this does not at once occur but is sure to come later. Not less than six different kinds of jackets (Figures 84–87) were formerly used in each large furnace; now the number has been reduced to three (Figures 92, 97), two (*A* and *B*) on the front one, and one (*E*) on the side, the opposite ones being duplicates. To these must be added the tapping-jacket (*n*, Figures 84, 90, 99). This reduction in number is made possible by giving the front of the furnace the same construction as the back, the opening required at the front being simply bricked up at the back. The length of the front jackets varies somewhat in different furnaces. Sometimes they reach to within 10 inches of the top of the crucible (Figures 99, 111), thus leaving open a 10-inch breast, which runs across the entire width of the furnace. It is usually closed by two small brick pillars (9 by 4$\frac{1}{2}$ inches) and three balls of clay, one in the middle, and one on the outer side of each pillar. In the

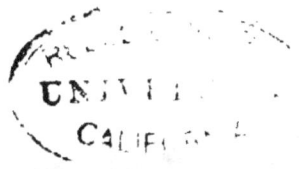

central clay-ball is placed the slag-tap. There is nothing in this arrangement to prevent a tapping-jacket from being put in, should the clay be eaten out by the slag. A second plan is shown in Figure 84. Here the front jackets reach at the sides down to the top of the crucible, and leave in the centre an opening, the upper part of which is closed by a separate jacket n, and the lower, the breast l, by a ball of clay, in which is placed the slag-tap. This plan is not a common one. A third modification (Figures 90, 98) is to fill the open place in front with a tapping-jacket to within $2\frac{1}{2}$ inches of the top. This space is left for convenience in taking out the tapping-jacket, and is closed by brick. The jacket (n, Figure 99), 26 by 14 inches, and $3\frac{1}{2}$ inches deep, has, $6\frac{1}{2}$ inches above its lower edge, a tap-hole m, which is $2\frac{1}{2}$ inches in diameter, and widens, after entering the jacket to the depth of one inch, to 5 inches on the inner side. The lower edge of the jacket is placed 4 inches beneath the upper edge of the crucible castings, and thus prevents at the front that leakage of lead from the crucible which is otherwise so difficult to stop. The last arrangement is the one now in general use at large smelting works. Having large ore-beds from which to make up the charges there is little probability of trouble in the furnace, and therefore a small opening closed by the tapping-jacket is sufficient for all practical purposes. At small smelters, where the charges are changed often, it is probably better to have the front jackets 10 inches smaller than the side jackets, in order to give room for working in the crucible, should it prove necessary.

The water-jackets have been joined in various ways by using wedges, screws, bands, etc. Now they are simply bolted to each other near the top and bottom, the bolts passing through lugs (y, Figures 89, 92, 96, 97) cast in the jackets.

Wrought-iron jackets have in many instances replaced those of cast iron. With circular furnaces they are exclusively used. They have no bosh, and are usually in two parts. They seldom have special water-feeders, like the cast-iron jackets, the water inlet pipe being usually near the bottom and the outlet pipe near the top. With this arrangement, it is important, if a complete filling of the jackets with water is to be made possible, to have one or two small pieces of pipe protrude upward and outward from the top of the jackets, and to have the water outlet pipe also bent upwards that it may discharge above the top of the jackets.

Oblong furnaces have only four wrought-iron jackets, one

on each side. Two kinds of jackets are in use. One is shown in Figures 99–101, and 112–115. The jacket is of the usual height; the inner wall forms a straight line slanting outward; the outer wall has a greater slant; thus the water-space grows wider toward the top, where the water outlets i are riveted to the outer wall, the inlets being at k''. The walls are stiffened by stay-bolts. The mud accumulating in the jackets is removed through the hand-holes q'. The details of the tuyere-box a' are discussed further on. The other form of jacket is like the cast iron one. The walls are vertical and parallel from the base to the bosh, and then slant outward; cast-iron water-feeders are bolted to the outer wall. If cast-iron and wrought-iron jackets be compared, there is no question that the latter last longer than the former, and that, although much more expensive at first, they are cheaper in the end. There may, however, be exceptions to this rule. If the water that is to cool the jackets be muddy or hard, and thus liable to form scale, a wrought-iron jacket will not last so much longer than one of cast iron. Again, if a foundry be not too distant from the smelter, the value of the leaking cast-iron jacket as old iron will contribute considerably toward the cost of a new one.

Water-jackets made of cast steel are strongly recommended by Terhune,[1] who introduced them at the Hanauer Works, where they gave great satisfaction. The walls are $\frac{4}{8}$ inch thick.

The cooling-water for the jackets is drawn from a wooden tank, the bottom of which should be at some distance above the water inlet, in order that there may be some pressure, as an extra amount of water is often needed in blowing in or blowing out. The main delivery-pipe starts from the water-tank and runs along the front or back of a row of furnaces. From it branch off separate supply-pipes ($x,'$ Figure 85), each of which ends in a pipe g (Figures 84, 85, 89, 90, 92, 96, 98–100), surrounding its own furnace. This supplies the small feed-pipes h, which deliver the water into the top of the feeders, the flow being regulated by a valve. The cold water, entering the jacket at the top, sinks down slowly and pushes upward the hot water, which runs off through the small pipe i below the inlet. This, the common arrangement, accomplishes on the whole its purpose, but if the temperatures of a jacket be compared at the top and the bottom, it will be found that the bottom is always hotter. There are two ways of equalizing this.

[1] *Trans. A. I. M. E.*, xvii., p. 131.

One is to attach a rubber hose to the feed-pipe *h*, thus letting the cool water come in contact with the hot water at about the middle of the jacket. The other is to have an extra supply-pipe through which a small stream of water runs in near the bottom of the jacket. If these two methods do not succeed in cooling the lower part of the jacket, it shows that it contains mud or scale, and requires cleaning, if it is not soon to burn through. The hot water from the jackets is discharged into a galvanized-iron trough (Figures 84, 85, 96, 98) *j*, surrounding the furnace, from which it passes off through a cast-iron stand-pipe *k* (Figures 85, 96) into a main underground. The troughs are often in the way when the furnace is running, and are very inconvenient sometimes, as, for example, when a cracked jacket has to be exchanged. To remedy this, the water is sometimes carried away from the jackets by long pieces of gas-pipe, terminating in the funnels of stand-pipes placed close to the supporting pillars and connected with the main underground. There are two or four of these stand-pipes. Thus, if a jacket has to be exchanged, it is only necessary to turn off the water-supply from it instead of from the whole side, as is ordinarily done, the trough being also removed. The amount of water required to cool the jackets varies with the size of the furnace and the slag that is being made. A furnace 36 by 92 inches at the tuyeres, making a siliceous calcareous slag, requires under normal conditions 11 gallons of water per minute. This is a good average figure. For blowing in or blowing out one must be prepared to use double this amount. It is sometimes said that cast-iron jackets require less water than wrought-iron ones. This remains to be proved.

g. *Blast.*—The machines that furnish the blast belong to the class of rotary positive pressure-blowers. Fans are used very rarely. A comparison of the two has been made by Howe.[1] The pressure-blowers at present in general use are the Baker and Root. The former was at one time almost exclusively used, but the latter is gaining ground, and is now found in a good many works. Its advantage is that it allows a greater difference in speed. Take, for example, a No. 7 Baker blower, giving a displacement of 60 cubic feet per revolution. Its greatest allowable variation in speed per minute is 35 revolutions; the No. 7 Root with 65 cubic feet displacement allows 100 revolutions. If a higher pressure is wanted in a furnace, the Root can therefore be more easily speeded up

---

[1] *Trans. A. I. M. E.*, x., p. 482.

than the Baker. A furnace of from 36 × 60 to 36 × 84 inches will require a Baker or Root No. 6 ; a furnace varying from 36 × 100 to 36 × 120 inches, a Baker or Root No. 7. These sizes give more blast than is actually necessary, but it is advisable to have a slight excess.

There are two ways of supplying the blast to a number of furnaces. Each furnace can have its own blower or several blowers deliver the compressed air into a blast-main, from which the single furnaces are supplied. The second method is the one generally accepted, as the plant is cheaper, the daily attendance easier, and the repair smaller. The pressure in the single furnaces is regulated by a gate in the branch-pipe. All blast-pipes are made of galvanized iron. The main blast-pipe or induction-pipe (s, Figure 81), which receives the wind from several blowers, has a diameter that is from $\frac{1}{6}$ to $\frac{3}{8}$ larger than the combined outlets of the blowers. It usually runs along back of the furnaces near the dust chamber, and is suspended eight or more feet above the furnace floor in a wooden frame. It has safety-valves, and is closed at both ends by blast-gates. From it branch-pipes, each with its own gate, furnish the blast to the bustle-pipes of the single furnaces (Figures 84, 85, 90, 96, 98–100). Back of the gate each branch-pipe has an opening, with a thimble, to be connected with the pressure-gauge.

The pressure of the blast is measured by different quicksilver or water gauges, 2 inches quicksilver or 28 inches of water equalling 1 pound or 16 ounces pressure per square inch. At some works colored glycerine is used instead of quicksilver or water.

From the bustle-pipe thimbles pass downward to be connected by the wind-bags q with the tuyere-pipes p. The wind-bags are of closely-woven canvas that has been soaked in water-glass or alum to prevent it from readily catching fire from a spark. The number, size, and form of tuyeres and tuyere-pipes vary a great deal. Hahn[1] reckons one tuyere with a 3-inch opening as being sufficient for 2 square feet hearth area. Thus a 3 × 5 foot furnace would require seven tuyeres ; one that is 3 × 6½ feet, nine tuyeres. They are so distributed that one tuyere enters at the back and three or four are inserted symmetrically on each side. At some works the diameter of the tuyeres has been reduced to 2½ inches ; at others it has been increased to 3¼ inches, but these figures are not common.

[1] "Mineral Resources of the United States," 1882, p. 336.

The tuyere at the back has been lately abolished at many works, while the tuyere at the front (Figures 84, 85) was discarded many years since as being very much in the way and of no special advantage. Blowing only from the sides has the effect of making the breast hard; blowing from the sides and ends of chilling in the centre. The former difficulty can be avoided by having the last tuyeres close to the ends, and also by the manner of feeding; the latter is not easily remedied. Hence it is advisable to leave out the end tuyeres.

The ordinary blast-pipe is made of galvanized iron. The horizontal arm, varying from 2 to 14 inches in length (Figures 84 and 90), is either slightly conical and fits into a brass nozzle p' (Figure 98) inserted into the tuyere-hole, or it is cylindrical and is soldered to the nozzle. The elbow which joins the other end to the wind-bag has a brass nipple soldered to it, which forms the eye or peep-hole s' (Figures 90, 96, 98). This is closed either by a slide or a cap having a glass or mica plate in the centre, or simply by a wooden plug. In the centre of the plug is left a small opening, the size of a pencil, to be closed by a small piece of wood. This has to be removed to observe the condition of the tuyere. To keep the blast-pipe in its normal position and to thus prevent it from delivering the blast upward, which is its natural tendency, an iron band, hooked by means of two springs to the jacket, is passed around the elbow, or an iron loop is soldered to its inner side, by means of which it is hooked to the jacket. In order to close the tuyere-hole, when the pipe has been temporarily removed, a tuyere-cup has in many works replaced the ball of clay commonly used. It is made of galvanized iron, has the form of a nozzle, and is closed at the back, to which is soldered a handle.

In this connection may be mentioned Werner's adjustable tuyere-pipe.[1] It consists of a cast-iron pipe to which a cast-iron elbow is fastened on the upper side by means of a hinge. The pipe is hooked with three springs to the jacket, and, working in a ball-and-socket joint, can be turned in any direction, and thus deliver the blast wherever desired, the springs keeping the pipe in position and very little air being lost by leakage. Other advantages are that should any slag run into the pipe it will not damage it and can be easily removed by raising the elbow. If the furnace is to be shut down for a short time, which usually necessitates the re-

---

[1] Emmons, "Geology and Mining Industry of Leadville," p. 682.

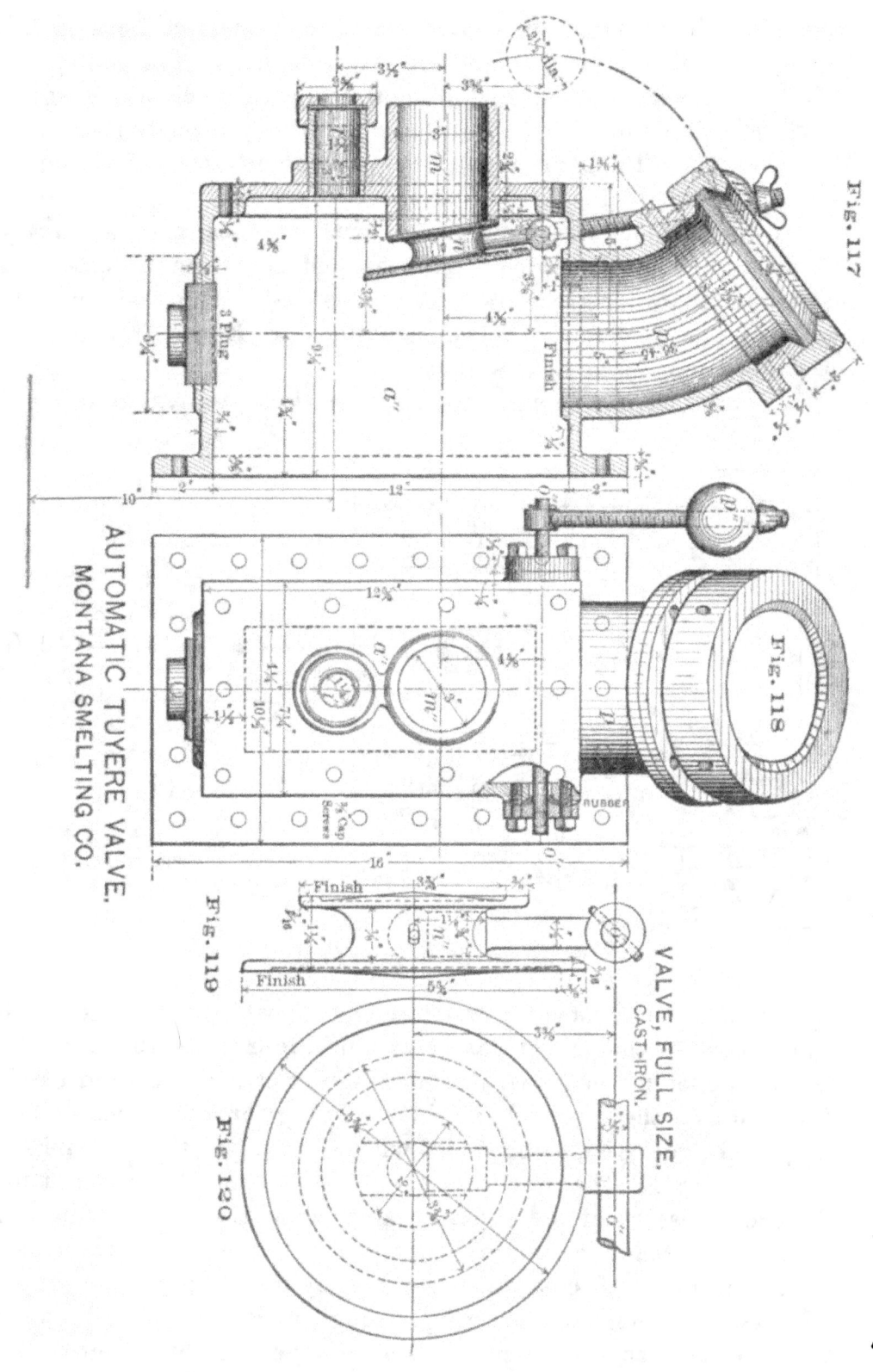

AUTOMATIC TUYERE VALVE.
MONTANA SMELTING CO.

Fig. 117

Fig. 118

Fig. 119

Fig. 120

VALVE, FULL SIZE.
CAST-IRON.

moval of the pipes, a piece of thin card-board inserted between the flanges at junction of pipe and elbow is sufficient. The writer has used this tuyere-pipe, and while there is not as much advantage as is claimed in being able to set the blast in any direction, its other features make it a good apparatus, preferable to the ordinary pipe of galvanized iron.

In the ordinary blast-pipe, even with the best care, a leakage of air cannot be prevented. This has led some works to adopt a cast-iron elbow or tuyere-box, which is fastened by an air-tight joint to the jackets. The constructions of Eilers, Murray, and Devereux may serve as examples.

Figures 117–120 represent the automatic tuyere-valve designed

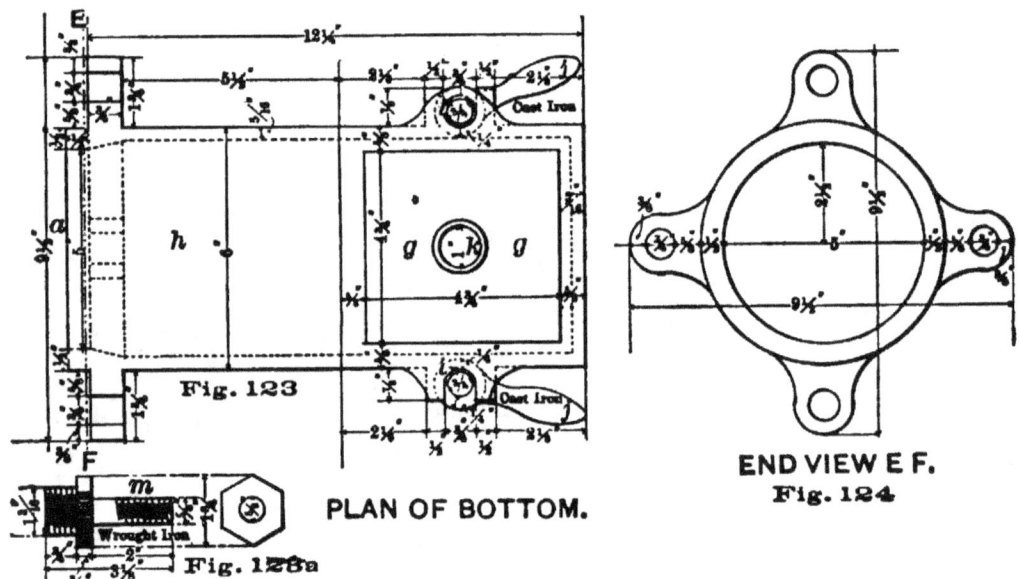

Fig. 123

PLAN OF BOTTOM.

Fig. 123a

END VIEW E F.
Fig. 124

THE MURRAY TUYERE-BOX.

by Eilers. It consists of a cast-iron box a'', which is fastened with cap-screws to the jackets, as shown in Figures 99, 100, 112–115. On the upper side is a cast-iron nipple p, to which the wind-bag is fastened; on the lower side a 3-inch opening through which chilled slag can be removed, to be closed with a plug. At the rear end are two openings: l'', the peep- and poking-hole, and m'', the outlet for the back pressure of the furnace gases when the blast is shut off. When the furnace is running, the valve n'', swinging on the cross-pin o'' and balanced outside of the box by the movable weight p'', is pressed by the blast entering through p against m'' and closes this; as soon as the blast is taken off, the weight p'' turns the valve, closes p, and opens m'', thus preventing the gases in the furnace from entering the blast-pipe and furnishing them an outlet into the open air.

Murray's tuyere-box is shown in Figures 121–124. Figure 121 shows an obtuse-angle elbow. The opening *a* through which the blast enters the jacket is rather large—5 inches in diameter. Opposite is the poking-hole *b* closed by the cap *c*, which contains the

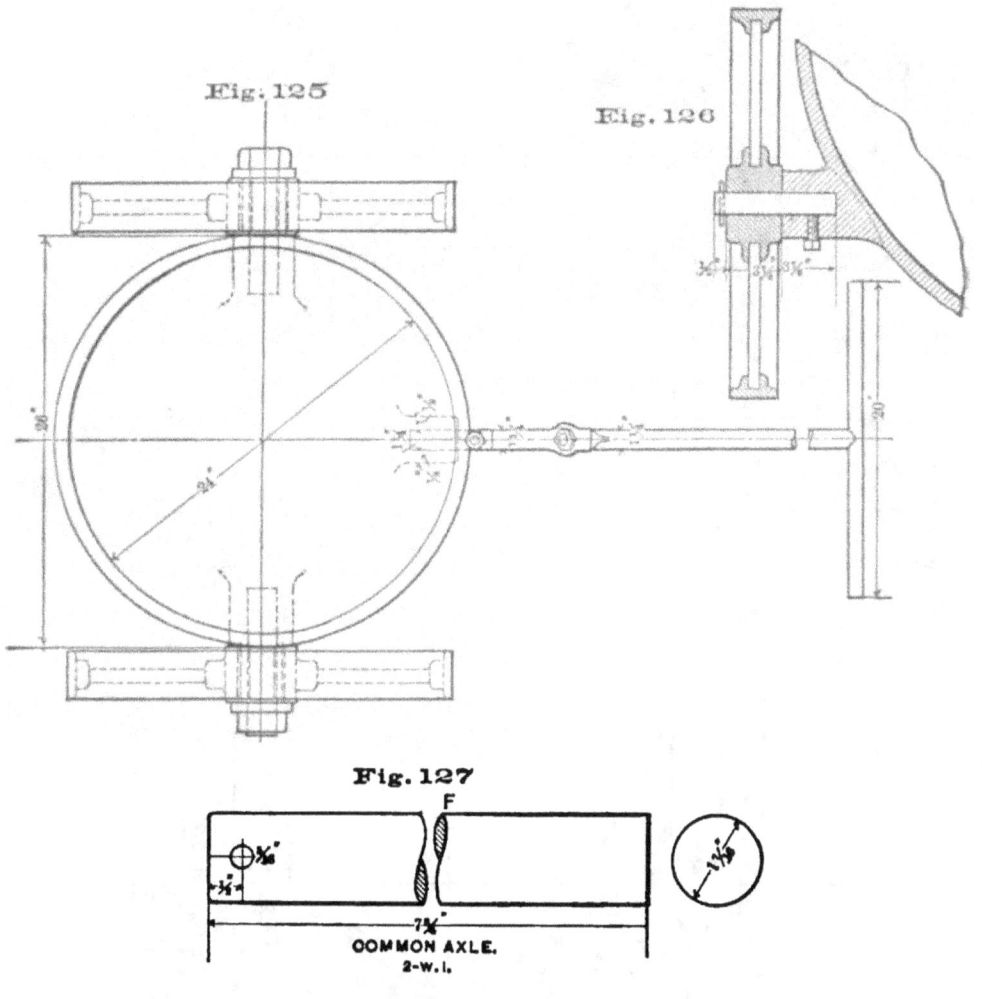

Fig. 125

Fig. 126

Fig. 127

COMMON AXLE.
2-w.l.

COMMON SLAG-POT.

eye-hole. Between the tuyere-pipe *d* and the cast-iron nipple *e*, over which passes the wind-bag, is the wind-gate *f*. Of special interest is the drop bottom *g* of the belly-pipe *h*, shown in Figures 122 and 123. It is held in place by the hinged bolts *i* and the crank-nuts *j*. The drop bottom has a fusible plug *k*, a thin disc of lead, which will melt as soon as any slag enters the belly-pipe, and thus call immediate attention to the accident. The elbow is fastened to the jacket by slipping it over four threaded bolts *m* that have been screwed into the jacket, and then tightening it with nuts.

The characteristic of the Devereux tuyere-box[1] is that the blast

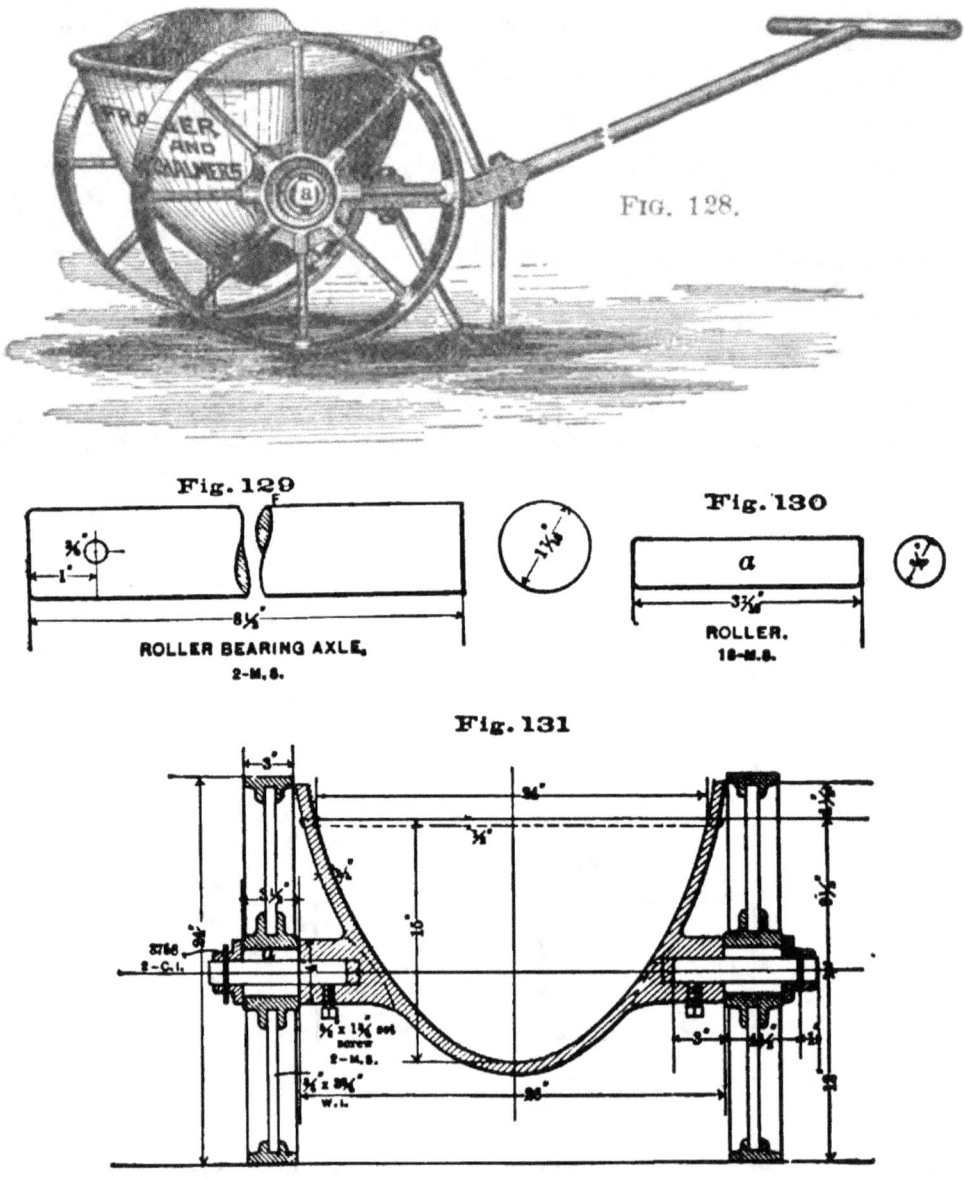

FIG. 128.

Fig. 129

ROLLER BEARING AXLE.
2-M.S.

Fig. 130

a

ROLLER.
18-M.S.

Fig. 131

SLAG-POT WITH ROLLER BEARINGS.

can be made to play in different directions. The tuyere-hole is lined with a cylindrical bronze tube in which can be revolved an iron sleeve having a diagonal bore. By turning this sleeve the

[1] Patent No. 318,604, May 26, 1885.

blast can be directed up and down to the right and left; it cannot, however, be directed centrally.

h. *Slag-Pots.*—Twelve ordinary slag-pots, 24 inches in diameter and 15 inches deep, are sufficient for a 33- by 100-inch furnace. Considerable improvements have been made in the construction, the aim being to make them light without diminishing their strength.

Figures 125 to 128[1] show the usual paraboloid form of a slag-pot with the ordinary compression-spoke wheel, the spokes being wrought iron, the axle machine steel. The length of the handle is 5 feet, the height of the cross-piece above the ground 2 feet 8 inches.

In order to lessen the friction at the hub, roller bearings (*a*, Figures 128–131[1]) have been introduced with satisfactory results.

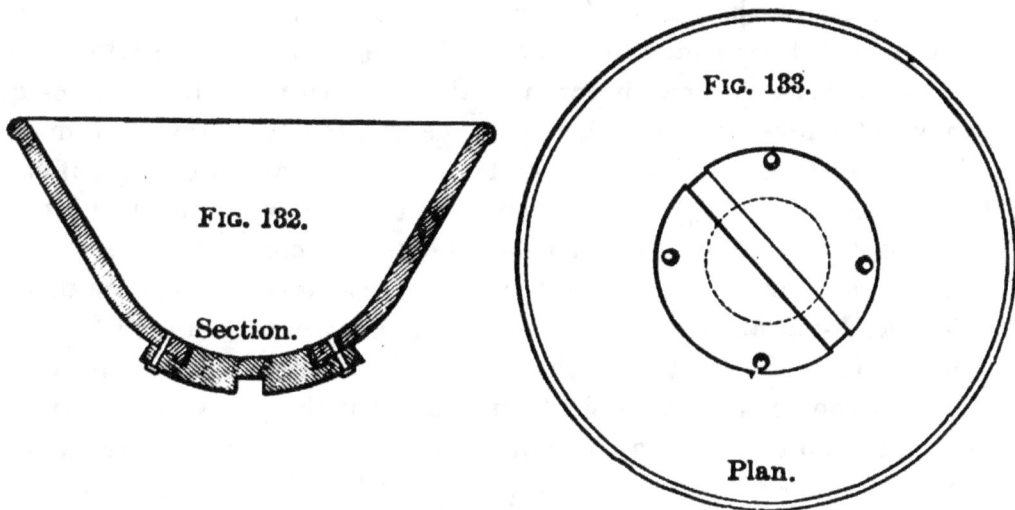

Fig. 182.

Section.

Fig. 183.

Plan.

SECTIONAL SLAG-POT.

Terhune,[2] Figures 132 and 133, has made the bottoms of slag-pots movable, so that they can be replaced when corroded or cracked.

When matte and slag are being tapped, there is danger that they may not separate well. To assist their doing so several slag-pots have been constructed.

The Iles-Keiper pot[3] is a large overflow-pot which retains the matte, while the slag runs over through a spout in the side into an ordinary slag-pot. To prevent the melted mass from solidifying the pot has a cast-iron cover. Overflow-pots have been used for a

[1] Taken from a drawing of Messrs. Fraser and Chalmers, Chicago, Ill.
[2] *Trans. A. I. M. E.*, xv., p. 92.
[3] Patent No. 835,224, February 2, 1886.

good many years. At some works the slag is allowed to harden
in them from the surface down to a depth of several inches, the inlet and overflow for the slag being kept open. Thus the hardened
slag, taking the place of the iron cover, prevents the liquid slag
below from cooling, and promotes a good separation of matte and
slag.

Werner[1] has patented a slag-pot which permits the separate
pouring off of the bulk of the slag. On the rim of the bowl opposite the handle a segmental cover is pivoted, which is provided
with a spout for the discharge of the slag. When the pot is filled,
the slag is allowed partly to solidify, is then broken, the pot tilted,
and a certain amount of the still liquid slag is allowed to run out,
the remainder solidifying with the matte and speise, and being
removed with them. The writer has never seen this pot in use.

A third method, and the one in common use for preventing a
loss of metal by shots of matte adhering to slag, is to allow the
matte to settle in a catch-pot and then to tap the still liquid slag
above the level of the matte. Thus the matte in the bottom of
the pots remains undisturbed, and the shell of chilled slag (that enclosed the liquid slag) is recovered and smelted over again, as it is the
only part of the entire slag that is liable to be rich. Several patents[2]
have been taken out for different applications of this method.
The catch-pot of Murray (not patented), shown in Figures 134–139,
will serve to illustrate the third method. The pot has a cast-iron
bowl of the usual paraboloid form, but $3\frac{1}{2}$ inches above the bottom
is the tap-hole a, through which the liquid slag is discharged after
the matte has settled out. To prevent injury to the pot at the tap-
hole, when this is opened with a steel bar, the casting is made thicker
by a ring b, ending in a rib c which reaches the top.

In this pot is seen a third class of wheel. The hub d has alternate spoke-sockets e and e', similar to those of the tension-spoke
wheel. This wheel is more common than the ordinary compression-
spoke wheel, as it is stronger.

The Nesmith Dumping Car, Figures 140–144.[3]—The object of
this car, with its two large tilting-pots A A, is to convey the waste
slag from a number of catch-pots near the furnace building to the
edge of the dump and to discharge it there. By this means the

[1] Patent No. 216,648, December 21, 1886.

[2] For instance, Devereux: No. 312,439, February 17, 1885 ; No. 335,114,
February 2, 1886.

[3] Patent No. 388,708, August 28, 1888.

distance that the pots receiving the slag from the furnace have to
be wheeled is shortened, and the disposal of the waste slag cheap-

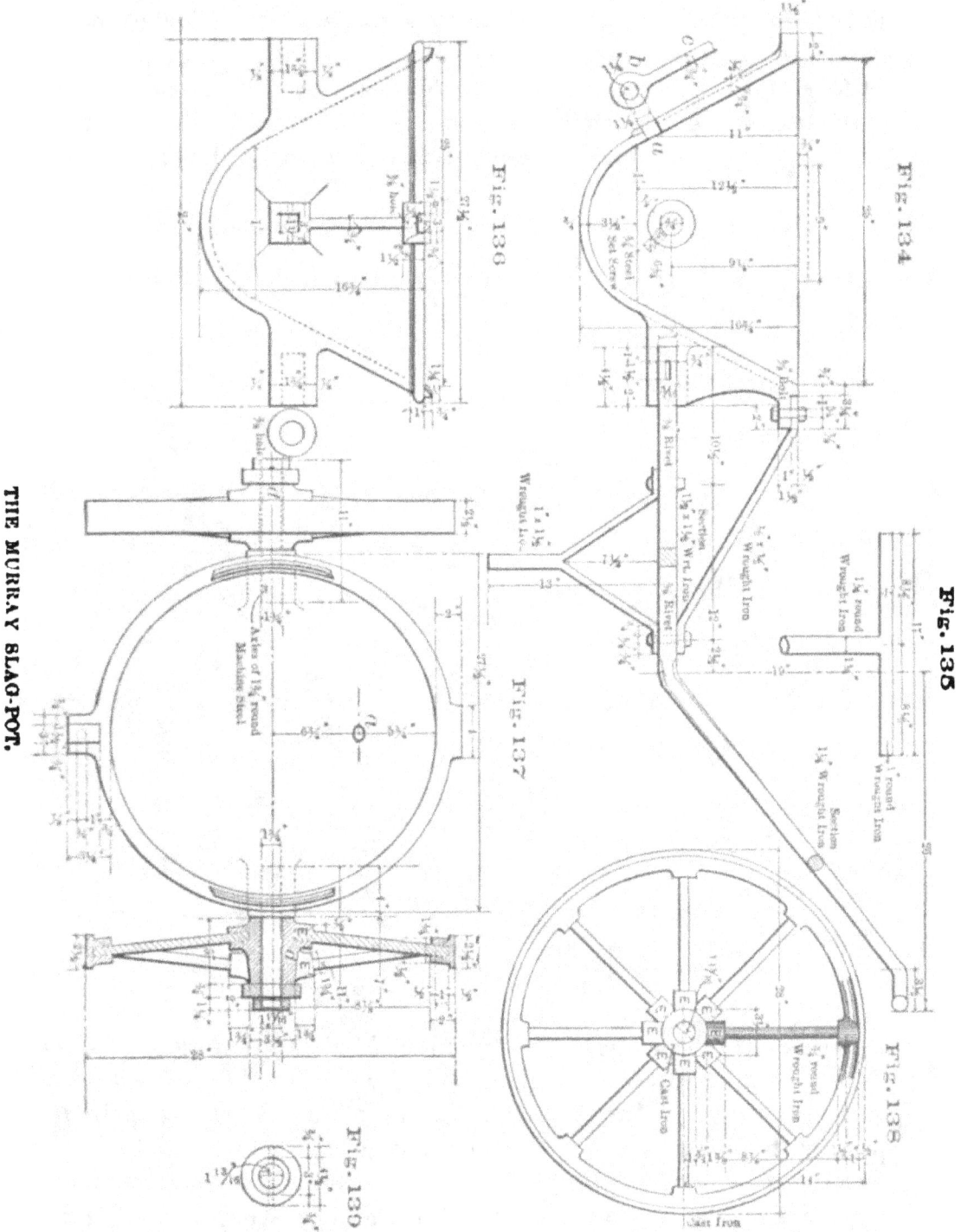

THE MURRAY SLAG-POT.

ened.   Each tilting-pot has a capacity of 7.38 cubic feet, and holds
about 1,280 pounds of slag. The car consists of a truck-frame *a*, with

platform *b*, brake *c*, and railing *d*, by which the driver holds on. The frame carries the boxes *e* for the axles *f* of the wheels *g*. Two bridge-beams *h*, lying transversely across the frame *a*, serve as support for the frame *h′*, which carries the central pin-socket *m*, in which the pin *n* is made fast by the nut *n′*. The swinging frame consists of the channel-irons *i i* (held apart by the central blocks *j j*, in which are the swivel-eye and the end-blocks *j′ j′*) and the beams *k k* (resting on the channel-irons); the latter have on their

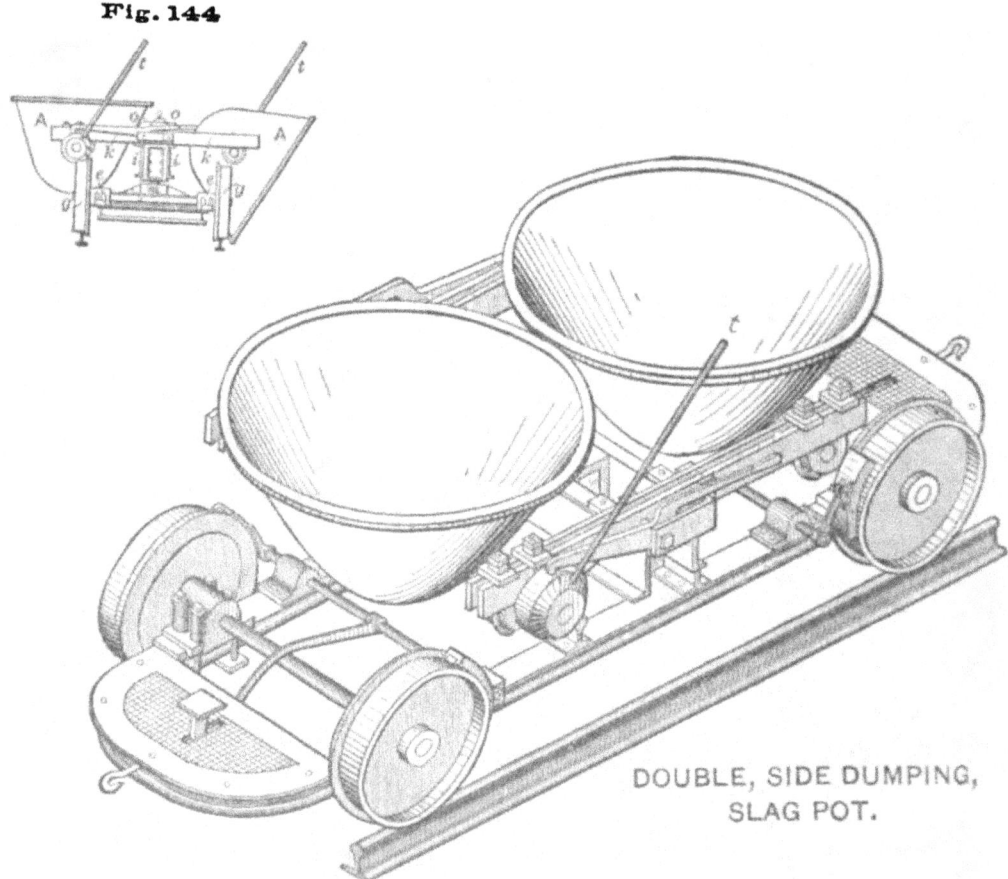

Fig. 144

DOUBLE, SIDE DUMPING,
SLAG POT.

under side the bearings of the trunnions *l l* of the tilting-pots *A A*. These are pivoted out of centre, so that when in their normal position they may lean against the stop-pieces *o*. They are held in position by the pawls *r*, with disengaging handles *r′*, which fasten into the teeth *q* of the projecting head or collar *p* of the trunnions. This collar also has the holes *s* for the operating-bar *t* (Figure 144), with which the pots are tilted. The weight complete of truck and pots is 5,000 pounds. When in their normal position, the pots are placed as shown in Figures 140 and 141. When they are to be emptied, the frame *iikk* is swung 90 degrees on the swivel (Figure 144), the pots are emptied together on both sides of the track; or, if the slag is to be discharged only on one side, the frame is returned to its normal place after one pot has been tilted and the other pot swung into position.

§ 60. **Chemistry of the Blast-Furnace** —The reactions of a lead blast-furnace have never been fully studied. Guyard[1] has a very interesting theoretical discussion of the subject in his paper, "Argentiferous Lead-Smelting at Leadville," but measurements of temperatures and analyses of gases are wanting. Schertel[2] has measured the melting temperatures of the slags and analyzed the gases of the Freiberg blast-furnaces, where slagged ores are smelted. With the aid of these two writers, whose work will be used to supplement the more general outline, the sequence of processes in the lead blast-furnace can be approximately given.

The main reactions that take place in the blast-furnace are reduction and precipitation; the latter has already been discussed (§§ 7 and 54, *l*). The principal reducing agents[3] are carbon and carbonic oxide.

a. Carbon acts on metallic oxides if its own affinity for oxygen is greater than that of the metal combined with the oxygen, and this affinity stands in direct proportion to the height of the temperature. The product of carbon and oxygen will be carbonic acid, if the carbon acts on readily reducible oxides (lead, copper). With oxides not easily reduced (iron, manganese) carbonic oxide is the principal product.

$$2PbO + C = Pb_2 + CO_2,$$
$$Fe_xO_{y+1} + C = Fe_xO_y + CO,$$
$$Fe_xO_y + yC = Fe_x + yCO.$$

b. Carbonic oxide is formed in the blast-furnace : first, by the direct combustion of carbon,

$$C + O = CO,$$
$$yC + Fe_xO_y = yCO + xFe;$$

[1] Emmons, "Geology and Mining Industry of Leadville," p. 731.
[2] *Wagner's Jahresberichte*, 1880, p. 188.
[3] Balling : "Compendium der metallurgischen Chemie," p. 63; Wright, in *Colliery Guardian*, January 3, 1890, or Roberts-Austen, "An Introduction to Metallurgy," Philadelphia, 1891, p. 194.

secondly, by carbonic acid being split into carbonic oxide and oxygen,

$$CO_2 = CO + O,$$
$$CO_2 + Fe_xO_{y-1} = CO + Fe_xO_y,$$
$$yCO_2 + xFe = yCO + Fe_xO_y;$$

thirdly, by carbonic acid combining with carbon,

$$CO_2 + C = 2CO.$$

The reducing power of carbonic oxide is on the whole favored by a high temperature. Its product of combustion is carbonic acid.

$$PbO + CO = Pb + CO_2,$$
$$Fe_xO_y + CO = Fe_xO_{y-1} + CO_2,$$
$$Fe_xO_y + yCO = xFe + yCO_2.$$

As carbonic acid begins to be decomposed into carbonic oxide and oxygen at 1,200° C., and cannot exist at 2,000° C., it follows that the reducing power of carbonic oxide is diminished as the temperature rises above 1,200° C. Hence the same mixture of carbonic oxide and carbonic acid may be a reducing agent at a low and an oxidizing agent at a high temperature.

c. Other sources of carbonic acid in furnace gases are :

First, the splitting of carbonic oxide into carbon and carbonic acid at a temperature of from 400 to 450° C. when in contact with iron oxide.

$$2CO = CO_2 + C.$$

Second, the decomposition of carbonates : lead carbonate, 170 –200° C.; magnesium carbonate, 650° C.; ferrous carbonate, about 800° C.; calcium carbonate, 850° C.

Third, its entrance into the blast-furnace in small quantities with the blast : 0.04 per cent. by volume.

The action of sulphur as a reducing agent has already been discussed (§ 9).

As has been seen, the carbon compounds in furnace gases are mainly carbonic acid and carbonic oxide. The oxygen of the air, on entering the furnace through the tuyeres, meets incandescent fuel, which is converted into carbonic acid or carbonic oxide according to the prevailing temperature. This depends on the melting-point of the slag.

Freiberg lead-slags melt, according to Schertel, at 1,030° C.;

other temperatures[1] given for lead slags are 1,220° C. and 1,273° C.; Guyard assumes 1,200° C.

In the products of combustion carbonic acid will therefore strongly prevail over carbonic oxide. In ascending, some of the carbonic acid will be reduced by carbon to carbonic oxide; to this will be added that which is produced by carbon acting on oxides of iron. These reactions decrease very quickly as the ascending gases expand at the boshes and come in contact with the cooler parts of the charge; here the percentage of carbonic acid again increases quickly through the reducing action of carbon and carbonic oxide on readily reducible oxides. Thus the percentage of carbonic acid in furnace gases increases, with one slight interruption, with their distance from the tuyeres, and that of carbonic oxide correspondingly decreases. The amount of carbonic acid will be further increased by the decomposition of carbonates in the charge.

Of the nineteen analyses of gases from Freiberg blast-furnaces made by Schertel,[2] a few, selected by Kerl,[3] may be quoted. The gases were taken from the throat of the furnaces. Table I. represents gases from smelting thoroughly slag-roasted ore to which oxidized lead from the refining works and roasted matte have been added; Table II., gases from resmelting the ore slag with roasted matte.

		TABLE I.			TABLE II.		
Volumes.		Ore Smelting.			Slag Smelting.		
In 100 volumes of furnace gas.	N...............	72.72	75.30	75.20	75.5	76.0	74.8
	CO₂............	16.26	17.80	17.20	16.6	17.4	18.5
	CO.............	10.06	5.20	5.40	5.9	4.8	3.5
	CH₄...........	0.36	0.10	0.70	2.0	1.8	0.3
	H..............	0.59	1.60	1.50	....	0.5	2.6
For every 100 volumes of N.	CO₂............	22.3	23.7	22.9	21.9	22.9	24.7
	CO.............	13.8	6.9	7.2	7.8	5.7	4.7
	C..............	18.1	15.3	15.0	14.8	14.2	14.7
	O..............	29.3	27.1	26.5	25.6	25.6	27.0
	Excess of O.........	+2.8	+0.6	....	−0.9	−0.9	+0.5

[1] Balling, "Metallhüttenkunde," p. 613.
[2] *Loc. cit.*
[3] *Berg- und Hüttenmännische Zeitung*, 1880, p. 85.

The presence of hydrogen is explained by Schertel through the decomposition of marsh gas. In none of the analyses is sulphur dioxide taken account of. The largest amount obtained from a number of determinations was 0.15 per cent. The ore being slag-roasted very little, if any, sulphur dioxide will be set free in the upper parts of the furnace; most of the little liberated in the lower parts is reduced to sulphur when passing through the incandescent fuel and forms a sulphide.

Schertel concludes from the analyses, in which the relation of carbonic acid to carbonic oxide by weight is as 4 : 1, that at the tuyeres most of the carbon is converted into carbonic acid. He says that atmospheric air contains, for every 100 vols. of nitrogen, 26.5 of oxygen. If the oxygen of 126.5 vols. of air is completely converted to carbonic acid, there will result 126.5 vols. of gas, containing 100 vols. nitrogen and 26.5 of carbonic acid, equal to 13.25 vols. of gaseous carbon. If on the other hand the oxygen is converted completely into carbonic oxide, the resulting gas will consist of 100 vols. nitrogen and 53 vols. of carbonic oxide, equal to 26.5 vols. of gaseous carbon. From the tables it is evident that the proportion of 100 vols. nitrogen to 13.25 vols. gaseous carbon corresponds more closely to the figures of gaseous carbon in the analyses than would 100 vols. nitrogen to 26.5 vols. gaseous carbon. Consequently the carbon burns before the tuyeres rather to carbonic acid than to carbonic oxide.

The slags produced in the furnaces from which Schertel made the gas analyses contain 4.75 per cent. lime and 0.54 per cent. magnesia. This would not increase the amount of carbonic acid to any extent.

The charge in passing from the throat of the furnace undergoes the following changes. It first loses its hygroscopic water, then that which is chemically combined. During the stages of incipient redness (525° C.) and dull redness (700° C.), oxide of lead will react on sulphides of lead, iron, and silver, unless already combined with silica or reduced to metal. Lead sulphate may also act on sulphide; its reduction to sulphide begins and that of porous ferric oxide (roasted matte). Carbonate of lead is decomposed, dolomite loses part of its carbonic acid, and the reduction of carbonic acid by carbon begins.

As the charge descends through incipient cherry-redness (800° C.) and cherry-redness (900° C.) the reactions begun in the upper zone are about completed, except the reduction of carbonic acid and that of porous iron oxides by carbonic oxide. The action of carbon on iron oxides now begins. The lead set free reacts on lead sulphate, arseniate, and antimoniate, if these have not been reduced to sulphide, arsenide, and antimonide by means of carbon and carbonic oxide. The lead also reacts on silver chloride and sulphide. The charge begins to become pasty; iron and lime carbonates are decomposed; any lead sulphate still existing is converted into silicate; the reduction of iron oxide by carbonic oxide decreases, that by carbon grows; sulphides and arsenides rich in lead form to a greater extent and are in part decomposed by metallic iron. Ascending vapors of metallic zinc are sulphurized and oxidized.

Passing through the stages of a clear cherry-red (1,000° C.) and a deep orange (1,100° C.) to a clear orange (1,200° C.), fusion is perfected. Ferrous oxide and lime combine with silica, setting free lead oxide, most of which is reduced by carbon to metal; what remains may oxidize metallic iron to magnetic oxide, sometimes found in matte and slag; metallic iron acts on sulphides and arsenides, setting free some lead, and forms the matte and speise, with prevailing iron, as tapped from the furnace. Lime and baryta may also act as a desulphurizing agent, carrying some calcium and barium sulphide into the slag; some zinc oxide will be reduced and vapors ascend in the furnace.

The melted masses separate according to their specific gravities, the lead collecting in the crucible and passing off through the lead-well; then come speise, matte, and slag, which are tapped at intervals into the slag-pot, where they separate again in horizontal layers.

§ 61. Calculation of Charge.

a. *Introductory Remarks.*—In calculating a charge for the lead blast-furnace, the typical slag best corresponding to the character of the ore is selected, and the necessary amount of fluxes and fuel then determined. Other considerations are the amount of lead that the charge will contain, the richness of the bullion to be produced, and the quantity of speise, matte, and slag that will ensue from the charge. A complete calculation will give full information on all these points.

To begin with the necessary amount of lead, which is expressed in percentage having reference to the sum of ore and fluxes: charges with as little as 6.5 per cent. of lead have been run successfully; the highest, with from 25 to 30 per cent., is hardly ever reached. With pure ores, containing little or no zinc, arsenic, or antimony, and not much sulphur, it is safe to go as low as 8 per cent.; if these impurities are present to any extent, the charge should contain not less than 12 per cent. Ordinarily the lead in charges ranges from 12 to 18 per cent. It is to be noted that more lead is lost by volatilization with a charge low in lead than with one that is high; the loss in silver depends mainly on the loss of lead and on the richness of the base bullion.

For calculating a charge, a carbonate ore containing some galena may serve as an example. Its composition is :

SiO₂.	FeO.	MnO.	CaO.	MgO.	BaO.	ZnO.	Al₂O₃.	S.	As.	Pb.	Cu.	Ag, ozs.	Au, ozs.
32.6	14.8	4.3	2.2	1.4	1.5	2.4	2.5	4.4	0.5	20.7	2.9	50.5	trace

The typical slag shall be :

$$30SiO_2 - 40FeO - 20CaO.$$

The charge shall weigh 1,000 pounds and contain 10 per cent. of slag; the fuel—coke—shall be 15 per cent. of the charge. The analysis of the iron ore shows :

SiO₂.	FeO	MnO.	CaO.
4.3	72.4	1.7	3.1;

that of the dolomitic limestone,

SiO₂.	FeO.	CaO.	MgO.
2.7	4.5	37.3	11.9.

The coke contains 10 per cent. of ash, consisting of

SiO₂.	FeO.	CaO.	MgO.	Al₂O₃.
40.3	26.5	6.9	2.4	20.4

Before beginning the calculation it is necessary to bring the different slag-forming components of ore, fluxes, and fuel under the three main heads of silica, ferrous oxide, and lime.

The atomic weights of iron and manganese being very nearly the same, 56 and 55, the two oxides are simply added. Both ferrous oxide and metallic iron have to be considered in the calculations:

$$FeO \times \frac{7}{9} = Fe \; ; \; Fe \times \frac{9}{7} = FeO.$$

It will also be necessary to express the equivalents of one component (ferrous oxide) in terms of the other two (silica and lime).

Let $SiO_2 = c$, $FeO = a$, $CaO = b$, and $a+b+c = 90$.

$FeO : SiO_2 : : a : c$, $FeO : CaO : : a : b$,

$FeO = \dfrac{a}{c} SiO_2$. $FeO = \dfrac{a}{b} CaO$.

Under the head of lime are to be classed magnesia and baryta. For instance:

$$CaO : MgO : : 56 : 40.$$
$$CaO = MgO \times 1.4.$$

In the same way $CaO = BaO \times 0.4$.

Some metallurgists bring also zinc oxide under the head of lime:

$$CaO = ZnO \times 0.7,$$

thus cutting down the lime of a slag with the increase of zinc oxide.

The analyses of ore, fluxes, and coke-ash, changed as indicated, are:

	SiO₂.	FeO.	CaO.	ZnO.	Al₂O₃.	S.	As.	Pb.	Cu.	Ag, ozs.	Au, ozs.
Lead ore....	32.6	19.1	10.16	2.4	2.5	4.4	0.5	20.7	2.9	50.5	trace.
Iron ore....	4.3	74.1	3.10	...	..	..	..	..	..	..	..
Limestone..	2.7	4.5	53.96	..	..	..	..	..	..	..	..
Coke-ash ...	40.3	26.5	10.26	..	20.4	..	..	..	..	..	..

In figuring the charge five calculations have to be made to find:

1. The amount of available ferrous oxide and metallic iron;

2. The amount of metallic iron required by the arsenic to form Fe_5As;

3. The amount of metallic iron required to combine to FeS with the sulphur not taken up by the copper as Cu_2S;

4. The amount of flux required for the 15 pounds of ash in the 150 pounds of coke;

5. The amount of flux required to slag the silica of the lead ore.

1. Available ferrous oxide and metallic iron in the iron ore, 100 lbs. In the slag, 30 SiO_2 require 40 FeO. In 100 lbs. iron ore there are 4.3 lbs. SiO_2. These require

$$SiO_2 : FeO : : 30 : 40 : : 4.3 : x;$$
$$x = 5.7 \ FeO.$$

The iron ore contains 74.1 per cent. FeO; deducting 5.7 gives 68.4 available FeO, or $\frac{7}{9}$ FeO = 53.2 available Fe.

2. Arsenic and iron. 100 lbs. lead ore contain 0.5 lbs. of As.

$$As : Fe_5 :: 75 : 280 :: 0.5 : x;$$

$x = 1.86$ Fe. How much iron ore is required?

$$\text{Iron ore}: \text{available Fe} :: 100 : 53.2 :: y : 1.86;$$

$$y = 3.5 \text{ lbs. iron ore.}$$

3. Copper, sulphur, iron. 100 lbs. of lead ore contain 2.9 lbs. Cu.

$$Cu_2 : S :: 126.8 : 32 :: 2.9 : x;$$

$$x = 0.73 \text{ S.}$$

Of the 4.4 lbs. S contained in the 100 lbs. of lead ore, 0.73 are required for the Cu; the difference, 3.67, must be combined with Fe:

$$S : Fe :: 32 : 56 :: 3.67 : y;$$

$y = 6.42$ Fe, which corresponds to 12 lbs. of iron ore, viz:

$$\text{Iron ore}: \text{Available Fe} :: 100 : 53.2 :: z : 6.42;$$

$$z = 12.$$

For the arsenic and sulphur of 100 pounds of lead ore 15 lbs. of iron ore are required. These have: 0.66 lbs. SiO_2, 0.48 lbs. CaO, and 0.88 lbs. FeO. Only the non-available FeO enters the slag according to 30 SiO_2 : 40 FeO; the rest, *i.e.*, the available FeO, combining as Fe with the As and S to form speise and matte. The 0.66 lbs. SiO_2 require

$$SiO_2 : CaO :: 30 : 20 :: 0.66 : x;$$

$x = 0.44$ CaO, which is balanced by the 0.48 CaO already present. If this were not the case, the 0.44 would have to be supplied by limestone:

$$\text{Limestone} : CaO :: 100 : 53.96 :: y : 0.44.$$

4. Coke-ash, 100 lbs.

b. *Murray's Method.*—The method used resembles very closely the one given by Murray[1] and elaborated by Newhouse.[2] The analyses show:

Desired Amount.	Material.	SiO_2.	FeO.	CaO.
100.	Coke-ash	40.3	26.5	10.26
x.	Iron ore...................	4.3	74.1	3.10
y.	Limestone	2.7	4.5	53.96

[1] *Engineering and Mining Journal*, August 13, 1887 ; March 5, 1892.
[2] *School of Mines Quarterly*, ix., p. 373.

Starting again with 100 lbs. of coke-ash, the necessary quantities of iron ore (x) and limestone (y) can be found by expressing the amounts of FeO first in terms of CaO, then in terms of SiO_2 (see above), and finally by putting these quantities equal to each other, when x and y can be easily calculated.

$$FeO = \frac{a}{b} \, CaO,$$

$$26.5 + 0.741x + 0.045y = \frac{40}{20}(10.26 + 0.031x + 0.539y),$$

$$x = 1.668y - 8.80.$$

$$FeO = \frac{a}{c} \, SiO_2,$$

$$26.5 + 0.741x + 0.045y = \frac{40}{30}(40.3 + 0.043x + 0.027y),$$

$$x = 39.6 - 0.001y,$$

$$1.668y - 8.80 = 39.6 - 0.001y,$$
$$y = 30 \text{ lbs. limestone,}$$
$$x = 39 \text{ lbs. iron ore.}$$

5. Lead ore, 100 lbs.   The analyses give :

Desired Amount.	Material.	SiO_2.	FeO.	CaO.
100	Lead ore.....................	32.6	19.1	10.16
x	Iron ore.....................	4.8	74.1	3.10
y	Limestone...............	2.7	4.5	53.9

$$FeO = \frac{a}{b} \, CaO,$$

$$19.1 + 0.741x + 0.045y = \frac{40}{20}(10.16 + 0.031x + 0.539y),$$

$$x = 1.8 + 1.52y.$$

$$FeO = \frac{a}{c} \, SiO_2,$$

$$19.1 + 0.741x + 0.045y = \frac{40}{30}(32.6 + 0.043x + 0.027y),$$

$$x = 35.51 - 0.001y.$$

$$1.8 + 1.52y = 35.51 - 0.001y,$$
$$y = 22 \text{ lbs. limestone,}$$
$$x = 35 \text{ lbs. iron ore.}$$

In summing up there are :

        15 lbs. coke-ash, requiring
         6 " iron ore.
         4 " limestone.   Then the charge contains
        100 " slag.
        ——
        125 lbs.

The difference from 1,000 lbs. = 875 lbs. is to be made up by ore and fluxes.   Now

100 lbs. of lead ore require to slag the SiO$_2$.............................	35 lbs. iron ore + 22 lbs. limestone.
100 lbs. of lead ore require to combine with As and S..................	15 " "
Total..................	172 lbs. of fluxes.

$172x = 875$; $x = 5.088$, gives as charge in round figures :

        Coke-ash.....................   15 (= 150 lbs. of coke).
        Slag.........................   100
        Lead ore.....................   510
        Iron ore for SiO$_2$.............   185
        Iron ore for As and S..........   75
        Limestone   115
                                        ——
                                        1,000

Figuring the pounds of each component of the charge and adding like to like must give the slag.

By $216.57x = 30$, the coefficient is obtained with which the totals of SiO$_2$, FeO, and CaO have to be multiplied to obtain the desired figures : 30, 40, 20.   The table shows that the calculation is correct.

By adding the different components that go to form the slag, 665 pounds are obtained, and any change that is to be made in the slag must be calculated as having reference to this figure.

For every 12.8 ounces of silver there are 102.5 pounds of lead ; the base bullion will therefore assay about 249 ounces silver to the ton.

There are 10 per cent. of lead in the charge.

In the charge are 2.5 pounds of arsenic, which, with 9.3 pounds of metallic iron, form about 12 pounds of speise.

The 14.8 pounds of copper, requiring 3.7 pounds of sulphur, will form 18.5 pounds of copper matte.   Deducting the 3.7 pounds of sulphur from the total sulphur leaves 19.7 pounds, which, with 34.5 pounds of metallic iron, give 54.2 pounds of iron matte ; the total matte formed will be about 72 pounds.

Material Name	Dry Weight lbs.	SiO₂ Per Cent	SiO₂ lbs.	FeO Per Cent	FeO lbs.	CaO Per Cent	CaO lbs.	ZnO Per Cent	ZnO lbs.	Al₂O₃ Per Cent	Al₂O₃ lbs.	Ag ozs. per ton	Ag ozs.	Pb Per Cent	Pb lbs.	As Per Cent	As lbs.	Cu Per Cent	Cu lbs.	S Per Cent	S lbs.
Coke-ash	15	40.3	6.04	26.5	3.97	10.26	1.54			20.4	3.06										
Slag	100	30.0	30.00	40.0	40.00	20.00	20.00														
Lead ore	510	32.6	166.26	19.1	97.41	10.16	51.82	2.4	12.2	2.5	12.75	50.5	12.8	20.7	102.5	0.5	2.5	2.9	14.8	4.4	22.4
Iron ore (SiO₂)	185	4.3	7.95	74.1	137.08	3.1	5.73														
Iron ore (As, S)	75	4.3	3.22	74.1	4.29	3.1	2.32														
Limestone	115	2.7	3.10	4.5	5.17	53.9	62.05														
Total	1000		216.57		287.92		143.46		12.2		15.81		12.8		102.5		2.5		14.8		22.4
Coefficient 0.1385			29.99		39.87		19.87		1.69		2.21										

249-oz. bullion. · 11.8 lbs. speise. · 72.7 lbs. matte.

The table further shows that there are 10 per cent. of slag and 15 per cent. of fuel to the charge, thus giving all the necessary data.

Before making up the charge as it goes to the blast-furnace, the moisture has still to be considered. If the lead ore contains, for instance, 5 per cent. of moisture, 535 pounds moist ore will have to be used to correspond to 510 pounds of dry ore:

$$\text{Moist Ore} : \text{Dry Ore} = 100 : 95 :: x : 510;$$
$$x = 537.7.$$

The same is the case with fluxes and fuel.

It is to be noted that figuring a charge according to Murray's formula has one great advantage over the method next to be described, viz., that it shows in what proportions any three classes of siliceous, ferruginous, and calcareous ores are best mixed so as become self-fluxing.

c. *Common Method.*—In the common method the same ores, fluxes, and fuel, and the same slag as before are taken as a basis.

The preliminary calculations, such as bringing the different components of ore, flux, and fuel under the heads of SiO_2, FeO, and CaO, are made in the same way. The total weight (1,000 pounds) that the charge is to have, and with it the percentage of fuel (15 per cent.) and slag (10 per cent.) to be added are fixed. The available FeO and Fe of the iron ore are determined as before.

Two calculations are now necessary to determine the amounts of iron ore and limestone required by the coke-ash and by the ore.

The analyses of the ash and the two fluxes, iron ore and limestone, are entered, as shown by the heavy type in the table below. The 150 pounds of coke contain 15 pounds of ash; for these the totals of SiO_2, Feo, and CaO are figured and entered in the table.

There are 6.04 lbs of SiO_2; how much FeO is required?

$$SiO_2 : FeO :: 30 : 40 :: 6.04 : x;$$
$$x = 8.05 \text{ lbs. FeO are necessary};$$
$$3.97 \text{ lbs. FeO are present.}$$

The difference, $y = 4.08$ lbs. FeO have to be added.

To find the necessary iron ore:

$$\text{Iron Ore} : \text{Available FeO} :: 100 : 68.4 :: z : 4.08,$$
$$z = 6 \text{ iron lbs. ore.}$$

They are entered in the table; their total pounds of SiO_2, FeO, and CaO are figured and also entered.

To the previous sum of 6.04 lbs. SiO₂ have been added, by the iron ore, 0.26 lbs. SiO₂, making the total SiO₂ for which limestone has to be provided, 6.30 lbs. How much CaO is required?

$$SiO_2 : CaO :: 30 : 20 :: 6.30 : u,$$
$$u = 4.20 \text{ lbs. CaO are necessary};$$
$$1.73 \text{ lbs. CaO are present.}$$

The difference, $v = 2.47$ lbs. CaO, have to be added.

To find the necessary limestone (leaving out the SiO₂ and FeO it contains):

$$Limestone : CaO :: 100 : 53.9 :: w : 2.47,$$
$$w = 4.5 \text{ lbs. limestone,}$$

which are entered with the pounds of CaO they bring to the slag.

To see if the calculaton is correct, the pounds of FeO and CaO are multiplied by a coefficient (4.76, from $6.30x = 30$, $x = 4.76$), which changes the pounds of SiO₂ to 30, the percentage of SiO₂ of the slag aimed at. The result will be 40 FeO and 20 CaO. The table gives 40.08 FeO and 19.75 CaO, showing the calculation to give sufficiently close results.

Material.		SiO₂.		FeO.		CaO.	
Name.	Dry Weight lbs.	Per cent.	lbs.	Per cent.	lbs.	Per cent.	lbs.
Coke-ash..............	15	40.80	6.04	26.50	3.97	10.26	1.54
Iron ore	6	4.80	0.26	74.10	4.45	8.10	0.19
Limestone	4.5	2.70		4.50		58.90	2.42
Total...............	25.5		6.30		8.42		4.15
Coefficient	4.76		29.98		40.08		19.75

The weights of iron ore (6 pounds) and limestone (4½ pounds) are practically the same as those found by using Murray's method.

Deducting 125 pounds (the sum of coke-ash, with its iron ore, limestone, and slag) from the total weight of the charge of 1,000 pounds, gives the same 875 pounds as before to be made up by the ore and its fluxes.

A table, like the one below, is laid out, the analytical data are entered, and the calculation is made on a basis of 100 pounds of ore.

1. The amounts of iron ore required (15 lbs.) by the As and S are calculated as shown on pages 211 and 212, and the results are entered in the table.

2. 100 lbs. of ore contain 32.6 lbs. SiO_2 for which the necessary iron has to be provided:

$$SiO_2 : FeO : : 30 : 40 ; : 32.6 : x ;$$
$$x = 43.46 \text{ lbs. FeO are necessary;}$$
$$19.10 \text{ lbs. FeO are present.}$$

The difference, $y = 24.36$ lbs. FeO have to be added.

To find the necessary iron ore:
$$\text{Iron ore : Available FeO : : 100 : 68.4 : : } z : 24.36 ;$$
$$z = 35 \text{ lbs. iron ore.}$$

3. To the 32.60 lbs. SiO_2 of the ore, have been added from the two additions of iron ore $0.66 + 1.50 = 2.16$ lbs. SiO_2 making the total of 34.76 lbs. of SiO_2 for which lime has to be provided:

$$SiO_2 : CaO : : 30 : 20 : : 34.76 : u ;$$
$$u = 23.17 \text{ lbs. CaO are necessary;}$$
$$11.17 \text{ lbs. CaO are present.}$$

The difference, $v = 11.45$ lbs. CaO have to be added.

To find the necessary limestone (leaving out the SiO_2 and FeO it contains):

$$\text{Limestone : CaO : : 100 : 53.90 : : } w : 11.45;$$
$$w = 21 \text{ lbs. limestone,}$$

which are entered upon the table.

Adding the pound-columns of SiO_2, FeO, and CaO, and multiplying by 0.86, proves that the calculation is correct.

Material.		SiO_2.		FeO.		CaO.	
Name.	Dry Weight lbs.	Per cent.	lbs.	Per cent.	lbs.	Per cent.	lbs.
Ore	100	32.60	32.60	19.10	19.10	10.16	10.16
Iron ore for As and S ..	15	4.80	0.66	74.10	0.88	8.10	0.48
Iron ore for SiO_2	35	4.80	1.50	74.10	25.93	8.10	1.08
Limestone.............	21	2.70		4.50		53.90	11.32
Total	171		34.76		45.91		23.04
Coefficient	0.86		29.89		39.58		19.81

The figure found for iron ore (35 lbs.) is the same as with Murray's formula; that for limestone is slightly lower (21 *vs.* 22 lbs.), as the SiO_2 and FeO contained in the limestone are left out, making the available CaO 53.90, which is slightly too high.

If the items of the ore charge are now multiplied by 5.088 (as on page 217) and those of the coke charge added, the sum of 1,000 pounds, the entire charge, will again be obtained.

GENERAL SMELTING OPERATIONS.

§ 62. **Blowing-in.**—This consists of three parts: Warming the crucible, filling the furnace, and starting the smelting.

The warming of the crucible in a new furnace must be done slowly and with great care, so as to raise the temperature gradually; otherwise the quickly escaping moisture will crack the masonry and allow the lead in the crucible to percolate. The practice of metallurgists differs in the manner of warming the crucible and the amount of time given to it. The writer has, whenever it was possible, taken forty-eight hours to warm the crucible, and has been accustomed to proceed in the following way: the water is turned into the jackets so as to fill them and have just a little overflow; then the flue leading to the dust chambers is closed, and the damper on top of the stationary stack raised, or the movable stack put in place, so that the gases may pass off into the open air.

With some of the latest furnaces the iron plate for closing the furnace is lowered, and the joint covered with fine ore, the gases formed, when blowing-in, being allowed to pass off into the dust chamber. This can only be done if the furnace is blown in with very little or no charcoal, which means having the crucible full of molten lead before the charge is introduced. If the furnace were filled to the top of the jackets with charcoal before adding the blowing-in charges, the gases, when the blast is let on, being rich in carbonic oxide, would be liable to ignite and damage the flue and dust chamber.

A wood fire is made in the bottom of the furnace that will not reach half-way up the crucible. If it is kept going for a few hours, always replenishing the wood, ashes will have collected in the crucible. These, being bad conductors of heat, have to be removed in order that the burning wood may be in contact with the bottom. When raked out by means of a hoe from the breast of the furnace, a new fire is kindled. After from three to four hours

too many ashes will have accumulated in the furnace for the heat
to have the desired effect, and the crucible is cleaned out again.
While the crucible is being dried and warmed, the lead-well is
filled with glowing charcoal, and the basin itself covered by a piece
of sheet iron, so as to admit only a little air, thus preventing the
charcoal from being burned quickly. Similarly the breast of the
furnace is closed with loosely set bricks, by which the draught is
checked and too quick combustion of fuel on the surface prevented.
The heating is continued for twenty-four hours, when the outside
of the crucible will feel warm to the touch. This shows that all
the moisture is expelled and that the crucible can stand a high
heat without endangering the brick-work.

The second warming has for its object the heating of the crucible to a good red-heat, and requires carbonized fuel and blast. A
new fire is kindled and some charcoal added ; the blower is started,
and all the tuyere-bags are tied up or wound up, or the blast-gates
are closed, except that of the tuyere nearest the breast to be used.
This is connected with an iron pipe inserted into the crucible from
the breast, and the blast allowed to play on the charcoal till it is
well ablaze. A second layer of charcoal is added and, when it is
fully aglow, a third one and so on till the crucible is filled to the
jackets. The iron pipe is then inserted deep into the glowing coal,
in order that the blast may reach the bottom of the crucible. Meanwhile the furnace-man works at intervals with an iron bar and
a hoe, turning the coal over and moving it from front to back and
vice versa, so as to get it all into a perfect glow. When this is
accomplished, the pipe, of which often a small part has been melted
off, is withdrawn, and the furnace is let alone for an hour or two.
The pipe is again introduced and the charcoal burned down, the
furnace-man stirring it to bring all parts into contact with the
blast. The ashes are removed, and the crucible and well thoroughly
cleaned. The heating is repeated from three to four times in
twenty-four hours. The outside of the crucible will then have become too hot to be touched with the hand.

The second step is the filling of the furnace. This must be done
with care, in order that a crust may not form later on the lead,
when the blast has been started, as it is very troublesome to remove, and sometimes remains in the furnace during an entire run.
To avoid a crust, it is essential to have a clean crucible entirely
filled with red-hot lead. If half filled with a mixture of charcoal-
ashes and small bits of charcoal, a dead layer will be formed be-

tween the lead and the slag which will soon harden to a crust and attract small obstacles that would otherwise be carried out by the slag. The proper mode of filling a blast-furnace depends a good deal on the kind of fuel, and is to some extent individual with the metallurgist. After supplying the crucible with the necessary lead, which is either charged from the top or melted down from the breast of the furnace, the first charges will consist of an easy-smelting slag with much fuel and the necessary fluxes. Ore-charge gradually replaces the slag-charge, and the high percentage of fuel will be cut down until finally the normal charge is reached. It is better for a blowing-in slag to be glassy than crystalline, as it melts easier, and a slag consisting of 30 SiO_2, 40 FeO, and 20 CaO is probably the best.

Two methods of blowing-in may be mentioned; they both require charcoal.

For the first, the thoroughly heated crucible is gradually filled with charcoal, the blast playing on it till it is all aglow; then the breast of the furnace is put in, the tuyere openings are closed, and the rest of the charcoal is added from the feed-floor, so as to reach about one foot above the tuyeres. Then follows a bed of coke one foot thick, which will reach about to the top of the jackets. Upon this are charged lead and coke, with some slag and the necessary fluxes. When all the lead has been given that is required to fill the crucible, the furnace is filled with alternate layers of fuel and half ore-charge and slag-charge, using the same amount of fuel for ore as for slag.

The necessary amount of iron ore and limestone are calculated in the same way as in making up the regular blast-furnace charge.

The tuyere-holes in the jackets are now cleared, the lead-well is partly uncovered, the tuyere-pipes are put in place, and the blower is started slowly, the number of revolutions as well as the pressure of the blast being noted.

Too much emphasis cannot be laid on starting the blast slowly. If the blower is allowed to make too many revolutions per minute, the result is that the fuel burns quickly, the heat rises, and is not concentrated at the tuyere-level, and the gases rush upward and do not thoroughly warm the charge; the pressure of the blast at the same time lowers the level of the lead in the crucible, with the charcoal ashes floating on top. When finally slag and matte come down, they become partially chilled, because they sink too much below the level of the tuyeres, and thus the dreaded blowing-in crust is formed. For this reason the blower should always be started up slowly, and the pressure not allowed to be much over one ounce until slag appears before the tuyeres. The normal pressure of the blast is not reached until seven or eight hours after the blower is started.

When the blast has been turned on a flame issues from the siphon-tap till the lower aperture is closed by the lead. Care must be taken that the passage of the well does not become choked with ashes, which are carried out by the flame in considerable quantities. The lead rises slowly as the bullion melts down in the furnace, and while often at first a little cool, it soon becomes hotter, and, when it reaches the top of the well, ought to be bright-red. Shortly after the blast has been let on, the flow of water into the jackets must be increased in order that it may not boil. Soon, however, when the slag charged begins to melt and come down, the jackets become coated and thus cooler. The large excess of water required just before is turned off, and less than the normal amount will be sufficient for an hour or two. On the feed-floor the charges will at first sink quickly when the blast has been turned on, and it may be sometimes advisable to continue feeding alternately ore-charge and slag-charge until the furnace has been entirely filled with this mixture, and only then to begin with normal ore-charge. This is, however, exceptional. The smoke issuing at first from the top of the charge will be black, and gradually become lighter, and decrease in amount. The furnace-man watches through the peep-hole in the tuyere-pipe to see when it is time to tap the slag. This will be indicated by the appearance of a little blue flame of burning carbonic oxide gas.

The ashes of charcoal have now to be removed from the furnace, the lead having lifted them up from the bottom to about the level of the tap-hole. They float on the lead, and are liable to prevent the necessary contact of lead and slag or lead and matte in the furnace. Generally the slag carries out the ashes, but sometimes this has to be assisted, which is done by entering with a rod through the tap-hole and loosening the ashes where they are inclined to accumulate and adhere, thus stirring them into the slag. It is often advisable after stirring with a rod to let the blast blow through the tap-hole, as it will blow out a good many ashes. If, nevertheless, a crust of ashes and slag should form on top of the lead, it is necessary to break it by thrusting bars through it and lifting it up while it is still soft. If it hardens, the furnace will have to be stopped, and the breast taken out before the crust can be removed (see § 66). The indications of a crust are that the lead in the well becomes dark and does not play freely with the blast. The first is easily seen; the second can be discovered by removing the plug or cap of the peep-hole in the tuyere-pipe just above the lead-well. If the lead plays freely, it will sink in the well; if not, it will remain immovable. By inserting a rod through the tap-hole the position of the crust can be felt.

If everything goes well, the furnace will have its full blast in about seven hours after starting, the slag its normal heat next day.

One of the main points in filling a furnace is to be quick about it; therefore as many men are put to work as is possible. A furnace 42 by 108 inches can be filled in from 30 to 40 minutes by two sets of feeders, of four each, helped by as many wheelers as may be required to bring the necessary materials to the furnace. If there is any delay in charging, the crucible may cool and a crust form. For this reason no weighing is done at the furnace, but the charge is measured by the number of scoopfuls and shovelfuls, the average weight of one having first been determined by weighing 10 scoopfuls of coke and 10 shovelfuls of iron ore and limestone.

With good charcoal this method of blowing in is quite satisfactory; a crust very rarely forms. With charcoal of an inferior grade the second method is used. It is as follows: when the crucible has been well heated up and cleaned out as described, the breast is put in, the tuyeres in the jackets are closed, and the lead-well is covered with sheet iron, its lower side being coated with a clay lute and the upper one weighted with a couple of bars of lead. Charcoal is fed from the feed-floor until it reaches the top of the jackets, and covered by a bed of coke about one foot in thickness. A man descends into the furnace and spreads out evenly the successive charges of bullion, coke, slag, iron ore, and limestone. Then follow the regular ore charges. When the furnace has been filled, which takes about two hours, the charcoal is kindled from the tuyeres and a gentle blast started. The work now proceeds in the usual way. When the lead, coming down in the furnace, has closed the lower opening of the well, the cover is removed. Whether this has occurred can be found by removing from time to time the iron rod which closes the tap-hole of the lead-well. The first slag is tapped about 25 minutes from the time that the lower opening of the well is closed by the lead, which shows how quickly the crucible fills.

The reason for the good results attained by this method may be explained in the following way : as the entire lower part of the furnace is closed during the filling, and the charcoal is not kindled until just before the blast is started, it cannot burn in the crucible and form the dreaded mixture of ash and fine coal. At the tuyeres much heat will be generated by the combustion of the charcoal. A slight effect may be communicated downward, but it proceeds principally upward. The first lead that melts and trickles down is very hot when at the tuyere-level, but cools somewhat below it. As more follows, the first becomes heated again, so that finally the bottom of the crucible is filled with liquid lead, although it is not yet very hot. The unburned charcoal is raised gradually towards the tuyeres, where it burns completely, leaving no mixture of ash and fine coal, and gives a very high temperature, which prevents the forming of a crust. The lead as it rises in the crucible becomes always hotter from the continual addition of red-hot lead from above and its approach to the tuyere level, as well as by its contact with the hot slag.

Both of these methods require passably good charcoal, if not the best. If this cannot be had, it is replaced by split wood which is thoroughly dry and cut into pieces of suitable length. When wood is used instead of charcoal, it is burned down in such a way as to make fresh charcoal in the crucible. Thus, when this is sufficiently heated and cleaned out, the last fire is made and the wood burned down quickly with the blast only so far as to be well charred, fresh wood being added and the burning down continued until the crucible is filled with charred sticks. The breast is put in place and a two-foot bed of coke given. Upon this come the regular blowing-in charges, the double amount of coke requiring an extra quantity of slag, iron ore, and limestone.

If neither charcoal nor suitable sticks of dry wood are available, as is often the case in Mexico, it will be necessary to melt down the lead in the crucible before filling the furnace, as filling the crucible with coke is nearly sure to form a very disagreeable crust, which it is difficult to remove afterwards from the breast. If neglected, it grows rapidly, and the result may be a crucible frozen up solid. The mode of procedure is to fill the well, after it and the crucible have been heated and cleaned with glowing coals, and to pile them high up over it, so that it shall keep hot. Then as many bars of lead as the crucible will hold are introduced from the breast by sliding them in on a board. A fire is kindled on top of the lead ; when that is melted down, the floating ashes are removed, a new fire is made, more lead melted down, and this repeated until the crucible is full of lead.   The last ashes are raked out, the breast is put in, and enough of the best brushwood that has been sorted out is charged from the feed-floor to reach well above the jackets. On top of this comes a two-foot bed of coke, to which eight or ten bars of lead are added, in order that this, coming down hot, may help to heat the lead in the crucible.   Then come the usual slag charges and slag-ore charges.

The bullion which used generally to be charged from the top is now more commonly introduced from the bottom.   The lead is melted down in the warmed crucible, as just described, the breast put in place, and charcoal fed to the top of the jackets ; then the coke-bed is given and the furnace filled, with the addition of a few bars of lead with the charge.

Some metallurgists avoid charcoal entirely.   The crucible is warmed with wood, the lead melted down with coke, all the clinkers are removed, a good coke fire is started on the lead, and the furnace filled with coke to the top of the jackets.   Then a few slag charges are given and the furnace is filled with an easy ore charge running high in lead.   This is soon changed for the regular charge.

The writer has never blown in a furnace without using lead. Hahn [1] says that, when no lead can be had, he closes the syphon-tap where it enters the crucible with a clay plug, removing it when the lead begins to flow out with the slag.   He adds that bottom crusts are unavoidable.   When sufficient lead has been produced to fill the crucible, the furnace is blown out and started up again in the normal way.

A method of blowing-in described by Henrich [2] refers rather to a furnace where metal is tapped from the bottom than to one having a lead-well.   It is given here because it describes the blowing-in of a blast-furnace used for concentrating matte, which is tapped from the bottom.   The crucible is warmed, filled with charcoal up to the level of the tuyeres, more fuel is given, and then blowing-in charges are fed, followed by ore charges.   When the furnace is filled to from one-half to two-thirds the distance between the tuyeres and the charging floor, the front is closed.

[1] " Mineral Resources of the United States," United States Geological Survey, 1882, p. 342.

[2] *Engineering and Mining Journal*, April 11, 1885.

All the tuyeres excepting one or two (according to the size of the furnace) opposite the metal tap (2 inches wide) are closed, the pipes inserted, and the blower is started. The flame passes through the tap-hole, blows out ashes, and heats the bottom. The tap is kept open by inserting a rod until slag begins to flow, when it is closed with a clay plug. The other tuyere pipes are now intro-

TABLE OF BLOWING IN CHARGES.

Charges, number.	Coke, scoops.	Charcoal, scoops.	Iron ore, shovels.	Lime-stone, shovels.	Bars of bullion, number.	Slag, shovels.	Size of furnace, inches.	Crucible filled from
1	20	....	3	2	6	....		
3	15	....	3	2	6	1		
5	10	....	3	2	8	2		
5	10	....	3	2	7	2	36 in diam	above.
2	10	....	3	2	5	5		
5	10	....	half ore-charge.		3	6		
1	10	....	half ore-charge.		....	6		
			regular ore-charge.					
1	10	....	....	....	6	....		
3	8	....	½	¼	6	1		
4	7	....	½	¼	6	2		
4	6	1	½	¼	6	2	36×84	above.
2	5	2	½	¼	5	2		
2	5	2	half ore-charge.			2		
4	4	3	half ore-charge.			2		
			regular ore-charge.					
1	20	....	3	2	6	....		
3	15	....	3	2	6	1		
10	15	....	3	2	6	2	36×84	above.
5	12	....	3	2	4	3		
4	6	....	half ore-charge.			4		
			regular ore-charge.					
1	25	....	1	1	15	12	42×120	above.
12	12	6	1	1	15	12		
			regular ore-charge.					
1	25	....	1	1	20	....	42×120	below.
15	9	....	1	1	2	33		
			regular ore-charge.					

duced at short intervals one after the other. Once or twice slag is tapped from the metal tap; it is then allowed to accumulate in the crucible and tapping is begun at the slag tap. In this way the crucible becomes thoroughly heated.

The only instance coming under the writer's observation of slag being taken out from the lead-well was at a desilverizing plant, when very rich bullion, obtained by smelting zinc-crusts in a circular 36-inch furnace (§ 112), had to be exchanged quickly for low-grade bullion. The furnace was fed with a number of slag charges to half fill it; then sufficient low-grade litharge and slag was added to fill the crucible again with lead. When the slag charges had about reached the tuyeres and much slag was being formed, the lead was all dipped out of the well and then the slag, until finally the lead of the litharge appeared and filled the well, when slag was tapped again in the usual way. The rich bullion was very little diluted and could go to the cupelling furnaces.

The making up of blowing-in charges varies a great deal. The preceding table gives a few examples. A scoop of coke weighs 12 pounds, one of charcoal 4, a shovel of iron ore 15 pounds, of slag 12 pounds, and of limestone 10 pounds.

§ 63. **Regular Work on the Charging Floor.**—The work on the charging floor consists in bringing ore, flux, and fuel from the bins to the scales, weighing out the required amounts, dumping them, and feeding into the furnace.

Ores and fluxes are brought to the scales either in trucks or wheelbarrows; the former are filled through chutes from the bins, the latter by shovelling; coke and charcoal are nearly always brought in wheelbarrows.

Everything that goes in the furnace must be accurately weighed. No thorough work can be done without. An approximate weight can be obtained by taking ten or twenty shovelfuls and reducing this weight to that of one, and then making up the charge by measure. But shovelfuls differ according to the men who make them, and the same man will not always shovel a uniform amount— say, at the beginning and toward the end of his shift. With the shovel system the furnace is always liable to get out of order, and the metallurgist has no means of accurately determining the cause. As seen from § 62 the shovel system is used in ore smelting in blowing in a furnace, where the filling with slag charges must be done as quickly as possible; as soon as ore is substituted it must be regularly weighed. In refining works, where certain lots of lead by-products are treated in a blast-furnace with slag alone, the shovel system is often found. This is permissible on account of the small amount of new slag that is being formed in the blast-furnace, but even then weighing proves more satisfactory.

The scales in use have a capacity of from 1,000 to 1,500 pounds. They are of two kinds: the single-beam scale, where the weights have to be shifted for every weight taken, and the multiple-beam, where by pulling a lever on the outside of the case the one of the four, six, or eight beams which bears the desired weight can be set free. By the multiple-beam errors are avoided, as the foreman sets the weights and locks the box surrounding them, the workman having only to pull the levers in their regular order.

In making up a charge and in distributing it in front of the feed door the fuel is always kept and fed separately, while ore, fluxes, and slag are sometimes mixed and sometimes kept separate. With small furnaces—say, 36 by 60 inches—ore, fluxes, and slag are

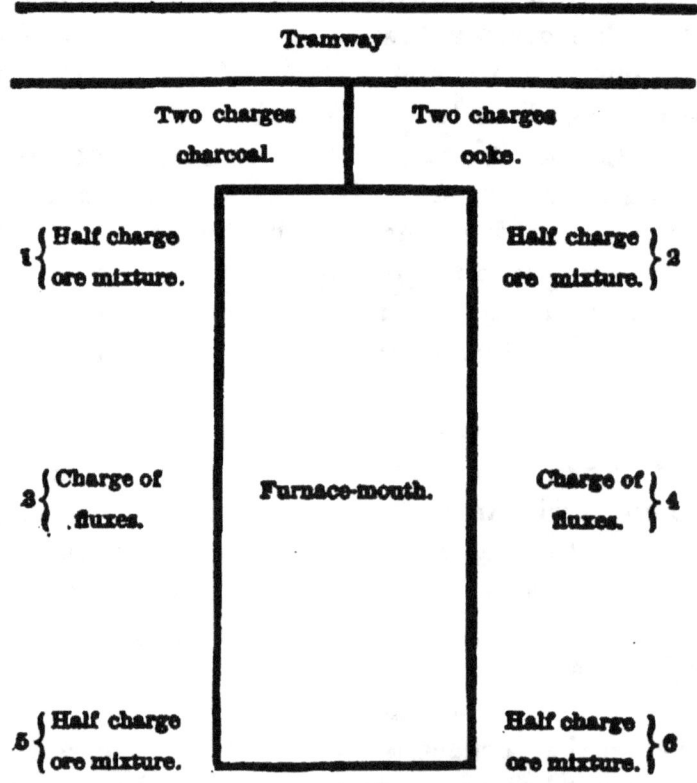

DISTRIBUTION OF CHARGE ON FEED-FLOOR.

usually dumped into one heap on the long side of the furnace, and the fuel has its place on the short side. With a large furnace (36 by 108 or 33 by 120 inches) ore and fluxes are sometimes distributed in separate heaps on the long sides of the furnace. Iles,[1] for instance, adopted the following disposition of the charge on the furnace floor (Figure 145). The numbers denote the order in which the materials are supplied to the furnace. The feeder first

[1] *Engineering and Mining Journal*, March 24, 1883.

charges one-half of the charcoal pile and on top of this the ore mixture and the fluxes 1, 3, and 5 on the same side of the furnace; then he gives one-half of the coke pile on the opposite side, which is followed by the rest of the ore mixture and the fluxes 2, 4, and 6.

The mode of distributing the charge in the furnace deserves great attention, as by wrong feeding a good charge can be spoiled and a furnace be put out of order. Fuel and ore charge go into the furnace in alternating layers, beginning always with the former. The fuel must never be exposed to the air, but must be well covered by the ore charge. In feeding, the fuel, especially charcoal, should be kept near the centre, and the charge distributed more towards the walls of the furnace, and, finally, the finer parts fed nearer the walls than the coarser ones. The reason for this is that in the descent of the charge the lighter fuel is liable to be pressed towards the walls of the furnace; further, as there is more friction between charge and furnace wall than between the parts of the charge itself, it will sink quicker in the middle while descending, and pack at the sides. If ascending gases are to pass evenly through the charge, their passage must be assisted in the centre, and retarded at the sides. Lastly, with carbonate ores an intimate contact of readily reducible lead oxide with the fuel in the upper part of the furnace must also be avoided, as it would assist in reducing it to metallic lead, and thus increase the loss by volatilization. Guyard[1] thinks that this distribution of fuel towards the centre and of charge towards the furnace walls favors the growth of accretions immediately above the jackets, and advises charging the fuel alternately towards the centre and the sides. He says the effect of this would be that the accretions would begin to form higher up in the furnace and could be removed more easily; but would not the top of the furnace easily become hot? In feeding a charge that contains much fine ore (sand carbonates) special precautions are taken to prevent this from trickling through the charge, as it fills up the pores of the coke, and, arriving in a crude state at the smelting zone, chills the furnace and often fills the tuyeres. Glenn[2] describes what he calls a filter charge. It is not uncommon with lead furnaces, and consists in making the charge large, and in feeding gradually on the heavy coke bed first the coarse parts, then

[1] Emmons, "Geology and Mining Industry of Leadville," United States Geological Survey, monograph xii., p. 665.

[2] *Engineering and Mining Journal*, July 19, 1884.

finer ones, and last of all the fines here and there, as the charge settles in the furnace. The finer the ore, the more uniform in size should the slag and fluxes be to make a good filter. Fine ores are often fed where the fumes are strongest, thus helping to equalize them. The distribution of ore and fluxes in a charge may effect its fusibility. Henrich[1] gives some very interesting experiences on this point, and calls attention to the fact that the formation of the slag, while the charge is descending in the furnace, is a gradual process. Pieces of ore and flux containing the necessary constitutents to form slag of a low-melting point may do this in the upper parts of the furnace if in close contact with one another. The liquid slag will then eliquate and take with it from the hotter zone below less fusible parts of the charge, thus forming prematurely the intended slag. If in charging, the ore is followed by a flux of similar composition, very liquid slag cannot form in the upper parts of the furnace, but slag requiring a higher temperature for its formation will be formed lower down, and will combine with the liquid slag at the tuyeres and form the normal slag. Thus, for instance, a siliceous ore should be followed by a siliceous iron flux, and the dissimilar flux—limestone—come last. In this way a higher temperature will prevail in the smelting zone than would if ore and fluxes were mixed and fed together into the furnace. Henrich explains the good effect large charges often have in furnaces as compared to small ones, by the separate charging of ore and fluxes with the former, while with the latter, ore and fluxes are usually dumped in one heap in front of the feed door and charged together into the furnace.

A furnace is in a good condition on top when this is cool, when the charges sink regularly and evenly, and the fumes ascend uniformly, or a little more densely at the sides than at the centre.

The labor required on the feed-floor varies with the size and number of the furnaces and the conveniences for handling materials. A furnace from 33 to 42 inches between the tuyeres and from 84 to 120 inches in length requires per shift one feeder and two wheelers; with ten furnaces under the same conditions are required ten feeders, but only sixteen wheelers. Both generally work twelve-hour shifts, although sometimes the feeders work only eight.

The tools required by the feeder are a square-pointed, long-handle shovel, and a scoop or fork for the fuel, a six-pound napping

[1] *Engineering and Mining Journal*, December 27, 1890; June 6, 1891.

hammer for breaking up the coke, and a stiff broom, several heavy 1½-inch steel bars of different lengths, the longest ones reaching from the top of the jackets up to 4 feet above the feed floor, and several ten-pound double-faced sledges. The bars and sledges are used in cutting out wall accretions (§ 66), and one set of tools is sufficient for a number of furnaces. The wheelers require several iron wheelbarrows for ore and fluxes and a wooden barrow or buggy for fuel, several round-pointed shovels, a scoop or fork, several coarse brooms, and generally a pick. A definite number for most of the tools cannot be given, as it varies too much with the general arrangement of the feed-floor.

§ 64. **Regular Work on the Furnace Floor.**—This consists in regulating the water supply, taking care of the tuyeres, tapping speise, matte, and slag together into the slag-pot and wheeling it to the dump, tapping the lead either first into the lead-pot or ladling it directly into the moulds, and piling the bars of base bullion upon the ground or on a truck.

The water of the jackets is kept at about 70° C., the usual test being that the hand can be quickly passed through the outflowing water without being scalded. Any irregularity in the temperature of the jackets indicates that the coating on the inside is thicker in some than in others, and thus that the smelting proceeds unevenly ; the slower descent of the charges on the cold side will corroborate this on the feed floor.

The pressure of the blast is constantly watched and all the changes in the gauge noted. It is regulated by means of a damper. If the furnace has its own blower, its revolutions are also counted.

The appearance of the tuyeres forms a good indication of the inner condition of the furnace. It is not necessary that the entire mouth of the tuyere should be bright ; it is often covered by a thin scale of slag, showing a star-like brightness in different places. If it becomes quite dark, a rod is inserted and the slag pierced to see if it has grown too thick. In this case repeated poking only aggravates matters, as a "nose" of chilled slag forms, reaching into the furnace. (For correction of this evil see § 66.)

The tapping of the slag is done with a square-pointed steel bar. The slag runs over the slag-spout into the slag-pot ; when full, the tap-hole is stopped with the stopping-rod. As regards liquidity, matte comes first, then speise, and last the slag. When flowing from the spout they can be readily distinguished. The clay stopper in the tap-hole ought to be sufficiently soft to be perforated by pricking it with the bar, but oftener it requires a few taps or strokes with a sledge. With a hard tap it is sometimes advisable to insert a small piece of charcoal into the clay stopper so that it protrudes slightly. On shutting off the slag, the charcoal enters the opening made and is followed by the clay. On tapping again very little sledging will be necessary. The use of this charcoal is kept up until the correction in the charge or the feeding has softened the tap. The taking of samples of slag will be discussed in § 81. It is a rule always to keep some slag in the furnace. At certain intervals, however, the lead and slag are allowed to rise until the little blue flame, seen through the tuyeres, indicates that the slag has nearly reached that level, when it is all tapped out to see how many pots of slag the furnace holds. If there is less slag than usual, it shows that the region of the tuyeres is not as free as it ought to be. It is generally desirable to accumulate the speise and matte in as few slag-pots as possible, as, when they run out with the slag, they are splashed, especially the matte, against the cool sides of the pot. The separation of matte and slag being complete only near the centre of the pot, where they remain liquid for some time, an outer ring of rich slag, assaying often 15 ounces and more to the ton, will be obtained when the bulk of the slag is poor. This occurs especially with acid slags; with slags that are very fluid the difference in outer and inner slag is not so great. If the furnace has a clay breast, the slag is tapped when it has accumulated in the furnace a little below the ordinary tap-hole. All the speise and matte, and perhaps a little lead, will run out and be collected in one pot, and all the slag of this pot will be saved. After a number of tappings from the ordinary tap-hole, regulated by the speise and matte the furnace is making, the tapping lower down is repeated. If the furnace has a water-cooled cinder-notch (tapping-jacket), the level to which the slag is allowed to rise is regulated by the lead alone, as a jacket with two tap-holes is not much used.

The loss of metal by imperfect separation of matte and slag is completely avoided by the use of an overflow-pot (§ 49, *h*) or a catch-pot having a tap-hole (§ 49, *h*).

It is of importance for the good running of a furnace that the level of the lead should be kept high. If too low, the speise and matte will be too far removed from the zone of fusion, and thus liable to chill and form a crust. The less lead there is in the charge the higher must the level of the lead in the crucible be kept, that it may not cool, as it is not frequently replaced by fresh lead that has just passed the hottest part of the furnace. With a charge containing from 7 to 9 per cent. of lead, the lead in the well is kept often as high as 5 inches above the slag-tap. It used formerly to be taken out altogether by dipping, but if this is not done with judgment, too much is apt to be removed, which causes trouble. The amount is to be regulated by the number of charges that have been fed into the furnace. In ladling the hot lead from the basin of the well into moulds, most of the dross held in solution rises to the surface when the bar cools, and is removed by skimming, but some always remains. This dross often causes dispute between buyer and seller as to the silver contents of a given shipment of bullion. The dross assays much lower in silver than the clean lead, and as it is not uniformly distributed in the bar, the sampling becomes irregular (§§ 71 and 72). Then it is the pride of the furnace-keeper to produce bars with a clean surface. This has led to pouring the mould one-quarter full of lead, then adding dross, and covering it up with a layer of clean lead. The trouble can be removed by allowing the lead to accumulate in the well, which is on purpose built up higher than the top of the crucible, and tapping it at intervals into a cast-iron pot. The lead not being dipped out, cannot sink lower than the level of the tapping-hole, which is on a line with the wall of the crucible, and can be kept as high as desired. In the cast-iron pot the dross is skimmed off, and the cooled lead is then ladled into a series of eight or ten moulds placed in a row alongside the furnace on two rails, supported by two wooden horses. Nevertheless, many metallurgists prefer to dip the bullion from the lead-well. The reason is that, if the lead is tapped at long intervals from the well, a large number of bars, say ten, will be taken out of the furnace at once, and the charge will sink suddenly to occupy the space set free by the lead, with the result that particles of unmelted charge pass too quickly through the melting-zone to be fused, and often form the nucleus for a crust. They therefore dip the lead from the well as fast as it forms; *i.e.*, the dipping is regulated by the number of charges fed into the furnace and the percentage of lead it contains. Dipping and tapping the lead are both used; the dipping may perhaps be preferable where charges run low in lead, say 8 per cent. Another reason for dipping the lead is that the refiner pays for the dross as lead. If the smelter skims the dross, he not only is not paid for it, but has to work it into copper-matte, and copper-refiners pay nothing for lead and only 93 per cent. for the silver of the matte. This is probably the main reason why many smelters prefer dipping to tapping.

When the bars have solidified they are marked with the running lot-number, removed from the moulds with a pick, and piled up to be sampled (§ 72), weighed, and shipped.

A furnace is in good working order below when the temperature of the jackets is uniformly high, the pressure of the blast does not fluctuate, and the tuyeres remain bright, having only short "noses." The furnace should within a given time produce always the same number of pots of slag, the tap-hole be neither too hard nor too soft, and the lead in the well be of a bright-red color, play with the blast, and sink slightly every time a pot of slag is tapped.

On the furnace floor are required for every furnace (from 33 to 42 by from 84 to 120 inches) one furnace-keeper, one tapper who looks after the lead, and from one and one-half to three pot-pullers, according to the size of the dump and the manner of disposing of the slag.

The tools required by the furnace-man are 2 tapping-bars (6 or 8 feet long, of $\frac{3}{4}$-inch steel), 2 stopping-rods (10 feet long, of $\frac{1}{2}$-inch iron, with a disc $2\frac{1}{2}$ inches in diameter for the clay plug), 2 iron rods (12 feet long, of $\frac{3}{4}$-inch iron) for special cases, 3 or 4 heavy (1–$1\frac{1}{4}$-inch) steel bars about 6 feet long, and 2 ten-pound double-faced sledges; a 2-inch steel bar 8 feet long is handy, if a crust is to be pried up. The tapper's tools are few: a $\frac{1}{2}$-inch steel bar, 3 feet long, a cast-iron ladle (6 inches in diameter, 3 inches deep), riveted to a 4-foot handle, a pick, a skimmer-rod to clean the well, a set of steel dies, a 4-pound hammer to number the bars with $\frac{3}{4}$-inch figures, and a chisel to trim them. The pot-pullers have to mop the pots with a clay wash; they require no special tools, excepting occasionally a steel bar and sledge to loosen a cake of matte or slag from the slag-pot, should it adhere firmly.

§ 65. **Work on the Dump.**—If the ordinary slag-pot is used, it is wheeled out on the dump and allowed to cool. When the contents have solidified the pot is tilted, and the cone of slag, with speise and matte adhering to the bottom, rolled out. This is broken up with a sledge, matte and speise are sorted out and piled up, and the clean slag is thrown over the dump. Sometimes the slag is loaded into flat cars in pieces as the result of breaking up the cones, or else the clean slag is first put through a crusher, thus preparing it for the direct use of the railroad as ballast.

When the overflow-pot is used which discharges slag free from shots of matte into an ordinary slag-pot, it is allowed to cool on the dump. The contents of the other pot are emptied while still liquid over the edge of the dump. The scale of slag forming on its sides is reserved, if required for the blast-furnace. With a catch-pot having a tap-hole, the pot is wheeled either to the edge of the dump and the liquid slag is discharged there, or to the large side-dumping pot, if this be in use. The tap-hole of the catch-pot is opened from the inside by thrusting a steel bar through the crust of the slag. The large tilting-pot is generally placed in a trench left open in the dump, and receives there from both sides the liquid slag. When filled, it is hauled to the edge of the dump by a horse or mule and emptied. Occasionally this receiving-pot is placed between two furnaces to receive their clean slag, and hauled, when filled, by a wire rope to the edge of the dump and there tilted.

All the slag remaining in the overflow-pot or in the catch-pot with tap-hole goes back to the blast-furnace.

The dragging-out of slag-pots to the edge of the dump is greatly facilitated by keeping its surface even and smooth. This is done by forming squares (about three feet) with rails or sand on top of the rough surface and filling them with liquid slag.

The number of men required on the dump varies with the size of the furnace and the manner of disposing of the slag. Two men in a ten-hour shift can clean up the slag and matte made by a furnace 42 by 120 inches at the tuyeres. This represents good work. The tools required are a steel bar, a pick, a sledge, a round-pointed long-handled shovel, and an iron wheelbarrow.

§ 66. **Irregularities in the Blast-Furnace.**—The disturbances that occur during the run of a blast-furnace, having different sources, are numerous. Some are caused by defective machinery and apparatus, others by refractory ores, faulty charges, or wrong manipulations. Some accidents due principally to the last-mentioned cause may be corrected in the following way.

When the charges do not descend as evenly as they should, one side sinking faster than the other, the jackets on the lower side being much hotter and the tuyeres brighter than those on the upper side, the first change is made in the manner of feeding. The fuel is placed more on the hanging side, and ore and fluxes on the quickly-descending one; then a full stream of water is turned into the hot jackets and that of the cool jackets reduced till there is just an overflow. By the combination of these remedies the smelting of the furnace on one side more than the other will be corrected and the charges will right themselves again after a few hours. Shaking up the charge with a long, heavy (1½-inch) steel bar, introduced through the feed-door into the hanging side, will often hasten matters. At some works that have to treat ores rich in zinc, it is the practice at the beginning of every day-shift to drive in a steel bar at the four corners of the furnace as far as the top of the jackets, thus loosening any wall accretions that are forming. By doing this, the number of times a furnace has to be barred down to remove accretions is greatly reduced and the whole running improved.

When the charges descend irregularly, it often happens that the fire creeps up and it becomes hot on the surface (over-fire, fire-tops, hot top). The furnace may then be fed down, which consists in adding only just enough fresh charge to keep the flame or heavy smoke from passing through. When the surface has thus been lowered for 2 or 3 feet, the furnace is filled up quickly again and the top thus cooled: Simple sprinkling of water on the top of the charge has only a temporary effect. This feeding down helps matters if the over-fire does not come from a crust in the crucible.

The cause of the irregular descent of the charges lies generally in the formation of wall accretions (§ 82), which begin on top of the jackets and grow thicker towards the feed-door. They assume different forms. The following figures, 146–149, by Iles[1] show some extremes. Figure 146 represents a more or less regular shape, and the smelting power of the furnace is not necessarily reduced. With irregular hangings, like those in Figures 147, 148, 149, the descent of the charges will be greatly obstructed and the amount of fluedust much increased. The charge will be tight at the narrow parts of the furnace, and the blast entering the tuyeres will be concentrated in a few places and cause "blow-holes." As soon as these wall accretions are discovered they have to be cut or barred down. In order to reach the lowest part, the furnace is fed down and the blast lessened at the same time, till the charge has reached about the top of the jackets. While it is being lowered no lead is removed from the well, in order that the crucible may be entirely full while the barring out is going on. The blast is stopped, the blast-pipes are removed, the tuyeres closed, all the slag in the furnace is tapped, and the flow of water into the jackets nearly shut off. When everything is ready a charge of coke is given, with some slag and flux, forming a bed for the accretions to fall on as they are chipped from the walls. The cutting-out is best begun

[1] *Engineering and Mining Journal*, February 6, 1886.

just above the jackets. A square-pointed steel bar about $1\frac{1}{4}$ inches in diameter and long enough (about 16 feet) to reach from the top of the jacket well into the opposite side on the feed-floor, is driven with a sledge into the crust above the jacket. If it does not yield, a rope tied around the head of the bar is thrown to the opposite side, where several men pull it and thus break off the crust. This is repeated until it has been removed in a number of layers, two sets of men working on opposite sides. While the crust is being barred off, fuel, slag, and fluxes are added that it may be smelted out when the furnace is started up again. The reason that the barring down is begun from below is that otherwise the broken crusts and small slag-charges that have been added might so fill

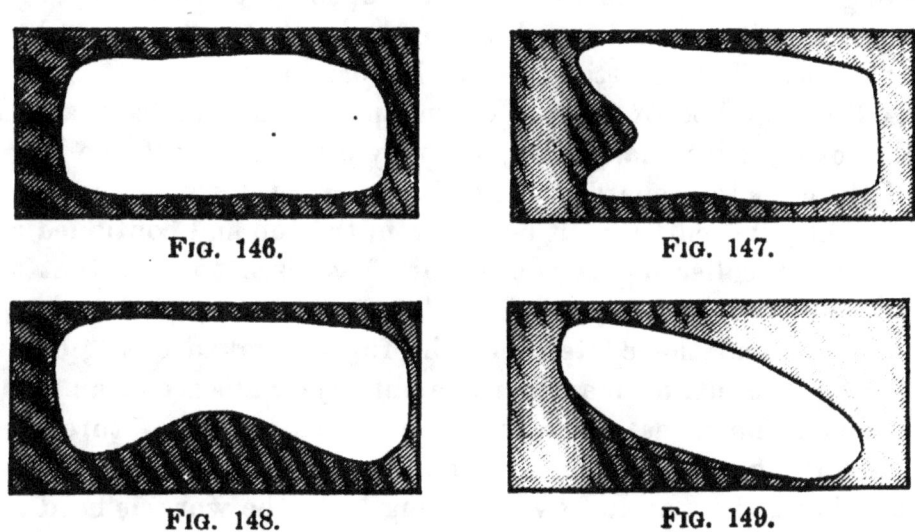

FIG. 146. FIG. 147.

FIG. 148. FIG. 149.

WALL-ACCRETIONS.

up the furnace that it would be impossible to reach the crust at the top of the jackets, and new accretions would form again quickly. Sometimes, however, if the crust is thick and hard, the barring-down is begun from the top, continuing until the accumulated crust and slag-charge meet the clean side-wall, when it is smelted out, keeping the charges low. The furnace is now let down again, and the second half of the barring-down is begun at the top of the jackets and continued upward.

When the sides of the furnace have been cleared, the tuyeres are cleaned out, the tuyere-pipes inserted, a weak blast, to be increased gradually, is turned on, and smelting resumed. Care must be taken about the water supply as the jackets grow hotter, and about the tapping of the slag, as the wall-accretions often melt very fast, and there is danger of the slag's entering the tuyeres. After starting the furnace it is sometimes found that a small crust has formed over the lead while the blast was shut off. This is perforated with a long iron rod or with a steel bar, if necessary, and will soon disappear, if the furnace was in good working order below before the cutting-out began.

A second method of barring down a furnace, said to work well, is to let down the charge to below the upper rim of the jackets, and give a bed of coke as previously described. In the meantime, the bricks between the jackets and the collar, on which the shaft rests, are removed for a distance of 1 or 1½ feet. The accretions are cut down, beginning from above, and raked out through this opening. When the shaft is clean, the opening is bricked up, light charges are given, the furnace is filled up with ore-charge, and the blast let on. The accretions go to the roasting furnaces.

A third method of barring down, applied in very bad cases, is the following; it is dangerous, but may be necessary if the life of the furnace is in jeopardy. The charge is let down to the top of the jackets, and cutting out begun from the top and continued until the crust collecting in the furnace has risen so high that the work cannot proceed any further. The two breast-jackets are now taken down and the contents of the furnace raked out (§ 67), in order that the cutting may continue until the walls are clean. This may take as much as eighteen hours. Two parties of three men each, working on opposite sides of the furnace, do the cutting, one man holding the bar and two sledging. As the work is hard and has to be done as quickly as possible, the regular hands have the constant assistance of furnace-men and helpers from other furnaces, who work half an hour at a time. While the accretions are being removed, a crust 6 inches thick or less will form on top of the lead. This is broken up, the breast-jackets are put back, and the lower front is closed. Charcoal is fed from above to the top of the jackets and the furnace blown in anew, adding twenty-five bars of lead to the first slag charges to heat up the lead in the crucible.

Guyard[1] suggests that by using caustic lime in the charge instead of limestone, these accretions might be made less troublesome, as lime has a decomposing action on the sulphides, of which they consist in great part. The writer is not aware that caustic lime is

---

[1] Emmons, "Geology and Mining Industry of Leadville," pp. 728 and 747.

used in any lead furnace. It would seem as if the lime, being exposed during its descent in the furnace to gases rich in carbonic acid, would be converted quickly into carbonate of lime in the upper part of the furnace and lose any decomposing power. Again, while accretions next to the furnace walls consist mostly of sulphides, they are usually covered by a thin crust as hard as flint; this is followed by a softer substance that is often a powder, and this again covered by a crust so tough that it is sometimes extremely difficult for a steel bar driven hard with a sledge to produce any effect on it. It does not seem probable, therefore, that caustic lime would have any important effect on these composite crusts, which must differ from sulphides as much in their chemical properties as they do in their physical.

It often happens, even when a furnace is otherwise doing well, that the tap-hole becomes hard and the tuyeres dark. The fuel is then distributed more over the cold places, additional fuel being given for a short time if necessary. One or two bars of lead are sometimes placed on top of the charge over the tap or the tuyere, but this cannot be commended, although it is often effectual. The change in feeding the fuel will generally soften the tap. To soften a crust in front of a tuyere, the bag or the gate should be closed, or nearly closed, as the blast playing on chilled slag can only have a bad effect. If it is turned off, the heat generated at the neighboring tuyeres will melt off the crust; then a little blast is allowed to pass through the tuyere, and gradually increased until the full blast can be turned on again.

This proceeding may also be necessary if some fine ore trickles through the coarser parts of the charge, appearing in a crude state at the mouth of the tuyere, and, as is sometimes the case, runs into the tuyere pipe. If fine ore appears at the same time at several tuyeres and too many would have to be tied up, the easiest remedy is to feed down the furnace and thus loosen the charge. If this does no good, a coarse ore-charge will have to be substituted for a short time, or if coarse ore is not to be had, a slag-charge.

The forming of hearth accretions (§ 83) is indicated on the feed-floor by the top becoming hot and the charges not sinking regularly, but in jerks, a foot at a time, and blazing up with every settling. On the furnace-floor the lead in the well becomes dark, and does not play with the blast. By inserting a rod through the tap-hole, the position and often the thickness of the crust can be ascertained. Sometimes it only forms a ridge across the furnace, and communication is open in front and at the back; more commonly the crust begins at the back and grows toward the front, gradually closing all communication. The lead produced in the furnace cannot get into the crucible, and runs out with the slag. A case like this generally needs a change in the composition of the charge; the slag analysis will show the defect and give the remedy. As it takes from six to eight hours before a new slag makes its appearance, holes are driven with a steel bar through the crust that the lead may find its way into the crucible. If the crust has not yet hardened, it is repeatedly perforated by introducing an iron rod and working this up and down and to both sides. The crust may be only a temporary affair, and can then be worked into the slag, which removes it. Cases do occur where the crust will not yield quickly enough to being fused out and has to be removed by force. This has been ironically spoken of as "muscular smelting," but it is sometimes unavoidable. Before beginning, all the slag in the furnace is tapped, the basin of the lead-well covered with glowing charcoal, the blast stopped, the tuyere-pipes drawn out, the tuyeres closed, and the flow of water into the jackets reduced. The breast of the furnace is now removed wholly or in part, any chilled slag in front is chipped off, and some of the loose material raked out into a wheelbarrow in front of the furnace and taken away. A heavy steel bar is passed through two opposite tuyeres nearest the breast to hold up the charge. Balls of loam or clay are tamped behind the lower part of the front jacket to prevent the charge from rolling down. The balls, placed on the end of a board, are slid below the front of the jacket and rammed upwards against the charge with a rod bent to a hook. When this is done, any lead that has accumulated in front is ladled out, and the crust thus laid bare. A hole is driven through it with a heavy steel bar. It is necessary to have a number of these ready, as the points soon become dull or bent. The hole is enlarged by driving the bar again close to it and breaking the crust towards it. When large enough to receive a 2-inch steel bar, this is warmed, inserted into the hole, the crust pried up, and the broken pieces raked out. If the crust will not yield, the hole is enlarged and the furnace started up again. The lead previously bailed out is returned, and, if necessary, fresh hot lead added, the clay balls are removed, the hollow space in front is filled with charcoal, the steel bars are withdrawn from the tuyeres, the breast is put in, the tuyeres are opened, the pipes inserted, and the blast is let on, but very gently at first. When the first two or three pots of slag have been tapped the rod is repeatedly inserted to keep the hole open until the new slag comes down. The lead will soon show the effect of having communication between it and the slag partly restored; it begins to play with the blast and becomes hotter, thus assisting the work of the new charge.

A crust is sometimes caused by a leaking jacket. This is first indicated by the appearance of moisture at the tuyere or the bottom of the jacket. The leak, if small, can be temporarily stopped by mixing corn-meal with hot water, pressing it with the hand into small balls, and throwing these into the water-feeder of the jacket. Soon, however, the jacket will have to be removed. For this purpose the crust. on the inside is first allowed to grow thick by cooling, which is done by turning in a full stream of water, and opening the discharge at the bottom. Two courses of brick are chiselled out above the jacket. When cool, the furnace is stopped, the cooling-water on the side of the injured jacket shut off, the water-trough removed, and the injured jacket unhinged, taken out, and a new one put in its place. The whole procedure need not take more than twenty minutes. Should the crust on the inside of the jacket prove too thin and break out, the opening is closed by the introduction of clay balls. The space where the new jacket is to be inserted must be absolutely clean, as any little pieces of brick or other hard matter will obstruct the placing of the new jacket and cause much delay. Of course the foregoing has no reference to a furnace with wrought-iron jackets extending its entire length.

The clogging up of the lead-well has still to be considered. In smelting sulphide ores rich in lead, sulphide of lead held in solution in the crucible often separates out when the lead ascends the channel toward the basin. A bent iron rod may be inserted to clean it out. This presence of sulphides is generally caused by an incomplete decomposition of galena in the furnace on account of lack of heat in the smelting zone. If the charge is rich in copper, this causes coppery lead gradually to close up the channel.

§ 67. **Blowing-Out.**—If the furnace needs to be repaired, or if an accident happens that cannot be remedied in a short time, say eighteen or even twenty-four hours, the furnace has to be blown out. This is done by stopping the ore-charges and substituting slag charges, until most of the ore has been smelted out. The charge is allowed to sink and the blast is gradually lowered. Soon volumes of dark smoke mixed with white lead-fumes will appear. When the charge has receded somewhat, and before a flame makes its appearance, the damper in the flue leading to the dust-chamber is closed, and the fumes are conducted into the air by opening the damper on the top of the furnace or by lowering the sheet-iron stack, or by whatever contrivance may be in use for the purpose. If this were not done, an explosion might occur in the dust-chamber. To check the flame and to reduce the temperature, water is often sprinkled over the charge, although its effect on the lining of the furnace cannot but be deleterious. When the charge has sunk as far as the top of the jackets, the blast is stopped and the tuyere-pipes are removed. All the liquid slag is tapped, the tapping-jacket removed, and the breast of the furnace is knocked in.

Sometimes the furnace is blown down, allowing the charge to sink only till heavy fumes, but no flames, appear and the entire contents then drawn. In this case there is no need of closing the damper to the dust-chamber; in fact, many furnaces have no damper at all.

The bulk of the slag remaining in the furnace is withdrawn with a hoe into iron wheelbarrows, which are emptied on the dump and then chilled with water. As it is important that there should be little delay in drawing the charge, four or five wheelbarrows are placed one behind the other near the front of the furnace. As soon as the first is filled with red-hot charge it is wheeled away and replaced by the second, the emptied wheelbarrow being put at the end of the line. When all the charge that can be easily reached with the hoe has been drawn out, the front jackets are taken down and the rest removed. Meanwhile a thin crust will have formed on top of the lead in the crucible. This is easily broken, and the lead is then ladled out into the moulds that have been moved from the lead-well to the front of the furnace.

§ 68. **Furnace Books.**—A daily record of the work done is kept for every furnace. The following is one of the many suitable skeletons (see Table A, page 243).

The time when any change in the charge is made, or when anything out of the regular way occurs is noted under the first column of "Remarks." The second column of "Remarks" refers to the shipment of the bullion.

§ 69. **Furnace-Assay Book.**—A separate book is kept to record the assays and analyses made of slag and matte (see Table B, page 243). Under the head of "Remarks" are brought the names of any of the other furnace-products that may be assayed now and then.

TABLE A.

1891, November.	Furnace, No.	Ore-bed, No.	Charges, No.	Composition of Charge, pounds.						Blower.		Remarks	Total pounds of Charge.						Output.						Remarks
				Ore.	Iron Ore.	Lime-stone	Slag.	Coke.	Char-coal.	Rev. No. per min.	Pres-sure, ozs.		Ore.	Iron Ore.	Lime-stone	Slag.	Coke.	Char-coal.	Pots of Slag, No.	Bars, No.	Lot, No.	Lbs.	Ag, oz. p. t.	Au, oz. p. t.	

TABLE B.

Date. 1891, November.	Furnace, No.	Slag. Per cent.							Ounces per ton		Matte. Per cent.		Ounces per ton.	Remarks.
		SiO₂.	FeO.	MnO.	CaO.	BaO.	MgO.	Al₂O₃.	Pb.	Ag.	Pb.	Cu.	Ag.	

FURNACE PRODUCTS.

§ 70. **Furnace Products.**—The products of a blast-furnace are base bullion, speise, matte, slag, wall accretions, hearth accretions, furnace cleanings, and flue-dust.

§ 81. **Base Bullion,** the commercial name for argentiferous lead, as distinguished from silver or gold bullion, contains in addition to the silver and gold small quantities of other metals, as seen by the following analyses:

	Clausthal.	Sophien-hütte.	Freiberg.	Mecher-nich.	Pribram.	Leadville.	Leadville.
Pb	98.80944	99.641096	95.088	99.5913	97.3597	99.0798210	98.492379
Ag	0.1412	0.000250	0.470	0.0215	0.4230	0.6112445	0.793417
Bi	0.0048	0.352053	0.019	....	0.0070	trace	0.011791
Cu	0.1862	0.000279	0.225	0.1332	0.1100	0.0479100	0.071450
Cd	trace	....	none	....	....	trace	trace
As	0.0664	....	1.826	....	0.2900	?	?
Sb......	0.7203	0.002872	0.958	0.2180	1.5340	0.2138940	0.347881
Sn......	....	....	1.354	....	0.2500	trace	0.000897
Au	....	....	....	....	....	0.000888	0.000891
Fe	0.0664	0.002877	0.007	0.0800	0.0036	0.0063000	0.012600
Zn	0.0028	0.000573	0.002	0.0060	0.0012	0.0016052	0.000232
Ni	0.0023	....	....	trace	0.0015	....	....
Co	0.00016	....	....	....	....	....	....
S........	....	....	0.051	....	0.0300	....	0.048984
Reference :	1.	2.	3.	4.	5.	6.	6.

1, *Zeitschrift für Berg-, Hütten- u. Salinen-Wesen in Preussen*, xviii., p. 203; 2, *Zeitschrift für Berg-, Hütten- u. Salinen-Wesen in Preussen*, xix., p. 169; 3, *Wagner's Jahresberichte*, 1887, p. 401; 4, *Berg- u. Hüttenm. Zeitung*, 1886, p. 434; 5, *Oesterreichisches Jahrbuch*, xxxix.. p. 52; 6, Emmons, "Geology and Mining Industry of Leadville," p. 694.

The impurities of a bar of lead, with the exception of silver and gold, always collect nearer the top than the bottom. This is illustrated by the analyses of Streng, made from non-argentiferous lead, and by Schertel, made from base bullion.

	Streng.[*]				Schertel.[†]	
	Top.	Bottom.	Top.	Bottom.	Top.	Bottom.
Ag............... ..	....	....	....	....	0.423	0.403
Bi...............	..	....	....	....	0.132	0.042
Cu	3.621	1.242	0.508	0.140	1.324	0.034
As	....	....	....	....	2.164	1.980
Sb	0.274	0.158	0.090	0.057	0.700	0.749
Sn.................	....	....	....	....	0.941	?
Fe	0.008	0.008	0.012	0 008	0.103	0.009
Zn.................	0.003	trace	0.002	trace	0.016	0.003
Ni	0.148	0.082	0.012	trace	0.029	....
S..................	....	....	....	....	0.500	....
Sp. gr.............	....	....	....	....	10.321	10.824

[*] *Berg- u. Hüttenm. Zeitung*, 1859, p. 14. [†] *Wagner's Jahresberichte*, 1887, p. 401.

Schertel's analyses would seem to show that silver has a similar tendency. He, however, took his sample from lead that had been kept for twenty-four hours above its melting-point in an iron cylinder, 3 feet $3\frac{3}{8}$ inches high, and thus created special conditions. In bars of base bullion which cool quickly just the reverse is the case. The following tests made by the writer in 1881 may substantiate this. (Figures 150 and 151.)

Bars of base bullion were sawed into three pieces and samples taken as shown by the numbers and assayed separately.

Fig. 150.
(1.) 149.7	(2.) 150.5	(3.) 146.0	(4.) 148.7	(5.) 145.0
(6.) 137.0	(7.) 152.0	(8.) 149.0	(9.) 149.0	(10.) 151.0
(11.) 148.0	(12.) 150.0	(13.) 150.5	(14.) 150.0	(15.) 152.0

Fig. 151.
(1.) 127.0	(2) 134.5	(3.) 128.5	(4.) 129.5	(5.) 125.0
(6.) 133.5	(7.) 124.0	(8.) 132.0	(9.) 126.5	(10.) 134.5
(11.) 129.0	(12.) 132.0	(13.) 133.0	(14.) 132.0	(15.) 134.0

The results (ounces per ton) clearly prove that the whole lower

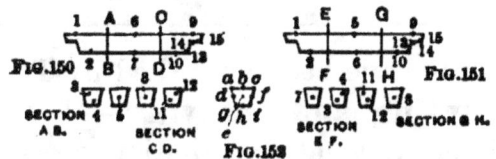

BARS OF BASE BULLION.

part of the bar is richer than the upper and that the centre of the upper surface is the poorest part. Similar results were obtained by Piquet.[1]

This concentration of the silver in the lower part of the bar can be explained [2] by the separation of the argentiferous lead, while cooling into crystals low in silver and liquid lead high in silver. When a bar has been moulded, the surface cools first, and crystals begin to form there; then the sides slowly solidify, and a mass of liquid lead and crystals remains in the centres. As the cooling proceeds from the top downward the crystals will continue to form on the cooler upper side and gradually force the richer liquid lead towards the bottom. The reason that the surface of a bar is so much poorer than the other parts is not only because of the crystallization, but also on account of the impurities that rise to the surface when a bar is cooling, forming the dross. They run much lower in silver than the pure lead. If bullion is so rich in

---

[1] Roswag, "Désargentation du plomb," Paris, 1884, p. 127.

[2] Roswag, *Op. cit.*, p. 126.

dross that the lead cannot eliquate freely, the natural distribution
of the silver will be much disturbed. This will account for the
fact that assays from the top of a bar run sometimes higher than
those of the bottom. For instance, coppery bullion from the
Ramshorn silver mine, Idaho, containing so much dross that, if
dropped on the floor, it would break, gave to Rhodes[1] the following results in ounces per ton (Figure 152):

$$a. \ 432 \qquad b. \ 432 \qquad c. \ 439$$
$$d. \ 450 \qquad e. \ 434 \qquad f. \ 441$$
$$g. \ 385 \qquad h. \ 397 \qquad i. \ 386$$

According to Kempf, Nenninger, & Co.,[2] gold also seems to be
concentrated with the silver near the bottom of the bar.

Sample from	Ag.	Au.	Ag.	Au.
Top............................	143.3	4.59	129.7	3.68
Bottom	148.4	4.76	134.3	3.82

Similar results have been published by Torrey and Eaton.[3]

§ 72. **Sampling and Assaying.**—The irregular manner of
sampling and the unequal distribution of the silver have caused
much trouble in estimating the value of base bullion. Samples
used to be taken, for instance, with a gouge from top and bottom
near the opposite ends of a bar. This method has two sources of
error, the form of the sample and the fact that they are taken only
at the ends. The sample obtained is a conical chip having its base
at the surface and its apex $\frac{1}{2}$ inch or less below. The incorrect
form has been remedied by using a punch which gives a cylindrical
chip about $\frac{3}{8}$ inch in diameter. If driven half way through the bar,
say 2 inches, it will represent a correct sample of that part of the
bar. Sampling, however, from the ends of the bar gives too high
a result, as can be seen by turning to Figures 150 and 151. The
poor parts of the bar, represented by No. 6 in Figure 150 and
No. 5 in Figure 151, are excluded. The only correct method which
is simple and quick is to take punch samples diagonally across a
row of, say, five bars by driving the punch every time deeper than
half way through, turning over the bars and repeating this on the
other side in the opposite diagonal. Three or four blows with a

---

[1] Private communication, July, 1891.
[2] *Engineering and Mining Journal*, July 1, 1882.
[3] *Ibid.*, December 25, 1886.

four- or five-pound hammer are sufficient to drive a stout punch having an $\frac{1}{2}$-inch opening about 2 inches deep; a few light taps on the sides loosen the punch and break the chip.

Austin[1] recommends removing from a bar 4 inches thick a chip a little longer than 2 inches, slipping it into a hole bored 2 inches deep into a block, and trimming off the projecting end with shears. This is too troublesome and is not followed. While in the method previously described a few chips may be a little longer than 2 inches, others are a little shorter, and the average chip will have the desired length.

Drilling one or more holes through every bar and taking the borings as sample has been suggested. The writer is not aware that this method is carried out anywhere. The large bulk of the sample and the liability to loss have probably prevented its use.

At some works a drill has lately been introduced which removes a cylindrical chip of the desired length from the bar. The operation requires no hard work and is quick; the bit has no complicated construction and outlasts many punches; the whole apparatus is simple and light, so as to be easily moved from one part of the works to the other.

The chips obtained by sampling a car-load (weighing from ten to fourteen tons) or a lot are collected in a wooden box, melted down, and cast into a bar. The writer[2] has always obtained satisfactory results by the following procedure. The chips are melted down quickly in a graphite crucible, which has first been heated for some time, so as to be dark-red near the bottom. When melted and heated till the dross is about redissolved, but before any cupelling has begun, the lead is well stirred with an iron rod for several minutes and then poured *with* the dross into a mould of, say, 3 by 7 inches. This sample-bar, representing from ten to fourteen tons of base bullion, weighs about eight pounds, rather more than less. Another form of mould is 4 by 12 inches, which gives a thinner bar. It has a ridge on the longer centre-line. The depression this makes in the bar marks the place where it is cut, if it is necessary to divide it in order that half may be sent to the buyer.

A different method of melting the chips is followed by Kempf, Nenninger & Co.[3] They call attention to the fact that the chips from the top of the bar are liable to be smaller than those from

[1] *Engineering and Mining Journal.* September 9, 1882.

[2] "Mineral Resources of the United States," 1883–84, p. 464.

[3] *Engineering and Mining Journal,* July 1, 1882.

the bottom, on account of the hardness of the dross and the unevenness of the surface. They melt down top and bottom chips separately, take from the resulting two bars an equal amount, and melt into a second bar, their final sample-bar.

At some smelting-works they hold back the poor dross (in pouring) with a rod, casting only the liquid lead, and use this as the sample. Refining-works, of course, object to this. If the dross is held back, as "Refiner"[1] suggests, and practises at his works, it must be assayed separately from the liquid lead. The final assay is calculated with reference to the weights of dross and lead and the result obtained is good :

$$\frac{A.a + B.b}{A + B.} = \text{Average assay.}$$

A. = weight of dross ; a = assay of dross.
B. = " " lead ; b = " " lead.

While at the works of "Refiner" the writer practised this method. Where a large quantity of base bullion is desilverized, it takes too much time. With bullion rich in dross "Refiner's" method is to be greatly recommended.

From the 3 by 7-inch sample-bar four samples are taken from the middle of the sides, cutting the entire bar through. From the 4 by 12-inch bar the samples are taken from half of the bar only: two on the outer side and two along the central depression. The writer prefers to cut out samples weighing a little over one assay-ton and to use four samples of one assay-ton each for the subsequent cupellation, as his results have been more accurate and uniform with silver than when half an assay-ton or less lead has been taken. That the results with gold would be more satisfactory from four one-assay ton assays than with a smaller amount, especially if there is little gold, as is generally the case, is clear from the advantage of the added weight, when the silver buttons have been dissolved in nitric acid.

The assay sample of bullion is usually cupelled without being scorified at as low a temperature as possible. Every cupel must show feather litharge, if the assay is to be accepted as definite. To regulate the temperature automatically Torrey[2] has constructed a heat-regulator, but generally the assayer uses his judgment as to degree of heat and draught of his muffle. If the temperature is

---

[1] *Engineering and Mining Journal,* **May 20, 1882.**
[2] *Engineering and Mining Journal,* **August 28, 1886.**

too low toward the end, lead is liable to remain with the silver, often also copper and bismuth. With impure bullion it is advisable to test the nitric acid solution, obtained in dissolving the silver buttons when determining the gold, for lead with sulphuric acid and for copper and bismuth with ammonia. Impure base bullion has to be scorified before it is cupelled, especially if it contains much dross (arsenic, antimony, and copper). Part of a table of the losses endured in cupelling argentiferous lead lately published by Rössler[1] is subjoined.

Material.	Ounces Ag per ton.	Charge.		Multiple of Pb.	Loss of Ag in percentage of amount present.
		Mgrs. Ag used.	Grs. Pb.		
Argentiferous copper......	43.75	10 mgrs. Ag + 10 grs. Cu...	200	12,000	8.3
Argentiferous copper......	583.32	200 mgrs. Ag + 10 grs. Cu...	160	800	4.5
Argentiferous lead........	43.75	150	100	600	2.5
Argentiferous lead.......	87.49	"	50	300	2.2
Argentiferous lead........	218.74	"	20	120	2.0
Argentiferous lead........	437.49	"	10	60	1.6
Argentiferous lead........	874.98	"	5	30	0.9
Argentiferous lead	4874.90	"	1	6	0.4

§ 73. **Speise.**—The speise obtained in lead-smelting is principally an arsenical speise. The predominant element is iron; this is somewhat replaced by nickel, cobalt, copper, and to a small extent by lead, bismuth, gold, and silver.

Guyard calls attention to the absence of cobalt in Leadville speise, which he found concentrated in the dross skimmed from the lead-well. He also found as much as 10 per cent. of grains of metallic iron free from arsenic in Leadville speise, which is uncommon. It has been already stated that speise always contains shots of lead, and that when coarsely-crystalline less is found than when it is fine-grained. As regards the presence of precious metals, the fact is to be noted that speise retains considerable amounts of

[1] *Berg- u. Hüttenm. Zeitung*, 1888, p. 480; *Engineering and Mining Journal,* July 26, 1890.

gold, while very little gold is found in matte; speise-assays show from a trace to 0.5 ounces gold per ton.

To treat speise so as to extract the silver, gold, and copper economically has always been a difficult problem. In a large number of works it is not treated at all, as only small quantities are produced. The cheapest and simplest way is to roast it in a heap of about fifty tons, which burns from two to four weeks. The imperfectly-roasted speise is sorted out, crushed and roasted in a calcining furnace. The whole is then smelted in the blast-furnace with pyrite or matte. The result will be base bullion and a matte rich in copper and silver, and perhaps a small amount of speise,

	Leadville.	Leadville.	Pueblo.	Pueblo.	Eureka	Pribram.	Pribram.
Ag	0.0085	0.0301	up to 0.016	up to 0.014	} 0.029	0.037	0.020
Au	trace	0.0009	....	....		....	....
Cu	0.3628	0.2566	2.09	5.06	1.06	1.956	0.409
Pb	1.4935	2.5030	1.87	0.69	2.18	1.752	8.245
Mo....	0.2110	0.2155	....	....	2.31	...	....
Fe	60.5780	70.4780	58.32	59.42	57.02	61.330	56.700
Zn	trace	trace	....	....	0.07	....	....
Ni	0.0876	0.0981	trace	trace	....	2.056	} 0.783
Co	....	....	trace	trace	....	0.194	
S	5.8191	4.4695	4.105	2.80	3.84	9.600	10.000
As	31.4725	21.8003	30.005	31.17	32.95	18.750	26.757
Sb	trace	0.1450	trace	trace	0.13	2.450	1.608
SiO₂	....	....	0.15	trace	0.23	....	....
CaO	....	....	....	....	0.34	Ca 0.500	Ca 0.535
Reference :	1.	1.	2.	2.	3.	4.	4.

1, Emmons, "Geology and Mining Industry of Leadville," p. 720; 2, Dewey, Bulletin No. 42, "United States National Museum," p. 52; 3, Curtis, "Silver-Lead Deposits of Eureka, Nev.," Monograph vii., United States Geological Survey, 1884, p. 160; 4, Balling, *Berg- u. Hüttenm. Ztg.*, 1867, p. 419.

in which any nickel and cobalt will be concentrated. This second speise goes to a new heap of first speise, as nickel and cobalt occur in such small quantities as not to call for any further attention.

Davies[1] invented a process for desilverizing speise which is said to give satisfaction at the Eureka Consolidated Works, Eureka, Nev. It consists in tapping 800 pounds speise into a small cylindrical iron converter lined with fire-brick (Figure 153), adding from 20 to 25 per cent. liquid lead and introducing from the bottom a blast of 17 ounces pressure through a ¼-inch pipe for 3 or 4 minutes. This stirs up the lead and speise and

[1] *Engineering and Mining Journal*, June 30, 1888.

burns off some arsenic. Most of the silver and gold is taken up by the lead; the liberated iron corrodes somewhat the lining, but eats through only very slowly. The converter is turned down and the contents are discharged into a cast-iron receiver having the form of a slag-pot, the bottom of which has a ½-inch hole for tapping the lead. The desilverized speise solidifies quickly, and when hard the still liquid lead is tapped. Best results are obtained when the lead assays 40 ounces to the ton. The desilverized speise is then a

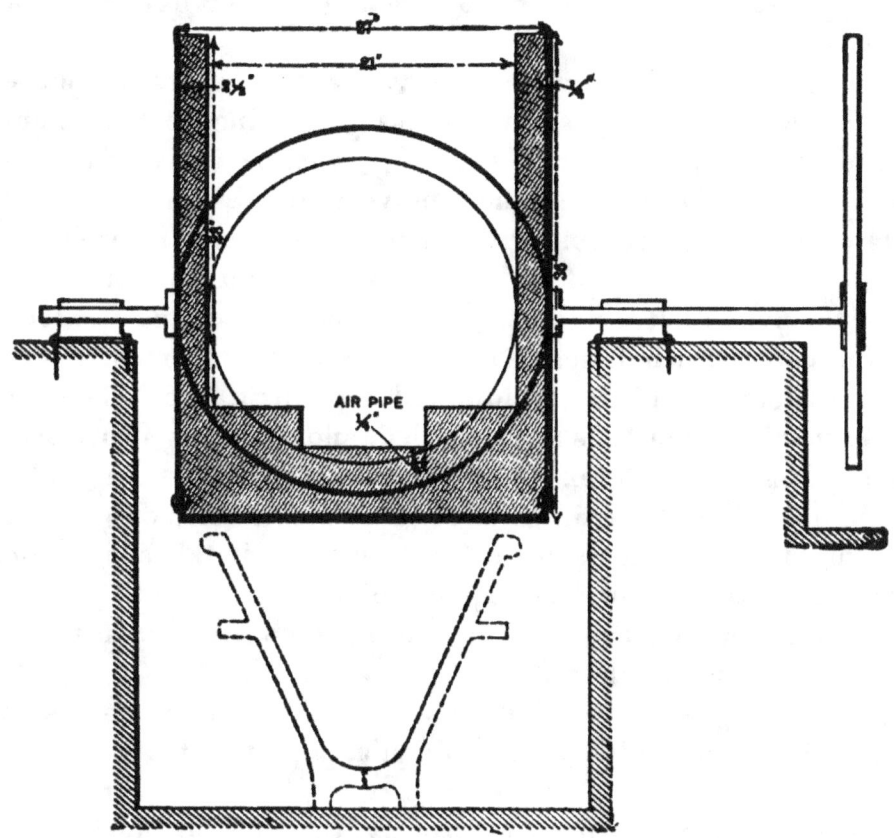

FIG. 153.—CONVERTER FOR DESILVERIZING SPEISE.

waste product. The claim is made that from a speise of the composition shown in the table, 85.5 per cent. of the silver and 89.28 per cent. of the gold are extracted, with a loss of lead varying from 5 to 8 per cent.

§ 74. **Analytical Determinations.**—The metals to be determined by assay in raw or roasted speise are silver and gold (§ 34), lead (§ 79), copper, arsenic, and iron (§ 81).

1. *Copper.*—In determining the copper by titrating with po-

tassium cyanide, the presence of arsenic [1] has an injurious influence. To counteract this, Blake [2] recommends two methods, both of which give satisfactory results. One of them only is given here. The weighed sample is heated with sulphuric acid (concentrated acid diluted with an equal volume of water); then nitric acid is added to complete the decomposition. The solution is evaporated and heated half an hour after the sulphuric acid fumes have appeared; it is then diluted, boiled, and the copper titrated after supersaturating with ammonia, either directly or after having been first precipitated with zinc and redissolved. (For details of the cyanide assay see § 69.)

2. *Arsenic.*—The quickest assay-method for arsenic is the one by Pearce.[3] Half a gramme or less (according to the amount of arsenic present, as 1 part $(Ag_3AsO_4)_2$ weighs nearly six times as much as 1 part As) of the finely-pulverized substance is melted in a large porcelain crucible with from 6 to 10 times its weight of a mixture of even parts of sodium carbonate and potassium nitrate kept for five minutes in a state of fusion, cooled, and treated with warm water, which dissolves the alkali arseniate formed. The solution is filtered, the filtrate, acidulated with nitric acid, boiled to drive off carbonic and nitrous acids, cooled, and carefully neutralized by first adding a slight excess of strong ammonia, then acidifying again by adding a few drops of concentrated nitric acid, and finally adding dilute ammonia drop by drop until it takes litmus-paper half a minute to show a change of color.

In determining the arsenic of ores, some alumina may be precipitated, which is filtered off.

Canby [4] adds an excess of an emulsion of zinc oxide instead of neutralizing carefully with ammonia, and titrates the silver as shown below.

The solution is stirred vigorously, the arsenic precipitated by adding an excess of a neutral solution of silver nitrate, as brick-red silver arseniate $(Ag_3AsO_4)_2$, the liquid is filtered, and the precipitate washed with cold water. To determine the silver, two methods may be employed : (*a*) The silver arseniate is scorified and the resulting lead button cupelled ; (*b*) the silver arseniate is dissolved in nitric acid and titrated with ammonium sulpho-cyanide, using

[1] Torrey and Eaton, *Engineering and Mining Journal*, June 6, 27, 1885.
[2] *Trans. A. I. M. E.*, ix., p. 317.
[3] *Engineering and Mining Journal*, May 5, 1883.
[4] *Trans. A. I. M. E.*, xvii., p. 77.

5 grs. NH₄SCN to the liter and 5 cc. of a saturated solution of ammonium ferric-sulphate as indicator.

1 part Ag = 0. 23148 parts As.

The results are accurate. Antimony is without effect, while molybdic and phosphoric acids interfere with the method.

§ 75. **Matte.**—The matte produced in lead smelting is principally a compound of iron and sulphur, in which the iron has been in part replaced by lead and copper and in a less degree by zinc, silver, nickel, cobalt, manganese, arsenic, antimony, calcium, barium, and magnesium.

	Clausthal, old.[1]	Clausthal, new.[2]	Pribram.[3]	Pueblo.[4]
Pb	41.50	11.399	11.16	10.72
Cu	0.36	5.976	1.59	0.61
As	....	...	0.55	0.56
Sb	0.66	0.212	0.93	none
Sn	....	....	trace	....
Ag	0.12	0.0311	0.105	0.084
Fe	34.05	51.963	41.31	52.27
Zn	....	1.963	11.55	4.27
Mn	....	0.300	....	...
MnO	....	....	1.40	....
Ni	....	0.112	trace	....
Co	....	....	trace	....
SiO₂	....	0.261	3.06	....
Ca	....	0.706	...	....
CaO	....	....	0.05	0.41
Mg	....	0.235	trace	0.47
Al₂O₃	....	....	trace	....
S	23.82	25.807	22.23	24.015
O	....	....	4.79	....

[1] Balling, "Metallhüttenkunde," p. 86.
[2] Private notes.
[3] *Oesterreichisches Jahrbuch*, xxxix., p. 24
[4] Dewey, Bulletin No. 42, "United States National Museum," p. 52.

The very high percentage of lead in the matte produced formerly (1867) in the Upper Hartz was connected with an enlargement of the furnace at the zone of fusion, made to lower the temperature and thus reduce the volatilization of lead and the corrosion of furnace-walls. This resulted, however, in an incomplete decomposition of galena. In modern furnaces that are contracted at the smelting-zone a matte contains from 8 to 12 per cent. of lead, if not otherwise enriched, *e.g.*, by the presence of zinc in the charge. It often happens that, if the constituents of a matte are figured as sulphides, the analysis does not show enough sulphur to combine with the metals as FeS, PbS, Cu$_2$S, NiS, etc.   The explanation given by Rammelsberg[1] for lead sulphide, and by Münster[2] and Schweder[3] for iron and nickel sulphide, is that, as subsulphides do not exist, lead, iron, and nickel are held in solution by their sulphides while liquid, and separate out during solidification.   Mackintosh[4] claims the existence of a subsulphide of nickel and iron. The value of a matte depends on the amount of silver, gold, lead, and copper, and (in exceptional cases) of nickel and cobalt it contains.   Matte does not retain much gold, but considerable silver (§ 7).   This is illustrated in a very interesting way by the fact observed by Carpenter,[5] that if a gold and silver bearing iron matte produced in pyritic smelting be added raw to the charge of a lead blast-furnace, all the gold, and from 60 to 70 per cent. of the silver, will be taken out.   The first matte produced in a lead furnace rarely contains more than 5 per cent. copper and from 1 to 3 per cent. nickel-cobalt.

The object aimed at in working the matte is to extract the silver and gold by means of lead, and to concentrate the copper into an enriched intermediary product, which is sold to copper works.   In exceptional cases it is sometimes worked at large refineries into metallic copper or blue vitriol.   The two operations by which matte is concentrated consist of roasting and smelting, with an acid flux to slag the iron and a lead ore to take up the silver and gold.

§ 76. **Roasting of Matte.**—Roasting is carried on in heaps, stalls, kilns, and reverberatory furnaces. The choice of method will depend greatly on the richness of the matte in silver.   In roasted matte, according to Plattner,[6] the exterior part of each piece or pellet contains more silver and lead than the interior.   Any dust that is made in handling the roasted matte, coming from the surface, will cause considerable loss in silver and lead ; again, if exposed to the rain, silver sulphate formed in roasting is liable to be lost by leaching ; finally, with rich matte the aim must be to roast as quickly as possible, so as not to lock up too much capital for months at a time.

[1] " Lehrbuch der chemischen Metallurgie," Berlin, 1865, pp. 48 and 232.
[2] *Berg- u. Hüttenm. Zeitung*, 1877, p. 195.
[3] *Ibid.*, 1879, p. 18 ; *Iron*, xiii., p. 292.
[4] *Trans. A. I. M. E.*, xvi., p. 117.
[5] Private Communication. May, 1892.
[6] " Die metallurgischen Röstprocesse," Freiberg, 1856, p. 205.

1. *Roasting in Heaps* has the advantage of cheapness and simplicity of plant; the matte need not be crushed fine, and most of the roasted matte is obtained in lump form, which is desirable for the blast-furnace. This is, however, more than made up for by the many disadvantages, as loss of metal by dusting and leaching, the consumption of fuel, slowness of roast, imperfect roasting, with consequent necessity of rehandling and reroasting, and obnoxious fumes. Therefore heap-roasting has not found much favor for matte. Where small quantities of matte are produced and where a calciner does not exist, heap-roasting is in place.

A heap for roasting matte should not be very high, as the fire then becomes too hot and melts the central part of the heap; on the other hand, if very low, the roast will be imperfect and the loss in silver and lead large. The sides of the heap are always covered with smalls, to prevent the fumes from passing off there. The height of a matte-heap varies from 5½ to 7 feet; the length and width have not much influence on the result.

The roast-heap is usually built on the level of the feed-floor, as the fumes will not enter the smelter-building, and it is important not to carry the roasted matte any further than necessary. The loss incurred in transporting the raw matte to the roast-heap is not worth taking into account. The ground for a heap should be even, hard, and slightly elevated. Soft ground is levelled by a plough or a scraper; hard ground is made even by filling; the elevation of 6 inches is obtained by making a bed of coarse slag, covering it with fine slag, and rolling it, if possible. The first matte, being rich in iron, oxidizes and crumbles when exposed to the air for a short time, so that the breaking-up into pieces of fist size does not cause much labor. In building the heap, a bed of light wood from 6 to 15 inches deep is made, leaving channels 10 inches wide, open every 6 feet, which are filled with kindling. Hard wood is not so good for the bed, as it makes too hot a fire, which is liable to fuse the lower part of the heap. Sometimes a chimney is placed in the centre, consisting of four boards nailed together and long enough to reach 2 feet above the heap. It helps to start the heap uniformly. In building the heap, first coarse matte is piled around the chimney, and then distributed over the bed of wood. The coarse matte is assorted from the finer matte with a sluice-fork. When the heap has been built in the form of a pyramid the edges of which reach to within a ½ foot of the border of the wood, the sides are covered with fine matte and sometimes also the top. Fine matte is, however, kept on hand, to be used when the heap has been started, to make the top layer thicker, and to check the draught in parts of the heap that are becoming too hot. In Utah[1] the pyramidal heaps are made 24 by 18 feet at the base and 6 feet high; they contain about 80 tons of matte, and burn from 30 to 40 days.

In the Hartz Mountains[2] matte-heaps are usually made 15 by 23 feet on a bed of wood 6 inches deep; they contain from 75 to 100 tons and burn 30 days. Quite a saving of cord-wood has lately been effected by placing pieces of wood, split from sound dry logs about 9 inches in diameter, end to end in parallel rows, about 12 inches apart, on the ground, and laying small sticks across them. On this bed are distributed in rows above the cord-wood large pieces of matte to the height of about 9 inches. Between these are piled faggots and soft coal. Looking at the heap from the top there will be seen alternate rows of coarse matte and faggots mixed with coal. On this bed the heap is built in the usual way.

The well-roasted part of a matte-heap amounts to about 50 per cent., if the roasting has been carefully conducted and the sulphur has been reduced to about 6 per cent. Analyses[3] of raw and roasted matte from Clausthal are subjoined:

Matte.	Pb.	Cu.	Sb.	Ag.	Fe.	Zn.	Mn.	Co.	Ni.	SiO_2	CaO	MgO	S.	O.	SO_3.
Raw.....	10.665	4.620	0.267	0 0299	53 112	2.110	0.385	0.215	0.097	0.510	0.383	0.054	26.877	.....	.....
Roasted.	10.492	4.123	0.128	0.0027	52.411	2.459	0.817	0.239	0.111	1.486	0.336	0.061	0.613	23.9663	4.225

To roast one ton of matte in 80 or 100-ton heaps, about 0.3 cords of wood are required and from 5 to 8 days' labor.

2. *Roasting in stalls* has many advantages over roasting in heaps. There is less loss of metal by dusting and none at all by leaching; little fuel is needed; the piles can be smaller; the roast is quicker and more perfect, requiring less matte to be roasted a second time, and the fumes are carried off. The only drawback is the cost of plant, and this is more than made up, especially if the stalls are built of slag-brick, by the advantages. That stalls have not superseded heaps is because, if a plant has to be built, it is

[1] Terhune, *Trans. A. I. M. E.*, xvi., p. 23.
[2] Private notes, 1890.
[3] Arche, "Die Gewinnung der Metalle," Leipsic, 1888, p. 72.

considered that a long-hearth reverberatory furnace may as well finish the roasting by one short operation. The writer, however, does not agree with this view, and thinks that stalls might in many cases be preferable.

The open stalls used at Pribram,[1] Figures 154–157, will serve as an illustration. The general arrangement is shown by Figures 156 and 157. There is a double row of stalls placed back to back with

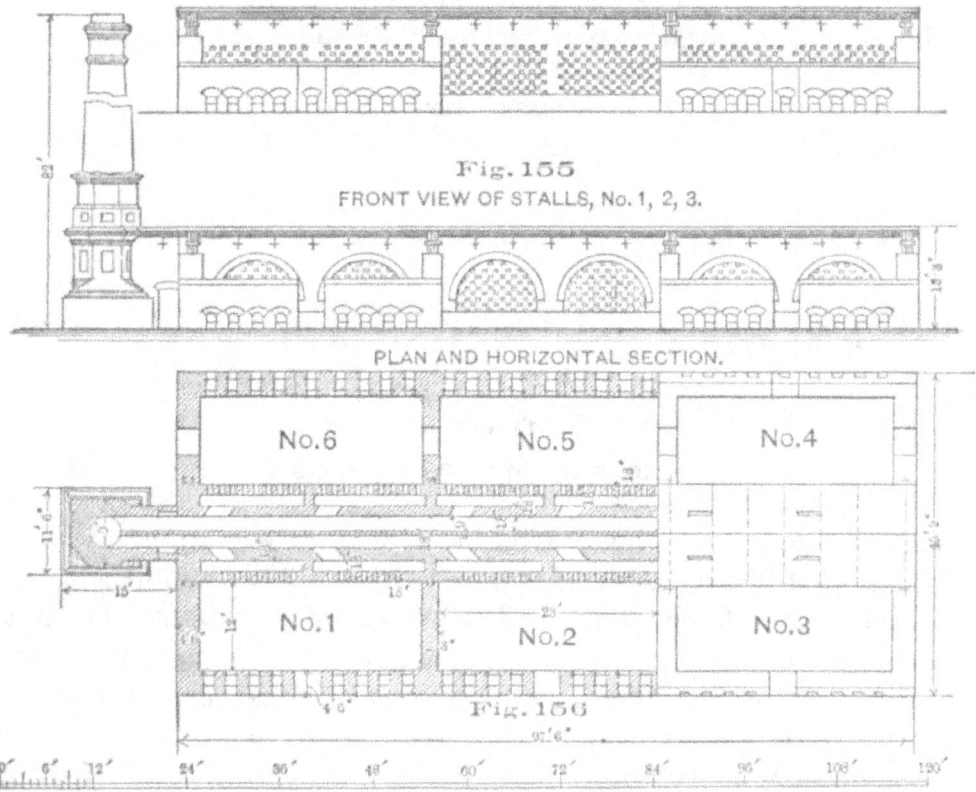

Fig. 154

ROAST STALLS FOR MATTE.

FRONT VIEW OF STALLS, No. 4, 5, 6.

Fig. 155

FRONT VIEW OF STALLS, No. 1, 2, 3.

PLAN AND HORIZONTAL SECTION.

No.6 No.5 No.4

No.1 No.2 No.3

Fig. 156

a flue between them, which is divided by a vertical wall into two parts, in order that each row of stalls may have an independent draught. The flue terminates in a chimney 82 feet high. The side-walls of a stall are built up solid; in the front wall on each side of the entrance are four fireplaces, from which the heap is kindled. The small flues in the back-wall open into two chambers, which con-

[1] Zdráhal, *Oesterreichisches Jahrbuch*, xxxix., p. 16, and Private notes, 1890.

nect separately with the main flue, and can be closed off by sliding dampers. The advantage of the divided flue is that in treating small quantities of matte one side only of the stall is filled, and when the roast is finished, the clinkered and insufficiently-roasted parts can be placed directly on the heap which is being erected on the other side. Stalls 1, 2, and 3 receive small charges of 65 tons, and stalls 4, 5, and 6 large charges of 140 tons. The stalls are built entirely of slag-brick. They are made by tapping the blast-furnace slag from a slag-pot like that represented by Figures 134–139, and letting it run into a cast-iron mould. The upper surface is made smooth by placing a heavy cast-iron plate on the mould when filled. If the slag is not sufficiently acid, it cannot be used alone for this purpose, as it cracks while cooling. It must then be

CROSS-SECTION.

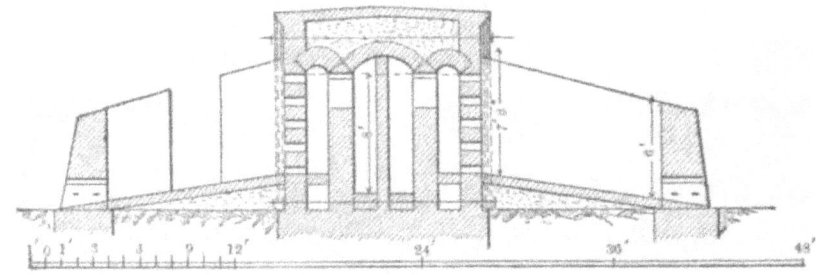

Fig. 157

ROASTING-STALLS OF PRIBRAM.

used to cement together clinkers, sharp-edged sand, pieces of brick, etc., placed in the mould. It is well to cover the mould with sand or ashes when filled, in order that the slag may cool slowly, as it is less liable to crack. Slag-bricks form an excellent building material for the stalls. They are laid with ordinary clay mortar; the walls are sufficiently thick not to require any binding.

Four sizes are made:

NO.	LENGTH.	WIDTH.	HEIGHT.
1.	17¼ inches.	8¼ inches.	8¼ inches.
2.	17¼ "	4⅛ "	8¼ "
3.	12½ "	12½ "	6¼ "
4.	12½ "	6¼ "	6¼ "

Bricks Nos. 1 and 2 are used for the side-walls; Nos. 3 and 4 for the bottom of the stall. Nos. 2 and 4, having half the width of Nos. 1 and 3, correspond to split brick, and are used as binders.

Two improvements in construction may be mentioned. The first consists in leaving open two horizontal air-passages, one above the other, in each side-wall, with openings into the stall. One may be placed at half the height of the side-wall, terminating at half its depth; the other at three-quarters the height, ending at three-quarters the depth. The second is to leave openings in the front-wall. This introduction of air gives better control of the roast.

In charging the stall, coarse matte is placed on a thin bed of wood arranged on the slag-bottom in the same way as when a heap is erected. Then follows a mixture of coarse and fine matte and the top is covered with sufficient fines to prevent the fumes from passing through the top, thus enabling them to be drawn off through the chimney. If only one-half of the stall is to be used, the open side is built up of coarse matte. The stalls are fired from the front with wood, and the fires kept going with bituminous coal and lignite until the heap is well ignited.

The time required for roasting is :

	65 tons.	140 tons.
Unroasted lead matte..................	48 days.	86 days.
Once-roasted lead matte...............	43 "	80 "
Twice-roasted lead matte.............	33 "	70 "

The well-roasted matte contains from 8 to 10 per cent. sulphur, one-third of the sulphur being present as undecomposed sulphide. It makes no difference in this result whether the stall is entirely filled or only half filled. The change in weight after roasting is insignificant.

The labor necessary per ton of matte is :

	65 tons.	140 tons.
For the first roast.	1.81 hours.	2.02 hours.
For the second roast.................	2.26 "	2.44 "
For the third roast..................	2.26 "	2.44 "
For pulling down and wheeling to the feed-floor........................	3.62 "	3.62 "

With four men working in the daytime only in ten-hour shifts, the number of days required to fill and empty the stalls will be:

	65 tons.	140 tons.
Filling with raw matte for first roast..	3 days.	7¼ days.
Pulling down and filling for second roast...........................	3¼ "	8¾ "
Pulling down and filling for third roast.	3¾ "	8¾ "
Pulling down and wheeling to feed-floor............................	6 "	13 "
Handling	16½ "	37¾ "

The fuel required per ton of matte is:

	65 tons.		140 tons.	
	Wood, cords.	Coal, pounds.	Wood, cords.	Coal, pounds.
For first roast...........	0.008	68.2	0.007	66.0
For second roast........	0.008	72.6	0.007	78.0
For third roast.........	0.008	71.6	0.007	84.4
Average	0.008	70.8	0.007	76.1

Analysis[1] of stall-roasted ~~ore~~ *matte* from the Colorado Smelting Co., Pueblo, Col.:

Fe_2O_3......................................	80.39
Mn_2O_3..	0.93
PbO..	7.91
ZnO..	0.98
CuO..	3.22
As_2O_3......................................	0.86
Sb	trace
S..	1.01
SiO_2..	3.21
CaO..	3.22
Ag..	0.0614

3. *Roasting in Kilns* is practised only in connection with the manufacture of sulphuric acid. In the Hartz Mountains and in Saxony[2] shaft-furnaces are used for roasting. At the Lautenthal Works[3] six Freiberg shaft-furnaces,[4] each having four doors, are in

[1] Dewey, Bulletin No. 42, "United States National Museum," p. 52.

[2] Merbach, "Freiberg's Berg- u. Hütten-Wesen," Freiberg, 1883, p. 251.

[3] Private notes, 1890.

[4] Drawings in *Zeitschrift für Berg-, Hütten-, und Salinen-Wesen in Preussen*, xix., plate viii.

use. The inside dimensions are: height, 10 feet 6 inches; width, 4 feet 2 inches; depth, 7 feet 5 inches. They hold about 300 cubic feet of ore. In twenty-four hours one furnace roasts in two charges 4,200 pounds of a mixture of 4 parts raw first matte (4 per cent. copper, 10 per cent. lead) and 3 parts half-roasted matte, or 5,400 pounds of a mixture of 5 parts raw second matte (12 per cent. copper, 7 per cent. lead) and 4 half-roasted matte. The roasted matte contains from 7 to 8 per cent. sulphur, and the gases 5.3 volumes sulphurous acid. The six furnaces are attended by eight men in twenty-four hours. Sufficient heat is generated by roasting to have a Glover tower; the sulphuric acid chambers have a capacity of 670,000 cubic feet, and terminate in a Gay-Lussac tower. In 1879, when the writer was at Lautenthal, there was a loss in the manufacture of sulphuric acid, and it was carried on only to make the sulphurous acid innocuous; now (1890) through the improvements in plant, acid is made with profit. Details about the management of furnaces and the manufacture of sulphuric acid are given in Lunge's excellent work.[1]

4. *Roasting in Reverberatory Furnaces.*—Of the different reverberatory furnaces only the long-bedded one-hearth calciner is in general use. Terhune[2] gives the results obtained at the Germania Works, Salt Lake City, in roasting matte in a Brückner cylinder 18 feet long and 7 feet in diameter. A charge of 8 tons is roasted in forty-eight hours down to 4 or 6 per cent. sulphur, requiring one man in a 12-hour shift and consuming 20 per cent. of Pleasant Valley coal. Two cylinders, 22 by 7 feet, were added later, and all the three roasters are run by two men in 12-hours shifts. From Plattner's experiments, already mentioned, which show that the silver and lead in roasted matte is concentrated toward the surface, it can be seen that the loss in these metals by dusting must be great with revolving furnaces, and this is probably the cause that they have not found more general application.

Terhune[3] gives some interesting figures on a short experiment in roasting matte in a Stetefeldt furnace.

The reverberatory furnaces used are very similar to those described for roasting ores (§ 56, *b*) except that they have no fuse-box and that the roasted matte is generally discharged through one or two openings in the hearth near the bridge into an iron car

[1] "Sulphuric Acid and Alkali," London, 1891, vol. i.
[2] *Trans. A. I. M. E.*, xvi., p. 19.
[3] *Trans. A. I. M. E.*, xvi., p. 22.

in the arched chamber below. Where ore-calciners are used, the matte is often roasted with the ore, if this is to be fritted or fused, furnishing the necessary iron flux. The details of manipulation, with a few slight exceptions, are the same as those given under the roasting of ore, and can be passed over. The hearth of the matte-calciner commonly used at present in Colorado has the same construction as that of the ore-calciner. It is 60 by 14 feet, and is divided into four separate hearths, each having two working-doors on either side. It receives a charge of 5,400 pounds, crushed through a 4-inch sieve, and lies three or more inches deep in the furnace. In twenty-four hours three charges are drawn (generally into slag-pots) by four men, and two tons of coal are consumed. Pyrite often replaces part of the matte in the charge.

The details of a matte-roasting furnace, which is also used for ore that is not to be fused, are given in Figures 158–163. It is one of the furnaces contained in the roaster building shown in Figure 82, and is chosen because it differs considerably in its constructive details from the type generally accepted in Colorado as standard. The hearth (14 by 40 feet) has about the same width as the Colorado furnace, but is considerably shorter. The reason is that, as the charge is not to be fused, the temperature is kept comparatively low at the bridge, which permits a 40-foot hearth to utilize the heat about as well as a 60-foot one. Special attention is called to the vaulted arches which support the hearth, the absence of off-sets (the hearth having a gentle rise from bridge to flue), the slope from the centre to the doors, the discharge-openings for roasted ore, and the damper in the flue. The distance between the working-doors is less than is usually found ; the width of the fireplace (21 inches, Figure 162) proving insufficient, it was increased to 36 inches (Figure 158), and to supply some of the room required for this, the fire-bridge (27 inches, Figure 163) was made narrower (22¼ inches, Figure 158) and received air-flues.

§ 77. **Smelting of Roasted Matte.**—The first matte obtained in ore-smelting (containing about 4 per cent. copper) is, after being roasted, always added as iron flux to the ore-charge in the blast-furnace. The resulting second matte, containing from 10 to 13 per cent. copper, is concentrated in the blast-furnace, in the reverberatory-furnace, or in the two successively.

1. In the Hartz Mountains the matte is concentrated exclusively in the blast-furnace to black copper (76 per cent. copper) and an enriched copper matte (67 per cent. copper). As many as four separate heap-roastings, each followed by a reducing smelting in the

blast-furnace, are carried on to obtain the final products that are sold to copper works.

The roasted matte is smelted with 147 per cent. slag and 15 per cent. coke in a low blast-furnace having a crucible that is partly internal and partly external, or only external.

One hundred parts of roasted second matte give 13.06 parts of impure base bullion and 35.42 parts of third matte. From this on no more lead can be extracted. The rate at which the matte decreases in quantity as its percentage of copper increases is shown by the following tables:

AMOUNT OF MATTE.

First matte 60.27
Second " 21.00
Third " 10.96
Fourth " 4.57
Fifth " 3.20

Total 100.00

ANALYSES OF MATTE.[1]

	Matte I.	Matte II.	Matte III.	Matte IV.	Matte V.
Pb	13.47	8.33	10.02	9.06	3.88
Ag	0.035	0.040	0.065	0.075	0.035
Cu	5.73	12.86	27.76	42.30	59.86
Fe	48.18	44.60	31.25	19.98	12.24
Zn	1.71	3.32	2.99	1.59	0.75
S	25.25	20.42	21.92	17.89	20.78

ANALYSES OF SLAGS.[1]

	Matte I.	Matte II.	Matte III.	Matte IV.	Matte V.
SiO_2	23.73	25.30	27.43	17.43	18.97
$BaSO_4$	....	0.36	1.06	trace	none
Pb	1.68	1.53	1.42	1.47	0.92
Cu	0.58	0.63	0.56	1.90	1.58
Ag	0.002	0.0019	0.0018	0.0015	....
Sb	0.14	0.05	0.14	0.09	0.20
FeO	48.16	51.99	49.67	62.14	61.84
Al_2O_3	8.12	3.86	5.28	5.87	6.47
P_2O	0.41	0.56	1.31	0.43	0.47
MnO	0.60	0.71	0.77	0.49	0.50
ZnO	8.11	6.69	6.38	4.30	2.90
Ni+Co	0.09	0.08	0.12	0.27	0.54
CaO	3.60	4.61	4.16	2.00	2.27
MgO	0.32	0.40	0.52	0.39	0.47
BaO	0.41	....	....	....	...
K_2O	0.42	0.47	0.64	0.57	0.77
Na_2O	0.36	0.33	0.33	0.33	0.53
S	3.95	3.49	2.39	2.64	1.48

[1] Private notes, 1890.

From the above analyses can be seen that the roasting of the second matte is imperfect; the results after smelting show an increase of 15 per cent. copper instead of 30 per cent., as might have been expected. Thus the fourth matte is only as high as the third should have been. While the amount of iron in matte V. has been somewhat reduced in comparison with that of matte IV., it is still very high, and can be removed only by smelting in the reverberatory furnace. It would seem, therefore, advisable to stop the concentration in the blast-furnace at 40 or 50 per cent. copper, and to treat the resulting matte in the reverberatory furnace.

2. The general practice of American smelters is to add the roasted first matte as iron flux to the ore-charge and to treat the resulting second matte in a separate furnace. This has the same general construction as the lead-furnace, with the exception that it has no crucible to speak of, the distance between centre of tuyere and slag-tap being ten or twelve inches. Often the crucible of a lead-furnace has been simply filled with well-beaten brasque and is thus changed into a matte-furnace, but special furnaces with a drop bottom, such as is common in copper-smelting, are often used.

The dimensions at the tuyere level vary from 36 by 60 inches to 33 by 100 inches, according to the required capacity, the latter smelting 100 tons charge in 24 hours. In separating matte from slag the practice varies.

At some works the tapping-jacket has two tap-holes, the lower one for lead and matte, the upper one for slag. The pot containing lead and matte with some slag is allowed to cool on the dump. When cold the button of lead is separated from the matte, and this from the slag, the latter being smelted over again. The slag from the upper tap-hole is also allowed to cool on the dump, and is examined for shots of matte before it is discarded.

At other works the entire contents of the furnace are tapped into an overflow-pot. This is to collect the lead, while matte and slag overflow into a catch-pot having a tap-hole. All the slag remaining in the second pot is smelted over again. The overflow-pot has a tap-hole placed on the side, as near the bottom as possible. Every six or twelve hours, according to the grade of matte that is being smelted, this pot is wheeled over a cast-iron basin let into the ground, and the lead tapped. As soon as matte appears the tap-hole is closed with a clay plug.

A third method of obtaining a good separation of slag and

matte is to have the bottom of the furnace inclined. Thus the distance between the centre of the tuyere and the slag-tap is 10 inches, and this is 8 inches higher than the matte-tap placed at the opposite end of the furnace. An excellent separation of matte and slag is effected.

The roasted matte is usually smelted with foul slag from the ore-furnace and with copper ore low in silver, if this be available. The slags made are similar to ore-furnace slags, with this difference, that they usually run lower in lime and are not quite so siliceous. Two special matte-slags may be quoted:

SiO_2.	FeO.	CaO.
32	36	14
32	40	16

The reason that the roasted matte is not smelted with dry silver ores is that only 93 per cent. of silver contained in the matte is paid for by copper refiners. The tendency is therefore not to enrich, but rather to reduce the silver contents of the matte. The first matte obtained in the blast-furnace runs from 24 to 30 per cent. copper; when this is smelted again, after having been roasted in a calcining furnace, its copper contents range from 40 to 60 per cent. when it goes to copper works.

3. The concentration of 12 per cent. matte in the reverberatory furnace alone is not practised. At Freiberg[1] it is brought forward to about 50 per cent. copper by two smeltings in the blast-furnace. Analyses of matte and slag by Schertel[2] are subjoined:

MATTE.		SLAG.	
Pb	16.86	SiO_2	82.71
Ag	0.29	PbO	8.16
Cu	55.43	CuO	1.40
Sb	0.22	FeO	38.01
Fe	3.50	(Ni, Co)O	——
Ni, Co	0.41	ZnO	9.14
Zn	3.44	Al_2O_3	5.30
S	20.12	BaO	2.88
		CaO	0.59
		MgO	——
		S	0.74

The matte is then roasted in a calcining-furnace and smelted in

---

[1] Grand, *Annales des mines*, 1875, vii., p. 314; Capacci, *Revue universelle des mines*, 1881, ix., p. 269.

[2] *Berg- u. Hüttenm. Zeitung*, 1888, p. 442.

a reverberatory furnace with barite and quartz as fluxes. The hearth of the furnace is 13 feet long, 4 feet 2 inches wide at the bridge, 8 feet at the middle, and 1 foot 2 inches at the flue; the grate, 4 feet 3 inches by 3 feet 4 inches, is 1 foot 3 inches below the top of the bridge, and this 1 foot 3 inches below the roof. The furnace has the usual two doors, one at the side and one beneath the flue. The bottom, which is built on an iron plate, consists first of a full course of fire-brick on which is rammed firmly a mixture of fire-clay and quartz, giving it the usual disc-like form, with the lowest point at the tap-hole. The thinnest part of this layer is 3 inches thick. The working-bottom, which follows, consists of 5,500 pounds of an intimate mixture of 5 parts of quartz and 1 part of matte slag. It is melted down, after being heated and patted into shape, in twelve hours, and lasts from 15 to 18 months. Five charges, each weighing 4,290 pounds, and consisting of

> 2,640 lbs. roasted matte,
> 440 " raw matte,
> 660 " barite,
> 550 " quartz,
> ———
> 4,290 lbs.

are treated in this furnace in twenty-four hours, 4¼ tons of bituminous coal being consumed. From these five charges 8,250 pounds of matte is produced. The composition of matte and slag as analyzed by Schertel[1] is given below.

MATTE.		SLAG.	
Pb	4.85	SiO_2	30.60
Ag	0.31	PbO	14.51
Cu	73.95	CuO	4.40
Bi	0.02	FeO	8.10
Sb	0.06	(NiCO)O	0.38
As	0.18	ZnO	1.95
Fe	0.13	Al_2O_3	1.80
NiCo	0.21	BaO	28.63
Zn	——	CaO	7.70
S	18.98	MgO	0.72
		S	0.18

In order to obtain matte running so low in iron, it is necessary to produce some bottoms. These go back to the next reverberatory charge or are added to the last blast-furnace charge to be again

[1] *Loc. cit.*

converted into matte. The matte forms the raw material for the manufacture of blue vitriol.

This Freiberg method of using barite in the final concentration in the reverberatory furnace will hardly recommend itself, as nobody will introduce baryta into a blast-furnace slag if it can be helped. By substituting gypsum, which, as shown by Schweder (§ 54, g), has the same effect, slags low in copper might be obtained, and the resulting slag, containing lime instead of baryta, would be welcome for the furnace charge. Thus it would be necessary for only a small amount of the copper from the concentrated matte to pass again through the lead blast-furnace and undergo the tedious process of matte concentration.

4. Treatment of Argentiferous Copper Matte containing Lead and Iron. The concentrated matte produced at lead works is usually shipped, as before stated, to copper works, and there treated by one of two methods: the first is to convert the matte into metallic copper and refine it, after which it is desilverized by electrolysis, vitriolization, or a modified Augustin process; the second is to dead-roast the matte, if it is sufficiently free from iron as at Freiberg (see *ante, a*), and then treat it with sulphuric acid to dissolve the copper, which is sold as blue vitriol, the residue going to a lead blast-furnace.

The aim at lead works for some time has been to work up the matte there. This is done in two ways—by the Crooke process and the Hunt & Douglas process, No. 2.

a. *The Crooke Process.*—Several years ago John J. Crooke invented[1] a process to desilverize copper matte and produce metallic copper. It is now in use at the works of the Pueblo Smelting and Refining Co., Pueblo, Col., but, as details are not published, only a short outline can be given.

The desilverization is based on the property lead possesses of extracting silver and gold from copper matte. If lead is kept at a red heat and crushed matte spread over it and allowed to remain in contact, from 75 to 80 per cent. of the silver and all the gold will be extracted. The lead, however, takes up copper, and if this goes too far it loses the power of desilverizing. Also, if it remains in contact with the matte beyond a certain time, the two are liable to combine into one mass. To prevent this, Crooke introduces into the lead bars of the best soft Swedish iron, placing them horizon-

[1] Patents No. 308,031, November 11, 1884; No, 354,182, December 14, 1886.

tally across the bottom of the reverberatory furnace and fastening them there. To extract the remainder of the silver, fresh lead is required.

In actual practice four reverberatory furnaces are used, each holding 25 tons of lead, arranged in terrace form, in order that the lead of the upper one can be discharged into the next one below. Into the highest furnace is charged lead free from or low in silver. This will be gradually enriched in its passage through the furnaces, until taken out at last as rich bullion, containing 2 or 3 per cent. copper and a little arsenic and antimony. The rich matte, charged into the lowest furnace, will be gradually desilverized in its passage upward over the four baths of lead, and will be drawn from the highest furnace as granular matte practically free from silver. About twenty-five tons of matte are thus passed in charges of one and a half tons each in twenty-four hours through each of the four reverberatory furnaces, taking about one and a quarter hours for the three operations: charging, treating, and withdrawing.

The second part of the process, the conversion of desilverized copper matte into metallic copper, requires four operations, which are carried on in four separate furnaces. The matte is first roasted in a calcining-furnace to reduce the sulphur contents down to 13 per cent. It then passes to a reverberatory furnace having a tuyere on either side of the bridge. Here it is subjected to a higher temperature, which is, however, kept below the melting-point of copper. By the action of heat and blast and by constant rabbling, the impurities of the matte will be oxidized, and the copper converted into moss-copper. When this has formed, the charge is transferred to a blister-furnace and melted down quickly, with an addition of quartz to slag the lead and any iron that may be present. The slags containing lead, iron, arsenic, and antimony are drawn; then follow coppery slags, and when the charge comes to work, the blister-copper, assaying 99 per cent. copper and over, is tapped into sand-moulds, and is then ready for the refining-furnace.

b. *The Hunt and Douglas Process. No. 2.*[1] This process, in-

[1] United States Patent, No. 227,902, May 25, 1880; Hunt, *Trans. A. I. M. E.*, x., p. 12; xvi., p. 80; Douglas, "Mineral Resources of the United States," 1882, p. 279; Howe, "Production of Gold and Silver in the United States, Report of the Director of the Mint," 1883, p. 798; Hunt and Douglas, *Engineering and Mining Journal*, October 3, 1885; Fulton, *Ibid.*, December 12, 1885; Howe, *Ibid.*, December 19, 1885.

vented over ten years ago, has lately been introduced at the lead-silver works of the Consolidated Kansas City Smelting and Refining Company, Argentine, Kan. It has not been running long enough to furnish final figures, although so far it appears to be doing satisfactory work. A general outline must therefore suffice for the present. The process consists in:

1. Dissolving the copper from roasted matte with dilute sulphuric acid containing some chloride, the residue, which goes to the blast-furnace, being iron, lead, and silver.

2. Chlorodizing the cupric sulphate with a solution of ferrous or calcium chloride which contains a slight excess of chloride over what is required to form cuprous chloride. If calcium chloride is used, nearly all the calcium will be precipitated as calcium sulphate, and have to be separated by an additional operation.

3. Reducing the cupric chloride formed to cuprous chloride, and precipitating this by pumping hot concentrated sulphurous acid, generated by burning pyrite in the Douglas "central-flue cylinder calciner,"[1] into the solution, an equivalent amount or more of sulphuric acid is formed. This is freed from sulphurous acid by injecting hot air, which also completely oxidizes any metallic salts held in solution. It is then ready to be used as solvent for roasted matte when it has cooled. Any iron, zinc, nickel, or cobalt salts gradually accumulating in the solvent have to be removed periodically by crystallization.

4. Converting the cuprous chloride by means of metallic iron into metallic copper, or by milk of lime into cuprous oxide, to be smelted for marketable copper. The ferrous or calcium chloride formed by the reaction serves as chlorodizing agent for the cupric sulphate of the next charge.

§ 78. **Nickel and Cobalt Matte.**—The behavior of nickel and cobalt in lead matte shows some points of interest. The non-argentiferous ores of Southeast Missouri (§ 19) carry small amounts of nickel and cobalt, which are concentrated in a blast-furnace with syphon-tap in a matte. At Mine La Motte[2] the first matte carries from 3 to 3.5 per cent. nickel-cobalt, from 0.5 to 1 per cent. copper, and from 20 to 25 per cent. lead. When this is roasted in a calcining furnace, there must remain, when it comes to be used as iron flux, from 5 to 6 per cent. sulphur in the matte, if the loss of nickel

---

[1] Douglas, *Engineering and Mining Journal,* August 14, 1886 ; Aaron, *Ibid.*, September 11, 1886.

[2] Neill, *Trans. A. I. M. E.,* xiii., p. 634.

and cobalt by slagging is to be avoided. In the same way the second matte (5–6 Ni-Co; 20–30 Pb; 1–2 Cu), when roasted, requires from 7 to 9 per cent. sulphur to reduce the loss by slagging; it is then smelted with siliceous matter to a third matte (12–17 Ni-Co; 35–40 Pb; 3–5 Cu), the slags nevertheless assaying 1.25 per cent. nickel-cobalt and from 2 to 2.5 per cent. lead. Attempts at further concentration in a reverberatory furnace proved unsuccessful, the slags carrying 5 per cent. nickel-cobalt and 5 per cent. lead. This illustrates the difficulty of concentrating nickel-cobalt in a matte when lead is present, which apparently drives it into the slag. The nickel can then be recovered only in part and the cobalt not at all. The way that remains is to introduce arsenic and form a speise. Neill used this method successfully. The experimental furnace he built was 30 inches wide at the tuyeres, and 48 inches at the charging-door; the height from tuyeres to charging-door was 6 feet 6 inches; the crucible, 18 inches deep, was partly internal, partly external, and had a syphon-tap. He obtained a speise with 22.53 per cent. nickel-cobalt ($\frac{1}{3}$ of which was cobalt), 6.4 per cent. lead, and 4.25 per cent. copper. The matte that formed contained 3.25 per cent. nickel-cobalt, 8 per cent. lead, and 7 per cent. copper to be retreated; the slag from 0.16 to 0.32 per cent. nickel-cobalt and from 0.6 to 0.8 per cent. lead.

§ 79. **Analytical Determinations.**—The bodies to be determined by assay in raw or roasted matte are silver and gold (§ 34), lead, copper, iron, and zinc (§ 81), and sulphur.

1. *Lead.*—The usual dry assay for lead (§ 34) gives unsatisfactory results, especially in raw matte. Only from 40 to 50 per cent. of the lead present is collected in the button. If the sulphur and metals other than lead are removed, the results obtained with the usual crucible-assay are accurate. To do this, 5 grammes of matte are dissolved in a small flask with nitro-hydrochloric acid; dilute sulphuric acid is added, and the whole evaporated to dryness and heated till sulphuric acid fumes are given off; when cool, hot water is added, the solution boiled, filtered, washed, the filter incinerated on a scorifier in the muffle, and the lead sulphate with any insoluble residue reduced in the crucible in the usual way.

Williams[1] recommends a wet method, devised by von Schulz and Low, of Denver, Col., which is suitable for ores as well as by-products. It is as follows. Dissolve 1 gr. ore in 10 cc. pure con-

[1] *Engineering and Mining Journal*, June 18, 1892.

centrated nitric acid, and 10 cc. pure concentrated sulphuric acid, heat till fumes of sulphuric acid are given off copiously, cool, add 10 cc. dilute sulphuric acid (1 acid : 9 water), then 2 grs. Rochelle salt, and when this is dissolved, 40 cc. water ; heat to a boil, allow to settle, filter and wash with dilute sulphuric acid. Dissolve the lead sulphate from the filter with a boiling concentrated solution of ammonium chloride held in a wash-bottle, and collect the filtrate in a flask in which have been placed three bits of aluminum foil (size $\frac{5}{8}$ by $1\frac{1}{4}$ inches, thickness $\frac{1}{16}$ inch.) Wash with the solvent, using little of it, boil the filtrate for five minutes, give the flask a rotary motion to collect the precipitated metallic lead, fill with cold water, invert in a casserole, discharge the contents, remove the aluminum foil, rubbing off any adhering lead with the fingers, wash with water by decantation and finally with alcohol. Collect and press the lead together with an agate pestle, warm gently to avoid oxidation, transfer to the scale-pan and, weigh. Assays made on ores by this method give results that are three per cent. higher than those obtained with the usual fire assays. The time required is less than forty minutes.

2. *Copper.*—The copper in the matte is determined by titrating with potassium cyanide or by electrolysis. The former method is universally used in estimating the copper in matte that is to be further concentrated; the latter is more common with enriched matte that is to be sold, although the cyanide method also finds much favor.

a. *The Cyanide-Assay.*[1]—From 2 to 5 grammes of matte are dissolved in nitro-hyrochloric acid and sulphuric acid, evaporated and filtered as above, the filtrate being caught in a porcelain casserole. The copper is precipitated with granulated zinc, care being taken that no zinc remains with the copper by dissolving the last particles with dilute sulphuric acid. The solution is poured off and the copper washed by decanting and transferred to a beaker. From 8 to 16 cc. nitric acid (1.2 spec. gr.), diluted with an equal amount of water, is added, the solution is slightly warmed, then

---

[1] Beringer, *Chemical News*, xlviii., p. 111 ; *Berg- u. Hütt. Ztg.*, 1884, p. 200 ; Brugman, *Engineering and Mining Journal*, May 18, 1889 ; Fessenden, *Ibid.*, May 31, 1890 ; Low, *Proceedings Col. Scientific Society*, vol. i., pp. 17, 69 ; *Berg- u. Hüttenm. Ztg.*, 1886, p. 53 ; Peters, "Modern Copper Smelting," New York, 1891, p. 33 ; Torrey and Eaton, *Engineering and Mining Journal*, May 9, June 6, 27, 1885 ; Wendt, *School of Mines Quarterly*, vii., p. 222 ; Eustis, *Trans. A. I. M. E.*, xi., p. 120.

cooled, made ammoniacal with 10 cc. of dilute ammonia (1 ammonia : 2 water), and titrated with potassium cyanide (20 grs. potassium cyanide to 1 liter water). To obtain satisfactory results with the assay the usual conditions of titration must be complied with, viz., that a given volume of solution (200 cc.) shall contain the same amount of acid and ammonia and the same temperature prevail as in standardizing the cyanide solution with pure copper. As regards the finishing point of the titration, it is customary to stop at a certain shade of pink, which each assayer fixes for himself. A more satisfactory way is to add cyanide until the solution becomes decolorized. This will be when a tube filled with the solution shows the same colorlessness as one filled with distilled water, when they are held together perpendicularly over a sheet of white paper and examined from above. Should the tube with the copper solution still show a slight shade of pink, it is returned to the main solution, a drop or two of cyanide is added, and the comparison repeated. In this way the slightest differences in color are easily detected.

In the decanted solution the iron can be titrated with potassium permanganate (§ 81). The residue from the solution can be weighed after the lead sulphate has been dissolved by ammonium acetate or citrate.

b. *The Battery-Assay.*[1]—The filtrate from the lead sulphate is neutralized with ammonia, 3 cc. of concentrated nitric acid are added, the liquid containing not more than 100 cc. A current that gives from 90 to 100 cc. of hydrogen and oxygen in 30 minutes from dilute sulphuric acid (1 : 12) will finish the precipitation in from 8 to 12 hours, the time required varying according to the amount of other salts present. The deposition can be hastened by removing the iron and other salts, *i.e.*, by precipitating the copper with zinc as in the cyanide-assay, and redissolving in nitric acid. To ascertain if all the copper has been precipitated, the solution is stirred and a little water added. If no copper is precipitated on the platinum dish or the platinum cone or cylinder in the beaker, the precipitation is complete. The other test is with hydrogen sulphide. If bismuth, arsenic, and silver are present, they are determined separately and deducted from the metal depos-

[1] Low, *Col. Scientific Society*, i., pp. 17, 69 ; *Berg- u. Hüttenm. Ztg.*, 1886, p. 53 ; Torrey and Eaton, *Engineering and Mining Journal*, May 9, June 6, 27, 1885 ; Peters, "Modern Copper Smelting," New York, 1891, p. 39 ; Eustis, *Trans. A. I. M. E.*, xi., p. 120 ; Glenn, *Trans. A. I. M. E.*, xvii., p. 406.

ited. The silver may be completely deposited with the copper if a few drops of tartaric acid be added to the solution of cupric nitrate.

3. *Sulphur.*—Two methods are used at lead works to determine the sulphur. Both aim to convert it into sulphuric acid, which is precipitated by barium chloride, and the resulting barium sulphate weighed. Before precipitating, both solutions should be brought almost to a boil and all the precipitant be added at once. In this way the barium sulphate will settle quickly and the filtrate be clear.

1 $BaSO_4$ contains 0.3433 SO_3 or 0.1373 S.

a. Fuse 1 gr. matte (ore, slag) with 10 grs. of a mixture of 2 potassium nitrate and 3 sodium carbonate in a porcelain crucible. Heating to quiet fusion is not necessary to oxidize all the sulphur. Boil the fused mass with water, introduce carbonic acid to precipitate any lead dissolved by the alkaline solution, filter, add hydrochloric acid, evaporate to dryness, take up with hydrochloric acid, filter the silica, boil, precipitate with boiling barium chloride. The evaporation to dryness is avoided by the second method.

b. Fuse[1] 1 gr. matte (ore, slag) with 25 grs. caustic potash in a silver or gold-lined platinum crucible for 20 minutes; cool, dissolve in water, filter, add 30 cc. bromine water, then hydrochloric acid; boil, filter, add barium chloride, etc.

§ 80. **Slag.**—The composition (§ 53) and the disposal (§ 65) of lead slags have already been fully discussed.

If slags are too rich in silver to be thrown away, and re-smelting is of no avail (*i.e.*, when the ore contains considerable amounts of zinc), they can be desilverized by smelting with copper-bearing pyritic ore or matte. Keller[2] smelted slag, estimated to run 5.3 ounces silver per ton, with 13 per cent. pyritic ore containing 10 per cent. copper and 11 ounces silver per ton in an oblong blast-furnace 36 by 80 inches. He put through in twenty-four hours 113 tons of charge with 10.9 per cent. coke, a little more than half the amount required to smelt ore-charges. There resulted 5.3 per cent. matte containing 20 per cent. copper and 92.7 ounces silver per ton, showing a saving of 80 per cent. of the silver contained in the original slag.

[1] Fahlberg-Iles, "Mineral Resources of the United States," 1883–84, p. 450.

[2] *Trans. A. I. M. E.,* xx.

§ 81. **Analytical Determinations.**[1]—Slags are assayed for silver, gold, and lead (§ 34), silica, iron, manganese, lime, baryta, magnesia, alumina, and zinc. The aim in making the determinations must be quickness, even if it be somewhat at the expense of accuracy. If a furnace is not working well, a satisfactory correction in the charge can only be made if the main constituents of the slag are known, and as it takes from six to eight hours for a new charge to have any effect on the outgoing slag, it is easy to see that the quicker the results of the analysis can be obtained the better for the furnace. For this reason it has been necessary to adapt the usual analytical methods in such a way that quick results may be obtained. The cardinal rule is to use as little of the reagents and wash water as possible.

An important feature is the complete decomposing of a slag without having resort to fusion with alkalies. All lead slags containing 35 and less per cent. silica are decomposed by hydrochloric acid if the slag sample is glassy and sufficiently pulverized. That glassy slags are more easily decomposed than cryptocrystalline can be explained by the same reasons Le Chatelier[2] gave for the fact that an iron blast-furnace slag, when chilled suddenly, shows the properties of a natural cement. If a slag is suddenly chilled and becomes vitreous, it still retains the heat of crystallization, which would be set free if the slag cooled slowly in the pot and became crystalline. The components are therefore in a less stable condition, and are easily decomposed by the acid. In taking the sample from a freshly-drawn pot of slag the hardened surface is perforated, and a clean steel bar inserted 3 inches into the liquid slag and quickly chilled by plunging into cold water. Sometimes slag is dipped out from the pot with a clean cold iron ladle, poured out again after a minute or two, and the thin shell of glassy slag adhering slightly to the ladle taken as sample. The former method, however, is preferable. The sample is crushed, cut down, and ground so as to pass a 120-mesh sieve. Care should be taken that the entire final sample should pass through the sieve. The part to be weighed out should be ground for a minute in an agate mortar before placing it on the tared watch-glass of the balance. This grinding is sometimes omitted, but it may save a great deal of trouble later on.

[1] Iles, "Mineral Resources of the United States," 1883–84, p. 449 ; Köhler, *Engineering and Mining Journal*, August 25, 1885.

[2] *Annales des mines*, 1889, xvi., p. 165.

1. *Silica.*—Add to 0.5 grs. slag in a 3½-inch porcelain casserole a "pinch" of finely-ground potassium chlorate, mix with a glass rod, add 2 or 3 drops of water, rub to a paste, add 5 cc. conc. hydrochloric acid, heat, add a few drops of nitric acid, evaporate to dryness over a lamp or on a sand-bath, stirring constantly. To prevent spattering, spread the mass over the sides of the casserole when nearly dry, and rub the residue down from the sides into a small heap. If heated too much, some iron becomes insoluble. Add a few drops of hydrochloric acid, evaporate again, then add 15 cc. conc. hydrochloric acid, heat, dilute with a little water, boil, filter, wash three times, dissolve any lead with ammonium acetate, wash three times, and discard the filtrate. Transfer the wet filter to a platinum crucible, ignite in the muffle, and weigh. This gives the silica and perhaps the insoluble silicate. To test for the latter, add twice hydrofluoric acid, evaporate over a lamp under a hood, ignite, and weigh again. The difference in the two weights is silica; the difference between the last weight and the weight of the crucible is the insoluble silicate.

Another way of eliminating lead sulphate is given by Iles.[1] When the hydrochloric acid solution has been evaporated to dryness, take up with hydrochloric acid, dilute, warm, add metallic zinc, which reduces to metal any lead, silver, and copper, and to a ferrous salt the ferric chloride. Filter and titrate the iron in the filtrate (§ 81, 2), dissolve the reduced metals, leaving as residue silica (and silicate).

If lead sulphate and undecomposed silicate need not be considered, transfer the wet filter to an annealing-cup, ignite in the muffle, empty the silica on a tared watch-glass, and weigh.

In an ore containing barite and a silicate that is not decomposed by acid, the insoluble residue will be treated as follows. Fuse with 4 parts of sodium carbonate, giving a cover of sodium carbonate, disintegrate in hot water, add hydrochloric acid, evaporate to dryness, take up with hydrochloric acid, dilute, boil, filter, wash, ignite, and weigh silica plus barium sulphate. Volatilize the silica as silicon fluoride with hydrofluoric acid and weigh again; the difference is barium sulphate. Or, fuse again with four parts sodium carbonate, disintegrate in water only, filter off the sodium silicate, wash, dissolve the barium carbonate from the filter with

[1] *School of Mines Quarterly*, iii., p. 223.

hydrochloric acid, wash, precipitate hot with hot dilute sulphuric acid, and weigh the barium sulphate.

Another method is given by Furman.[1]

Iles[2] recommends carrying on the fusion in a silver crucible with caustic potash. Fuse 40 or 50 grs. caustic potash in a silver crucible (holding 70 cc. water) to quiet fusion, cool, add 1 gr. pulverized substance, heat for 30 minutes. The caustic alkali acts in the same way as the carbonate. The advantages of this method are quickness of fusion at a low temperature and cheapness of crucible.

2. *Iron.*—The iron is titrated either with potassium bichromate (5 grs. of the powdered and dried salt to 1 liter of water), using potassium ferrocyanide (0.1 gr. in 100 cc. water) as an indicator on a dry white porcelain plate, or with potassium permanganate (5.686 grs. to 1 liter of water). Both solutions are standardized with 0.2 grs. iron wire (weight × 0.997 = Fe).

a. *Bichromate Method.*—Decompose the slag as in determining silica, but do not filter or use the filtrate of the silica if this has been determined. Add to the hot solution in the beaker stannous chloride (20 grs. stannous chloride and 20 cc. dilute hydrochloric acid to 1 liter) drop after drop until colorless, cool in water, add a 2½-per cent. solution of mercuric chloride in excess, stirring vigorously (the precipitate formed should have a silky lustre), and titrate.

b. *Permanganate Method.*—Decompose with hydrochloric acid as before, and add small pieces of zinc; when the ferric chloride is reduced, add cold water, transfer to beaker, dilute to 500 cc., add 25 cc. conc. sulphuric acid, and titrate. Another method: Decompose in a casserole with hydrochloric and sulphuric acids, heat until sulphuric acid fumes are given off ; cool, dilute slightly, reduce with zinc, dilute, and proceed as before.

3. *Manganese.*[3]—Manganese is usually titrated with potassium permanganate. Decompose the slag with hydrochloric acid in a casserole, add a few drops of nitric acid and 5 cc. of sulphuric acid, evaporate till sulphuric acid fumes are given off ; cool, transfer to a beaker, dilute to 200 cc., boil, precipitate the iron with a cream-like emulsion of zinc oxide in large excess, filter into a flask, and wash.

---

[1] *School of Mines Quarterly*, vi., p. 163.

[2] *Engineering and Mining Journal*, January 22, 1881.

[3] Iles, *School of Mines Quarterly*, v., p. 223.

Now either boil and titrate, shaking the flask, that the precipitate may settle and the finishing point be easily recognized, or filter into a ½-liter flask, fill up to the mark, take out 100 cc., transfer to a casserole, heat to a boil, and titrate.

The Fe indicated by the cc. of the permanganate multiplied by 0.2946 = Mn.

4. *Lime.*—This is titrated with potassium permanganate. Decompose the slag in a casserole as in determining silica, add to the hot silica filtrate a small excess of ammonia, dissolve the precipitated iron, etc., in a hot solution of oxalic acid, using a slight excess; repeat the precipitation and solution, avoiding an excess of oxalic acid, boil for a few minutes, allow to stand 20 minutes in a warm place, filter, and wash with hot water till the filtrate does not taste of oxalic acid. Perforate the filter, wash the precipitate with hot water into a beaker, then dissolve with hot dilute sulphuric acid (some chemists use hydrochloric acid) any precipitate that may adhere to the filter, and titrate hot with potassium permanganate.

The Fe indicated by the cc. of the permanganate divided by 2 = CaO.

5. *Baryta, Magnesia and Alumina* are determined by the usual methods.

6. *Zinc.*—This is determined by titrating with potassium ferrocyanide (41.25 grs. to 1 liter), using uranium acetate (0.2 grs. in 100 cc. water) as an indicator. It is standardized with zinc oxide.

a. *Headden's Method.*[1]—The ferrocyanide is standardized by dissolving pure zinc oxide in hydrochloric acid. Dissolve 1 gr. ore or slag in a casserole in hydrochloric acid, add nitric and sulphuric acids, evaporate till sulphuric acid fumes come off; cool, dilute precipitate once[2] with ammonia, filter, wash, add hyrochloric acid (the solution should not exceed 50 cc.), precipitate the copper with aluminum foil, add from 8 to 10 cc. hydrochloric acid, dilute to 225 cc., and titrate at 40 or 50° C.

b. *v. Shulz and Low's Method.*[3]—The ferrocyanide is standardized as follows. Dissolve 200 mgrs. recently ignited zinc oxide in

[1] *Proceedings of the Colorado Scientific Society.*

[2] Coda, *Oesterreichische Zeitschrift für Berg- u. Hütten-Wesen*, 1890, p. 123 ; *Engineering and Mining Journal*, March 14, 1891.

[3] *Proceedings of the Colorado Scientific Society.*

a mixture of 5 grs. ammonium chloride, 15 cc. concentrated ammonia, and 25 cc. hot water; add 25 cc. concentrated hydrochloric acid, 1 gr. sodium sulphite and from 50 to 75 cc. hot water. The solvent for an ore (slag) is a saturated solution of potassium chlorate in nitric acid. It keeps well; the flask is best kept loosely covered. The mode of operating is: Dissolve 1 gr. of the substance in 25 cc. of the solvent in a 3½-inch casserole, keep this uncovered, warm gently till no more greenish vapors come off, cover with a watch-glass, and evaporate to dryness ; cool, add 5 grs. ammonium chloride, 15 cc. concentrated ammonia, and 25 cc. hot water, boil one minute, break up any lumps with a stout glass rod, filter, and wash with hot water. If the filtrate be blue, precipitate the copper by adding 30 grs. of test-lead to the required 25 cc. hydrochloric acid and shake until it becomes colorless. Add 1 gr. sodium sulphite and shake the solution (any sulphur that may be precipitated does not interfere with the result). In titrating, diluting the solution to one liter and taking a measured quantity are avoided as follows : Pour two-thirds of the solution into a beaker and titrate, passing the end point ; then add the rest of it, excepting a few cubic centimeters, and titrate again, using one cc. at a time until the end point is again passed ; finally rinse out the vessel holding the solution into the beaker, and finish the titration, testing after every 2 or 3 drops of the standardized solution. A determination takes from 20 to 30 minutes. Gold, silver, lead, manganese, iron, sulphur, or the other constitutents do not interfere with the results.

§ 82. **Wall-Accretions.**—These accretions begin in lead furnaces just above the water-jackets and reach up to the feed-door. They sometimes come from the galena in the charge, or from galena formed artificially during the descent of the charge which adheres to and filters into the brick walls of the furnace; but their principal origin is in the volatilization of lead, zinc, and their compounds, which takes place in the lower parts of the furnace. These fumes condense in the upper cooler parts. During their ascent they are partly oxidized, and the oxides may act chemically on the unaltered parts. Thus in wall-accretions may be found volatilized metal, metallic sulphides, arsenides, antimonides with their oxides, and secondary products. Any insoluble silicates found in the accretions come from parts of the furnace-lining or from fine particles of the charge.

	Lead-ville.[1]	Lead-ville.[1]	Lead-ville.[1]	Pueblo.[2]	Pueblo.[2]	Tarno-witz (sul-phide).[3]	Tarno-witz (ox-ide).[3]	Claus-thal.[4]
S............	8.291	2 600	2.725	25.98	19.74	13–9	....	14.67
Pb	47.491	70 631	25.953	1.48	37.48	20–32	24	82.71
PbO	0.405	....	....	....	....	....	....	....
Bi	trace	....	....	....	....	....	....	....
Ag........	0.0754	0.297	0.0944	0.0188	0.0824	0.005	0.005	....
Au	trace	0.0007	....	....	....	....	....	....
Cu	trace	....	....	none	none	....	....	....
Zn	6.977	trace	53.392	45.68	38.99	30–27	60	trace
ZnO	0.300	....	....	....	....	C. 2.	....	....
Cd..........	trace	....	....	....	....	....	....	....
Fe	6.754	0 028	....	6.71	0.94	9–6	....	2.03
Fe_2O_3......	1.100	3.456	1.381	....	....	....	....	....
Mn_2O_3.....	2.887	....	....	0.54 Mn.	none	....	....	....
As..........	0.039	5 009	....	none	none	....	....	....
As_2O_3	....	....	0.071	....	....	....	....	....
Sb..........	trace	2.697	....	trace	0.12	....	....	1.79
Sb_2O_3......	....	....	0.056	....	....	....	....	....
Sn..........	....	5.593	....	....	....	....	....	....
SO_3.........	0.965	....	0.287	....	....	....	....	....
P_2O_5	2.166	....	....	....	....	....	....	....
Cl. Br	....	....	0.035	....	....	....	....	....
CO_2..........	trace	....	....	....	....	....	....	....
CaO	5.361	....	0.200	6.20	0.30	....	....	....
MgO.......	4.297	....	0.159	....	....	....	....	....
Al_2O_3......	1.672	0.600	0.328	....	....	⎰ 18	....	....
SiO_2	10.100	2.830	1.577	8.13	0.57	⎱	....	....
O	....	6.256	13.751	....	....	....	....	....

[1] Emmons, "Geology and Mining Industry of Leadville," p. 727.
[2] Dewey, Bulletin No. 42, "United States National Museum," p. 54.
[3] Dobers and Dziegiecki, *Z. f. B.- H.-, u. S.- W. i. P.,* xxxii., p. 102.
[4] Mezger, *Berg- und Hüttenm. Ztg.,* 1853, p. 52.

The Clausthal analysis, where galena is smelted raw, represents a crystallized wall-accretion, consisting principally of galena. At Tarnowitz, where slag-roasted galena rich in blende is smelted with the gray from the reverberatory furnaces (§ 44), two kinds of accretions form, one a sulphide principally black, the other an oxidized compound having a greenish color. In the three analyses by Guyard, of Leadville, accretions where carbonate ores formed, if not the whole, at least the major part of the charge, shows a great variety in composition. The lead, for instance, is present as metal, as sulphide, and as oxide; zinc, arsenic, and antimony as sulphides and as oxides.

The method of removing these accretions while the furnace is running and their treatment have already been discussed (§ 66).

§ 83. **Hearth-Accretions or Sows.**—In a furnace with
an Arents' syphon tap these unwelcome products form on top of
the lead below the tuyeres; in furnaces where lead, speise, and
matte are tapped from the bottom, they form there. They result
from a faulty charge or from a lack of fuel, and are mixtures of
slag, speise, matte, metallic iron, metallic lead, coke, and charcoal.
The metallic iron results from the reduction of ferric oxide, some
of it being held in solution by melted matte and dropped when
this cools. The iron of a sow is generally carbonized and contains
silicon and phosphorus.

	Leadville.[1]	Pribram.[2]	Pribram.[2]
Fe	72.828	48.685	38.875
Pb	18.793	7.309	20.140
Ag	0.1149	0.060	0.160
Au	0.00003	....	....
Cu	trace	0.199	1.107
As	5.083	8.446	2.889
Sb	trace	0.450	0.735
Mo	0.161	....	....
Ni	0.045	trace	trace
Co	trace	trace	trace
Zn	trace	8.610	5.417
Mn	0.015	....	....
S	0.650	13.760	18.466
P	0.109	....	....
Graphite	0.750	....	....
Combined carbon	0.550	....	....
Si, slag loss	0.900	....	....
CaO	....	0.166	0.780
SiO$_2$	....	15.850	9.300

[1] Emmons, *Op. cit.*, p. 723.
[2] Balling, *Berg- u. Hüttenm. Ztg.*, 1867, p. 419.

It does not usually pay to work up a hearth-accretion; it is
thrown over the dump or buried, being an eye-sore. Flechner[1]
suggests several methods of working furnace-sows. The following,
used at the nickel works of Schwerte in Westphalia (Prussia), is
of interest. Furnace-sows containing from 75 to 85 per cent. iron,
5 to 8 copper, 3 to 6 molybdenum, 2 to 4 nickel-cobalt, and weighing from 500 to 600 pounds apiece, are gradually melted down
with coke on the bottom of a blast-furnace, the melted parts running out continuously. In this way a large crust is easily reduced
in size, and can then be added again to the ore-charge, where it will
be taken up by the speise and the matte. Any mechanical means

[1] *Oesterreichische Zeitschrift für Berg- u. Hütten-Wesen*, 1889, p. 196.

of breaking up a hearth-accretion is sure to cost more than will be recovered from re-smelting afterward.

§ 84. **Furnace Cleanings and Furnace Refuse.**—Furnace cleanings and refuse are a mixture of fire-brick, metal-bearing compounds, fuel, etc., obtained in cleaning out a blast-furnace when blown down. They are assorted; the waste goes to the slag-heap, the valuable part is added to the ore-bed.

§ 85. **Flue-Dust or Chamber-Dust.**—This product is as important as it is unwelcome. It consists of fine particles of the

	Wyandotte, Mich		Pueblo.	Hartz Mts.	Ems.		Freiberg.		Sheffield.	
Pb.....	....	....	....	84.8	60.48	67.04	85.2	27.90	......	PbS2.25
PbO.....	19.91	23.77	87.65	18.0	.....	.....	.....	.....	44.80	68.35
Pb₂SiO₄	.....	.....	.....	2.9	.....	.....	.....	.....	.....	.....
Zn.....	.....	.....	.....	1.0	3 17	4.22	5 28	49.50	.....	.....
ZnO....	0.09	trace	5.32	1.5	.....	.....	.....	.....	4 80	1.80
Cu.....	.....	.....	trace	.....	trace	trace	trace	.....	+Bi 1.52*	.....
CdO....	.....	.....	.....	.....	.....	.....	1 30	.....	.....	.....
As.....	.....	.....	.....	} 3.0	0.24	0.16	28.3	1.60	} 3.03*	.....
Sb.....	.....	.....	.....		0.42	0.31	.....	.....		.....
Ag.....	0.286	0.292	0.04	.....	0.003	0.003	.....	.....	.....	.....
Au.....										
Fe.....	.....	.....	.....	1.0						
Fe₂O₃..	} 14.43	} 18.54	24.98	4.5	} 2.12	} 1.00	1.57	.....	trace	.....
Al₂O₃..			1.31	.....			.....	.....	10.00	5 40
NiO....	0.08	trace	.....							
CoO....	0.09	trace								
CaO....	8.74	6.62	5.26	.....	1.15	0.61	1.01	7.00	.....	2.63
MgO....	3.66	.....	none	.....	.....	.....	0 25	.....		
SiO₂....	16.11	13.06	8.63	12.3	.....	.....	6.19	... ..	9.00	2.25
S.....	.....	.....	2.53	7.8	6 22	5.42	.....	.....		
SO₃.....	9.30	8.85	1.61	2.8	14.78	14.07	3.88	13.00	28.81	16.84
H₂O.... CO₂....	} 1.28	} 3.76	} 11.20	.....	.....	.....	.....	.....	.....	.....
C.....	19.23	22.14		2.5	8.00	5.80	1.17	.....		
Reference :	1.	1.	2.	3.	4.	4.	5.	5.	6.	6.

* As oxides.

(1) Curtis, *Trans. A. I. M. E.,* ii., p 95 ; (2) Dewey, Bulletin No. 42, "United States National Museum," p. 53 ; (3) Balling, *Metallhüttenkunde,* p. 87 ; (4) Freudenberg, *Op cit.,* p. 19 ; (5) Hering, *Op. cit.,* p. 34 ; (6) French, *Engineering and Mining Journal,* January 17, 1880.

charge carried out of the furnace by the ascending gases, and of metals and their compounds that have been volatilized in the lower parts of the furnace and not been again condensed in it, but settled out only while passing through condensing flues or chambers.

The flue-dust from blast-furnaces has generally a dark color, which is caused by the admixture of finely divided fuel. It is black when charcoal forms even a small part of the fuel. The amount of flue-dust formed depends upon a variety of causes. Fine ores or fluxes are carried away easily by the current of gases; charcoal, being friable, makes dust; soft coke is broken up to some extent, and causes mechanical losses; and the manipulation of the furnace, affecting the descent of the charge, has a very great influence. Thus careful feeding and cutting out of wall-accretions will reduce the forming of flue-dust to a great extent. Then, a high temperature in the smelting zone causes much volatilization; in the same way a high blast will cause much vapor to be carried out of the furnace if the quick ascent of the gases be not checked by the form of the furnace (by having boshes). Extreme figures of the amount of flue-dust formed are 0.8 and 15 per cent. of the weight of the ore charged.[1] An average figure is 5 per cent.

The different methods of condensing flue-dust[2] may be classed as wet and dry. With the former the gases are either drawn through water alone,[3] or they are pressed through one or more horizontal filters,[4] through which fine sprays of water trickle, carrying with them the condensed vapors and filtered solid particles. Water condensation, while still found in England,[5] is not much used elsewhere, as it is on the whole imperfect, and the apparatus and their maintenance costly. Dry condensation is the almost universal method used. It consists in cooling the gases and retarding the velocity of the current. Dry filtering is found in a few instances and electrical condensation has been experimented with.

The cooling of gases is essential to an effective condensation, as the single particles, being brought closer together, unite more easily into flaky masses, which then settle out. While the gases from a lead blast-furnace require very little cooling, the vapors and dust coming from reverberatory furnaces used in smelting (§ 39) and slag-roasting (§ 56), and in desilverizing base bullion (§ 97 and 105), are apt to be lost, if their temperature is not reduced.

An arrangement whereby the gases are cooled by air alone is shown in the following sketch,[6] Figure 164. It represents a horizontal sheet-iron flue, with small sliding doors at intervals of two

[1] Hahn, "Mineral Resources of the United States." 1882, p. 344.

[2] Hering, C. A., "Die Verdichtung des Hüttenrauches," Stuttgart, 1888.

[3] Eilers, *Trans. A. I. M. E.*, iii., p. 310.

[4] Percy, "Metallurgy of Lead," p. 442.

[5] Rösing, *Z. f. B., H. u. S. W. i. Preussen*, xxxvi., p. 103.

[6] Eilers, *Trans. A. I. M. E.*, iii., p. 309.

feet, suspended by means of iron rods from wooden trestles. It was used at the Richmond Works, Nevada, was 800 feet long, and ended in a wooden stack 40 feet high.

A sheet-iron flue of a different form but filling the same purpose, used at the Ems[1] Smelting Works (Prussia), is shown in Figures 165 and 166. The flues proper, 1 by 0.75 meters (3 feet 3⅜ inches by 2 feet 5½ inches) and 1 meter square (3 feet 3⅜ inches), made of

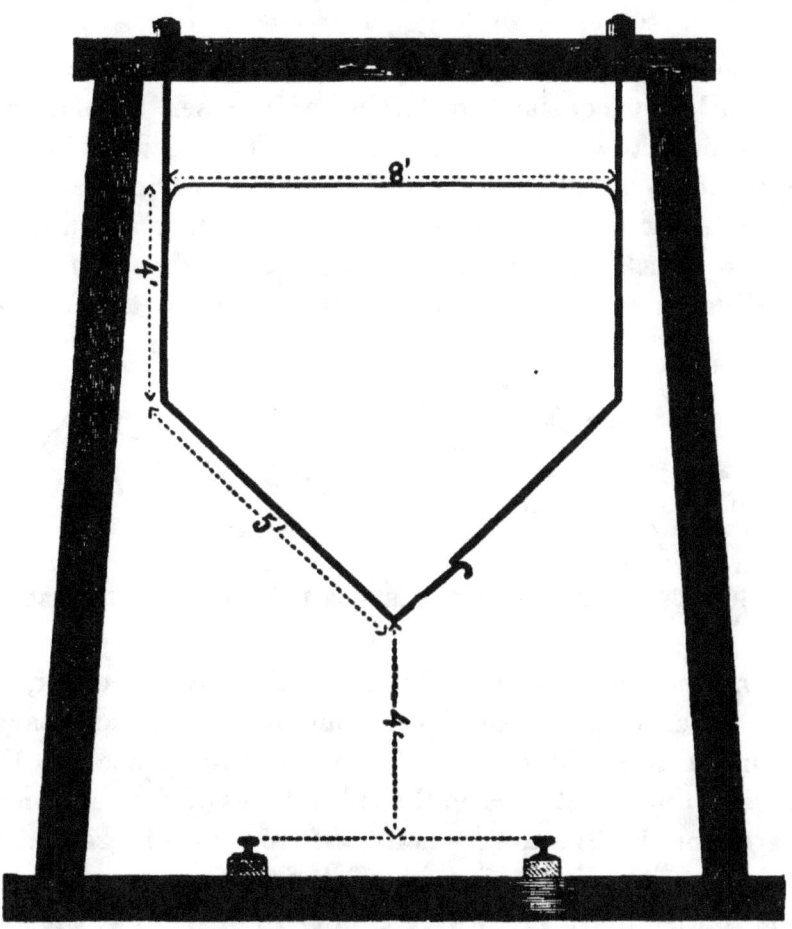

FIG. 164.—SUSPENDED SHEET IRON FLUE.

1/16-inch iron, have triangular projections, the base of the triangle being 6½ feet long and the height 3 feet 3⅜ inches.

To make the cooling more effective and thus shorten the flue, water has been used in different ways. In 1877 Hagen[2] introduced in Freiberg his water-cooled flues made of sheet-lead (Figures 167

[1] Egleston, *Ibid.*, xi., p. 410, plate ii.

[2] *Freiberger Jahrbuch*, 1879, p. 151; Hering, *Op. cit.*, p. 20, and Private notes, 1890.

and 168).   The roof (5 feet 3 inches) and sides (8 feet 2¾ inches) are suspended from a wooden framework in the same way as with a sulphuric-acid chamber.   The top *c* is cooled by a slight flow of water *d.*   This circulates along the sides through elliptical pipes *b,* soldered to the flue and to each other.   Figure 167, a vertical longitudinal section through the pipes *b,* shows the circulation of the water.   This arrangement permits cooling with less water than would be required if it overflowed from the roof down the sides.

Another arrangement has lately been introduced at Freiberg by Richter.[1]   It consists in letting the gases ascend in a set (five) of wooden towers, lined with ⅛-inch sheet-lead.   These are 18½ feet high and 6½ by 10 feet in the clear.   Water trickles down in the towers, each of which contains ten rows of small roofs, supported by lead-coated rails.   Each roof consists of three sheets of hard lead, 3 feet 3½ inches long and ₁₆⁵ inches thick, bent to the form of a V,

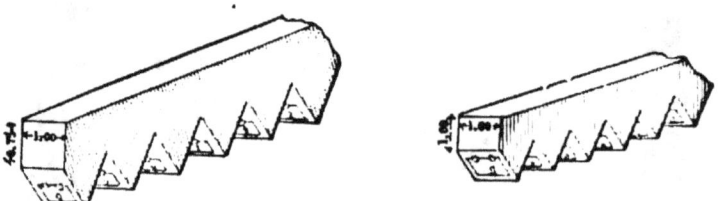

FIGS. 165–166.—SHEET-IRON FLUES WITH TRIANGULAR PROJECTIONS.

and having incisions 1⅛ inches deep at the lower border, in order that the water may run off freely and leave the necessary interstices for the ascending gases.   The roofs are placed so that the discharge of one set strikes on the ridge-poles of the next one below. The water for the five towers is drawn from two large tanks, and a short distance above the lowest row of roofs is a coarse grating, which is to distribute the water evenly in order that the gases, on striking the first roof, may be cooled as much as possible.   The temperature of the gases on entering the towers is from 100 to 115° C., and on leaving from 40 to 60° C.; they then pass through a leaden flue 1,673 feet in length before entering the main stack, which is 453 feet high.   Each tower can be separately shut out from the main flue, which is done at certain intervals, to remove,

[1] *Freiberger Jahrbuch*, 1889, p. 57; *Berg- u. Hüttenm. Zeitung*, 1890, p. 129; *Engineering and Mining Journal*, Feb. 15, 1890, and Private notes, 1890.

by means of a stream of water, the flue-dust that collects beneath the roofs.

The main effect of this apparatus is due not so much to the larger surface it offers to the gas-current as to the fact that it cools the gases. Five times as much flue-dust is collected with it as before.

Other arrangements for cooling gases are those by Schlösser and Ernst,[1] who let cooled solutions circulate in coils of iron pipe placed at certain intervals in the main gas-flue. Another plan[2] is to make gases pass through a circular tower in which iron pipes,

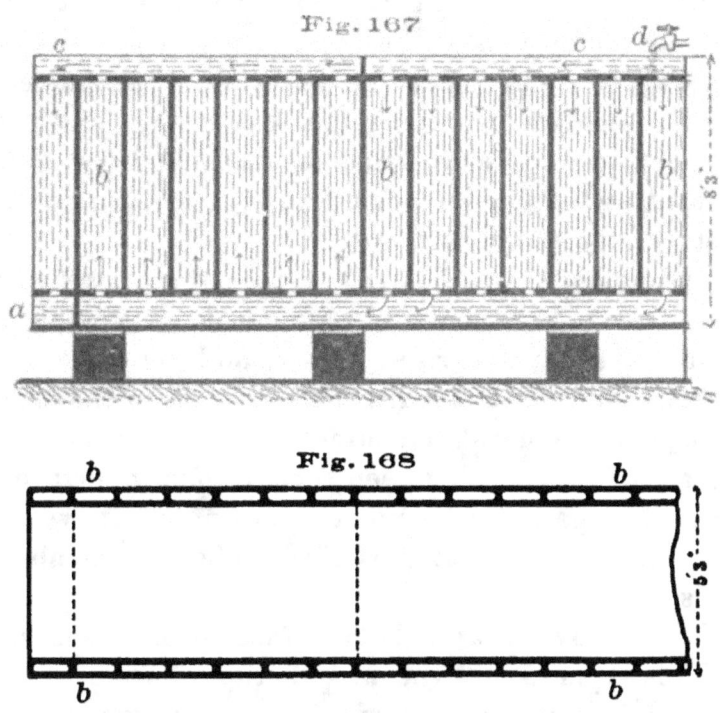

Fig. 167

Fig. 168

WATER-COOLED FLUE OF SHEET-LEAD.

with water circulating in them, are suspended. Any dust adhering to the pipes is removed periodically by a jet of water, introduced through a movable central pipe having nipples at certain intervals.

In settling out flue-dust the most important thing after cooling is the retarding of the velocity of the air-current and the exposing of a large condensation surface to it. It has been found that the point of least friction is on the centre line of a flue at $\frac{7}{10}$ of its

[1] *Berg- u. Hüttenm. Ztg.*, 1885, p. 464; 1887, p. 134.

[2] *Wagner's Jahresberichte*, 1889, p. 300.

height. Here the air has its greatest velocity. This diminishes very little upward, but so much downward that there may even be a small counter-current at the bottom of a flue. The speed of the air-current can be retarded by increasing the volume of the flue, by changing its direction or by increasing its surface.

Increasing the size of flues through which the gases pass has always been considered a very effective means of settling out flue-dust. On entering an enlarged chamber the velocity of the current is gradually slackened to the point where the draught near the exit begins to show its effect; then there is a gradual increase of velocity. On this account a comparatively small part of the enlarged flue or chamber is really useful in the settling out of the dust.

The reason that a change in the direction of the current is also not so effective as might be expected, is that, though at the turning-point the speed of the current may be slackened for a short distance, the pressure of the air behind and the draught in front will quickly restore it to the normal rate.

It is Freudenberg's[1] great merit to have discovered that an increase of surface is the most effective means of settling particles held in suspension by a current of air, and that the amount of flue-dust settled out stands in direct proportion to the area of surface with which it comes in contact. This increase of surface means increase of friction between the stationary surface and the moving current of gas, a consequent retarding of its velocity, and with it a settling out of dust particles. The surface also attracts these, if it is cold.

Aitken[2] has shown that a hot surface repels them, especially if moist. This explains how the introduction of vaporized or finely divided water[3] into the air-current has not proved so effective as was anticipated.

In order to increase the surface, Freudenberg suspended thin sheet-iron plates parallel with the air-current; and to prevent the

---

[1] Freudenberg, "Die auf der Bleihütte bei Ems zur Gewinnung des Flugstaubes getroffenen Einrichtungen," Ems, 1882. Abstract, *Engineering and Mining Journal*, July 1, 1882. Egleston, *Trans. A. I. M. E.*, xi., p. 379. Stetefeldt, "Comment on Freudenberg's Plates," *Engineering and Mining Journal*, July 28, 1883.

[2] *Proceedings of the Royal Society of Edinburgh*, xxxii., p. 239; *Wagner's Jahresberichte*, 1884, p. 1307.

[3] Iles and Keiper, *Engineering and Mining Journal*, February 27, 1886.

dust from being carried off, after it had once settled out, he placed partitions across the bottom, reaching nearly to the hanging plates. In Figures 169 and 170, representing a vertical and a horizontal section of a flue, are seen the sheet-iron plates B, $3\frac{7}{8}$ inches apart, having pieces of iron D, bent to the form of a hook, riveted to their ends. By these they are suspended from pins passing through rectangular cross-bars L, which, twisted flat at both ends, reach $1\frac{1}{4}$ inches into the side-walls A of the flue. The flue-dust collecting on the plates falls off when it has grown to a certain thickness. Every 16 or 20 feet cross-partitions E, $7\frac{3}{4}$ inches high, protect the

FIG. 169.   FIG. 170.

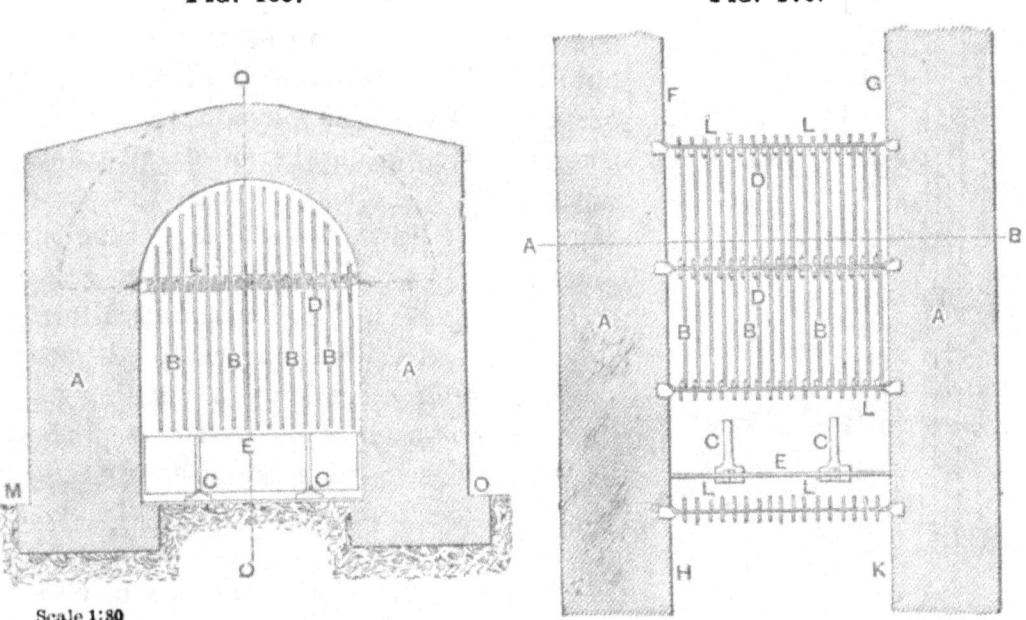

Scale 1:80

FLUE WITH FREUDENBERG PLATES.

lower part of the flue from the air-current. They are bolted to cast-iron supports C.

The investigations of Freudenberg form the basis for the construction of most modern dust-chambers. Rösing[1] has substituted iron wires for Freudenberg plates. By a shaking arrangement the dust can be more easily removed from the wire than from the plates. In Hering's book, already quoted, are given numerous sketches of condensation chambers that have been projected or built since Freudenberg's results became generally known.

Filtering is not much used, owing probably to the fact that

[1] Patent No. 432,440, July 15, 1890.

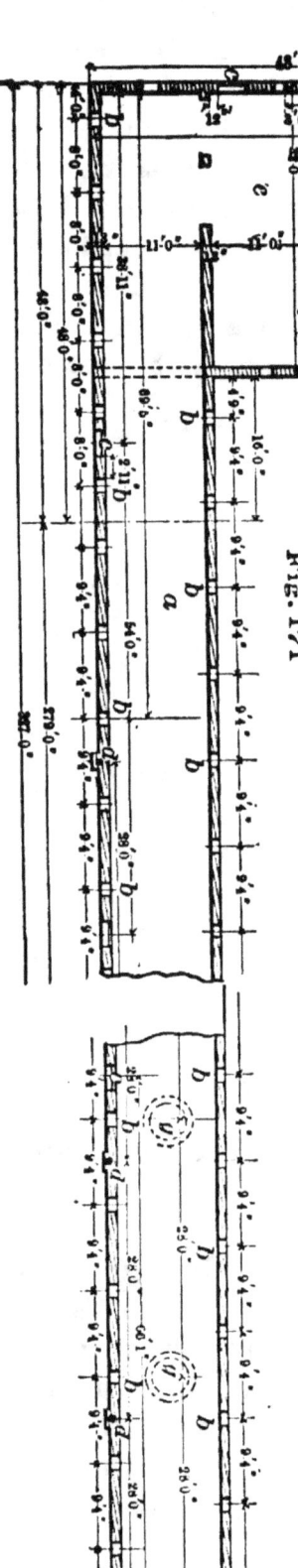

Fig. 171

The partition wall over each of the four openings of the Dust Chamber is supported with three rails laid in the wall and resting in the centre of their length on a brick wall.

nearly all the dust can be recovered without it. The effectiveness of filtering is certain. Whether it pays to draw off all the gases from a number of blast-furnaces with fans will be known when the results obtained by Iles at the Globe Smelting and Refining Co.'s Works, at Denver, become public. The Lewis-Bartlett filtering process has already been described (§ 51).

The use of electricity[1] in settling flue-dust has not been so successful as the first experiments seemed to promise. In closed chambers static electricity quickly clears dust-laden air; when, however, the air is in motion, electricity produces no effect.

As regards the arrangement of the plant, it is important to have a dust-chamber near the furnace, as most of the heavy dust is settled out quickly. The chamber will lead into a flue (or flues) with Freudenberg plates, where the gases will drop most of their dust, and then pass off through a chimney. Sometimes it will be necessary to have a fireplace at the

[1] Hutchings, *Berg- u. Hüttenm. Ztg.*, 1885, p. 253; *Engineering and Mining Journal*, May 8, 1886. Rösing, *Berg- u. Hüttenm. Ztg.*, 1885, p. 290. Bartlett, *Engineering and Mining Journal*, March 13, 1886.

foot of the chimney to create the required draught. Sometimes a fan in the flue may be needed to suck and push onward the gases.

The dust-chamber of the Montana Smelting Co.'s Works, at Great Falls, Mont., shown in Figures 171–177, may serve as an ex-

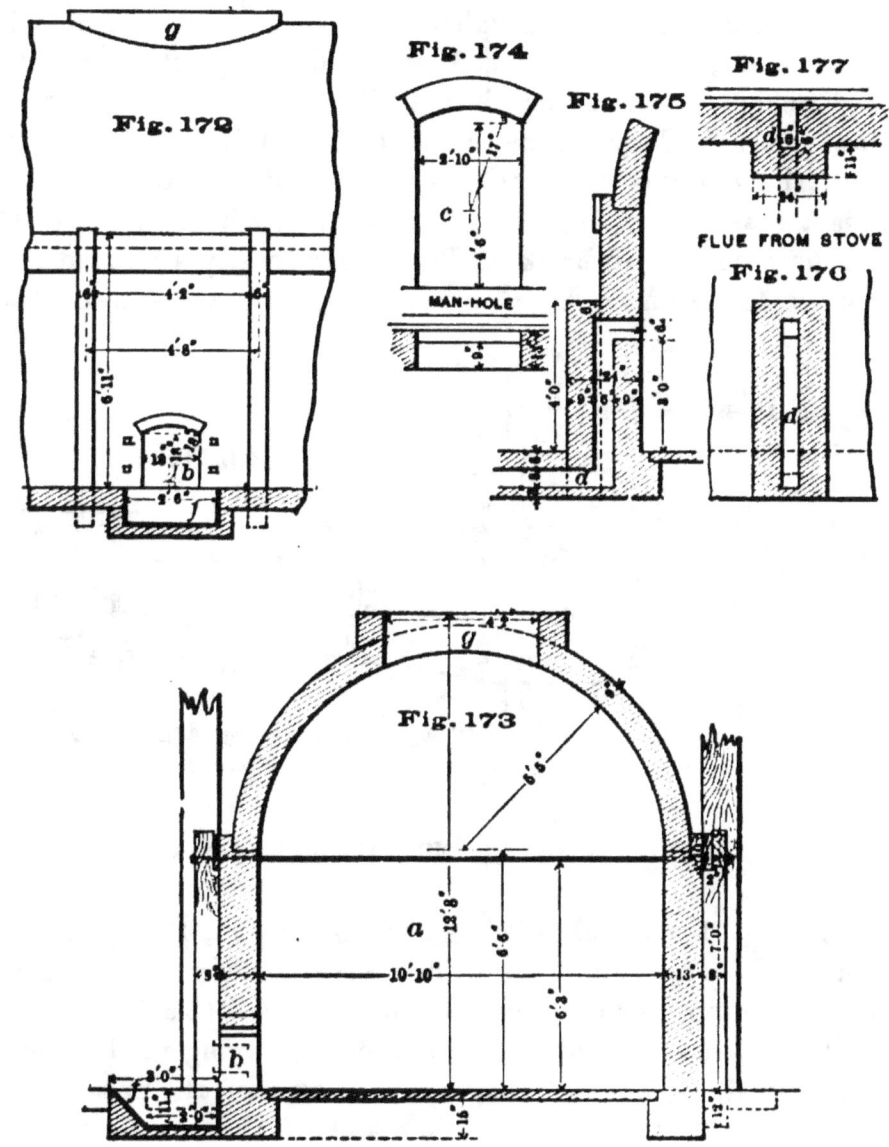

FLUE AND DUST-CHAMBER OF THE MONTANA SMELTING CO.'S WORKS.

ample of the manner in which the dust is now collected at most lead smelters. A general plan and elevation of the works has been given in Figures 82 and 83. In Figure 171 is shown a horizontal section of the main flue *a*, which runs along the back of the furnace floor. It receives the dust-laden gases from the blast-furnaces through openings *g* in the roof. The gases pass from the flue to the chambers *e*, and then into the stack.

If Freudenberg plates were used, they would be suspended in that part of the flue reaching from the chamber *e* to the first blast-furnace *g*.

The main flue has at intervals of 9 feet 4 inches small openings *b* (Figures 171, 172, 173), placed diagonally opposite each other, through which the flue-dust is raked out periodically into the shallow pits *f*. They are closed with hinged iron doors. Every 56 feet (the distance between the centres of two blast-furnaces) there is a manhole *c* (Figures 171 and 174), which is bricked up with a half course of bricks. These are easily removed when the flue is to be entered. For each blast-furnace will be found

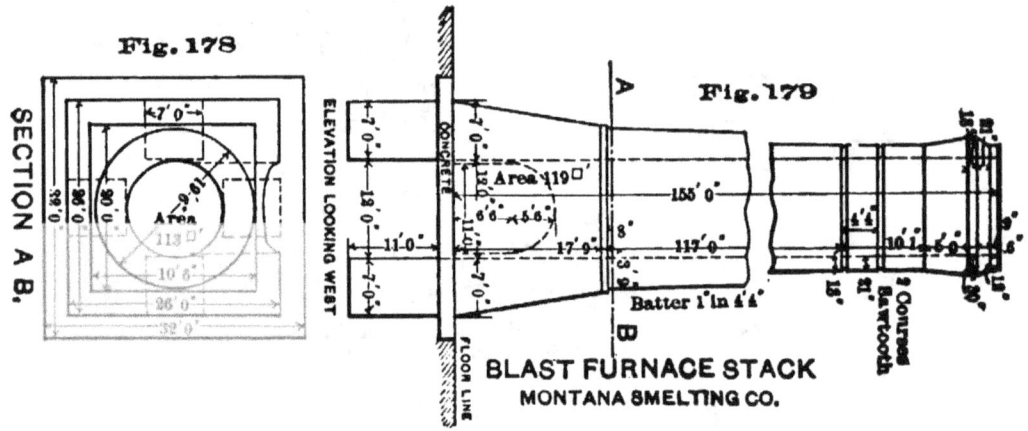

BLAST-FURNACE STACK OF THE MONTANA SMELTING CO.'S WORKS.

in the side wall a small opening *d* (Figures 171, 175–177). This is connected by an underground flue with the stove that heats the cast-iron lead-pot placed next to the lead-well (§ 64). From the flue *a* the gases pass four dust-chambers *e*, having vaulted arches running parallel to that of flue *a*. The area of the passage from the main flue to the first chamber and of the one between the single chambers is somewhat contracted. The gases must thus pass the openings in the partition walls with an increased velocity, which diminishes again as soon as they enter the next chamber. The last chamber ends in the stack, of which Figures 178 and 179 give the details.

Dust-flues and chambers are usually built of brick, and require strong walls and close binding to keep them air-tight. They are therefore expensive, and are often made too small and too short to serve their purpose satisfactorily. A lightly constructed and relatively cheap dust-flue, 6 feet 6¾ inches by 9 feet 10⅛ inches and 1,640 feet long, was erected two years ago at the Victor-Friedrich Smelting Works [1] in the Hartz Mountains, according to the Monier [2] system. It has so far given complete satisfaction. The principle of the Monier system is to reinforce the defective tensile strength of cement concrete by incorporating in it a coarse network of iron wire. It is still in the process of development, but would seem to be well adapted for dust-flues and dust-chambers, as the thinner walls must materially assist the cooling of the gases and thus promote the condensing of vapors and the settling out of dust particles.

The success of the new flues and chambers just erected at the works of the Grant and Omaha Smelting and Refining Company, Denver, Col.,[3] will be of great interest. They are built underground, the walls being made of common hollow brick opening at one end into the air and at the other into a stack, which draws fresh air continually through them and thus keeps them cool.

§ 86. **Treatment of Flue-Dust.**[4]—The question how to treat flue-dust with a minimum of loss is a difficult one. Many suggestions have been made and various methods tried.

To wet down flue-dust and put it back into the furnace is of no use, as when dry it will simply be blown out again.

An improvement on this is to mix the flue-dust with from 8 to 10 per cent. of slacked lime and form it into bricks. These, when hardened, are added to the furnace charge. Cahen [5] gives an example of mixing flue-dust (with 44 per cent. lead) and slacked lime. The bricks formed were smelted alone in the blast-furnace with 74 per cent. of slag, and only 9.60 per cent. of the lead charged was lost. If the bricks are to stand any rough handling, or the weight of the charge in the blast-furnace, they must not be made in a pug-mill, but in a machine and under considerable pressure. They must then be dried slowly with artificial heat. Harbordt,[6] having

---

[1] *Berg- und Hüttenmännische Zeitung*, 1891, pp. 175, 439; 1892, p. 66.

[2] *Engineering and Mining Journal*, June 6, 1891; *Berg- und Hüttenmännische Zeitung*, 1891, p. 175.

[3] *Iron Age*, May 19, 1892.

[4] Iles, *Eng. and Min. Journal*, January 30, February 6, 27, 1886.

[5] "Métallurgie du plomb," Liége, 1863, p. 102.

[6] Private communication, July, 1891.

found bricking with lime unsatisfactory, substituted clay. The bricks, simply air-dried, can stand considerable handling, and do not crumble in the blast-furnace. Many different ways of bricking fine ore and flue-dust have been tried. Hahn[1] found that flue-dust mixed with a solution of ferrous sulphate and formed into bricks (by hand) became so hard when sun-dried that it would stand much handling. Perhaps an addition of slacked lime might prove effective, as it precipitates the ferrous oxide, which, on being exposed to the air, becomes oxidized and forms hard lumps.

The writer observed, while roasting blende-bearing galena to obtain zinc sulphate for leaching, that, after removing as much of the zinc as was possible, the residual ore, when dry, became so hard that it required strong blows with a sledge to break it.

Church[2] made brick of fine ore and flue-dust in Tombstone, Arizona, by using pan-slimes, settled in tanks, as binding material. The bricks were made without pressure, and sun-dried. These slimes contained 85 per cent. quartz and from 2 to 3 per cent. clay, the remainder being calcite, manganese, iron oxides, sulphides, and lead carbonate. Church attributes the binding quality of the slimes entirely to their fineness. It is a well-known fact that material when ground very fine will often show adhesiveness it did not possess before.

Peters[3] used at the Parrott Works, Butte, Montana, a brick-machine that exerted a pressure of four tons to the square inch, and was capable of making forty bricks a minute, each weighing five pounds when dry. The ore to be compressed was roasted pyrite, and no binding material was used. The bricks had to be dried slowly by artificial heat to be suitable for the blast-furnace. He states that with an addition of from 2 to 4 per cent. lime or clay, slow drying could be dispensed with.

It is to be noted that in all bricking experiments the results obtained with fine ore cannot be simply adopted for flue-dust, as this contains carbonaceous matter which counteracts some of the binding property of adhesive substances. The bricking of flue-dust, even if imperfectly done, has always the advantage that the bricks will be carried some distance down with the descending charge before they crumble. The major part will thus be held back

---

[1] "Mineral Resources of the United States," 1882, p. 344.

[2] *Trans. A. I. M. E.*, xv., p. 611; *Engineering and Mining Journal*, August 22, 1885.

[3] "Modern Copper Smelting," 1891, p. 279.

by the charge above, and prevented from passing off again into the dust-chamber. The disadvantage of bricking is that, oxidized lead and carbonaceous matter being in intimate contact, finely divided metallic lead is formed in the upper part of the furnace and causes loss by volatilization.

For this reason, as well as on account of the cost of making bricks, it is more common to melt the flue-dust in a special reverberatory furnace (see *ante*) or in the fuse-box of the roasting furnace, where it is added in quantities of, say, 100 pounds to the roasted ore. It is charged before the ore is drawn from the roasting-hearth, and is thus covered by the roasted ore, so that little metal is carried off mechanically. It cannot, however, be denied that slagging flue-dust causes some loss in lead by volatilization; the loss in silver is inconsiderable.

§ 87. **Analytical Determinations.**—The assay of flue-dust requires no special discussion, as it contains nothing which cannot be found under the heads of speise (§ 74), matte (§ 79), and slag (§ 81).

§ 88. **Losses in Smelting.**—The losses in smelting are due to slagging or to particles of metal being carried off in by-products and not recovered. As has already been stated (§ 53), ore slags should not contain over $\frac{3}{4}$ per cent. lead and $\frac{1}{2}$-ounce silver to the ton with 300-ounce bullion; but they often contain over 1 per cent. lead and about 1 ounce silver, and are considered satisfactory. Special causes, such as the presence of foreign substances having a deleterious effect (§ 54), may make slags run still higher. The quantity of by-products (speise, matte, flue-dust) formed has an important influence on the output of lead and silver, as they have to be roasted and resmelted several times, and each of these operations causes an unavoidable loss in metal. It is difficult, therefore, to give an average figure of the lead and silver recovered in smelting in the blast-furnace. It may be said that a total loss of 7 per cent. of lead represents very good work, the lead assays being made in the dry way. Losses of twice the amount are, however, not uncommon. With silver the output is generally over 95 per cent., the silver being also determined in the dry way. If a smelter can afford to pay (§ 35) 95 per cent. of the assay value of the silver, 5 per cent. or less must be the maximum he endures with the ordinary run of ores.

§ 89. **Cost of Smelting.**—The cost of smelting a self-fluxing ore at the large smelters of Denver and Pueblo is $4.50 per ton, as previously stated (§ 35). It may go as high as $5.00, but not higher. The figure includes all incidental expenses; in fact, is the total cost.

The following estimate is by C. Henrich, who has often been quoted in the preceding pages. It is for a single furnace, 33 by 84 inches, based on Colorado data, and assumes that the furnace puts through 48 tons of ore in 24 hours and requires 16 tons of flux.

MATERIALS :

16 tons flux @ $3.00	$48.00
11$\frac{x}{11}$ tons coke (18 %) @ $12.00	132.77
3 cords wood @ $6.00	18.00
Supplies, etc.	12.50
	$211.27

PAY-ROLL :

Superintendent	$11.70	
Assayer	5.00	
Foreman	5.00	
Weighmaster	2.50	
2 engine-men @ $3.00	6.00	
3 furnace-men @ $3.00	9.00	
4 slag-men @ $2.50	10.00	
3 feeders @ $3.00	9.00	
4 charge-wheelers @ $2.50	10.00	
8 laborers @ $2.00	16.00	
2 inside laborers @ $2.25	4.50	
1 sampler	2.50	
2 bullion-men @ $2.00	4.00	
		95 20
		$306.47
For unforeseen expenses, 5 per cent... ⎱ Delays in repairing furnace, 5 per cent. ⎰		30.65
Cost of smelting 48 tons of ore		$337.12
Cost of smelting one ton of ore		7.03

It will be seen that he assumes 8-hour shifts for furnace-men and feeders, 10-hour shifts for day laborers, and 12-hour shifts for weighmaster and engineers; 10 per cent. is added to the cost for unforeseen expenses and repairs of the furnace. This is as accurate a general statement as can be made.

PART III.

DESILVERIZATION OF BASE BULLION.

CHAPTER IX.

PATTINSON'S PROCESS.

§ 91. **Introductory Remarks.**—The process is based on the fact discovered by Pattinson in 1833, that, if silver-bearing lead is melted and cooled down almost to its fusing-point, crystals of lead will separate which are much poorer in silver than the original lead. If they are removed and the process is repeated, always adding fresh lead of the same tenor in silver, a large quantity of market lead low in silver will result, and a small amount of enriched lead ready to be cupelled. By the repeated meltings and crystallizations many of the impurities will also have been collected in drosses, and the market lead become purified.

The fact that lead low in silver solidifies before the enriched lead still lacks a satisfactory explanation. Pattinson[1] observed that on heating carefully a bar of lead that ran low in silver, until a few drops of metal oozed out, these were richer in silver than the residual lead, while with a bar running high in silver just the reverse was the case. Reich[2] measured the melting-points of different lead-silver alloys, and found that lead with 1.89 ounces silver per ton melts at 321° C.; with 139.03 ounces, at 309°; and that with 656.23 ounces silver the lowest melting-point was reached. If, however, the lead contains much silver, for instance 33 or 50 per cent., its melting-point is much higher than that of lead free from silver. The inferences to be drawn from this are that Pattinson's process is adapted only for low-grade bullion, and that the enriching of the liquid lead can be carried only to a certain degree. The following table, by Reich,[3] shows how far the silver can be concentrated in the lead :

[1] Percy, "Metallurgy of Lead," London, 1870, p. 137.
[2] *Berg- und Hüttenm. Ztg.*, 1862, p. 251.
[3] *Ibid.*

at the Ems Smelting and Refining Works, Prussia, where the three processes were used one after another. To these have been added the amount of lead to be cupelled and the traces of gold.

	Cupellation Process.	Pattinson's Process.	Parkes' Process.
Cost.............................	3	1½	1
Amount of lead to be cupelled...........	100	13	5
Loss in lead and silver..................	6	2	1
Impurities remaining in the lead, per cent.	0.2	0.05	0.015
Silver remaining in the lead, ounces per ton	0.73	0.58	0.17
Traces of gold.........................	lost	lost	recovered

The task of modern desilverizing works (or refining works, as they are also called), is not only to separate effectively and cheaply the precious metals from lead, but also to make out of a base bullion containing from 95 to 98 per cent. of lead, a refined lead of not less than 99.9 per cent. of lead and salable products of the impurities contained in it, such as copper, tin, arsenic, and antimony.

In the following chapters, cupellation, being now only an auxiliary to Pattinson's and Parkes' processes in desilverizing base bullion, although still an independent process in the manufacture of litharge, will be treated last. The three processes will be discussed in the following order :

Pattinson's Process.
Parkes' Process.
Cupellation.

PART III.

DESILVERIZATION OF BASE BULLION.

§ 90. **Introductory.**—The final separation of silver and lead is universally accomplished by the process of cupellation. Keith[1] [2] desilverized base bullion on a working scale by means of electricity, but did not make a pecuniary success of it. Roesing[2] experimented at Tarnowitz, Silesia, in oxidizing lead in a Bessemer converter lined with basic refractory material. He worked with charges weighing 13,200 pounds, and enriched lead from 12.4 to 1863. ounces silver per ton, the fumes collected assaying 75 per cent. lead and 2.5 ounces silver per ton. He also refined, in a few minutes, desilverized zinc-bearing lead. In what way this new idea can be applied to desilverization so as to compete with the present methods remains to be seen. Formerly all argentiferous lead was cupelled, but this was found to have disadvantages, prominent among which are the cost and the loss in metal, the limit being very soon reached, where the separation of silver from lead ceases to pay. This is with base bullion assaying about 30 ounces silver to the ton. Below this the silver recovered will hardly pay for the labor, fuel, and material required, the loss in metal, and the impurity of the lead obtained from the reduction of the litharge. It becomes, therefore, necessary to concentrate the silver in a smaller amount of lead before cupelling. The processes of Pattinson and Parkes do this successfully.

The progress made in desilverizing during the last sixty years is well illustrated by the following table. The figures published by Hermann[4] are derived from the actual working-results obtained

---

[1] *Engineering and Mining Journal*, July 13, 27, 1878 ; June 8, 1882 ; December 15, 1883, p. 372. *Trans. A. I. M. E.*, x., p. 312 ; xiii., p. 310.

[2] Hampe's criticism : *Zeitschrift fur Berg-, Hutten-, u. Salinen-Wesen im Preussen*, xxx., p. 91. *Engineering and Mining Journal*, March 18, 1882.

[3] *Revue universelle des mines*, 1892, xvii., p. 110 ; *Iron*, March 25, 1892 ; *Berg- und Hüttenm. Ztg.*, 1892, p. 102 ; *Engineering and Mining Journal*, April 16, 1892 ; *Stahl und Eisen*, 1892, p. 870.

[4] *Berg- und Hüttenm. Ztg.*, 1883, p. 382.

OUNCES SILVER PER TON.		
In the molten lead before crystallization.	In the crystals.	In the liquid lead. [*]
205.33	113.74–135.91	298.95
213.49	92.75–109.08	313.83
281.34	119.58–198.33	422.91
288.16	113.74–181.99	446.24
420.57	198.91	560.57
609.57	586.53	659.15
615.15	503.99–646.31	655.65
643.40	645.15	660.32

The process has to stop when the liquid lead assays from 600 to 650 ounces silver per ton. In practice the concentration is stopped when the liquid lead assays from 450 to 500 ounces, as the nearer the silver-contents approach the 650 ounces silver per ton, the smaller become the crystals, and the more difficult is it to drain off the liquid lead, especially as this also tends to solidify at the same time.

The process of concentrating the silver in a small amount of lead may be conducted according to two systems, called the method by thirds and the method by eighths. In the first of these systems $\frac{2}{3}$ of the lead contained in the kettle is withdrawn in the form of crystals, while $\frac{1}{3}$ remains behind as liquid lead. The crystals will then be half as rich, and the liquid lead twice as rich, as the original bullion. In the second system the bullion in the kettle is divided into $\frac{7}{8}$ crystals and $\frac{1}{8}$ liquid lead, and the silver-contents of the crystals is $\frac{1}{4}$ as much, and of the liquid lead 3 times as much, as that of the original bullion. The latter method is, therefore, to be applied to low-grade bullion. Stetefeldt [1] tried to find a general mathematical formula which would show the proportions in which leads of different contents in silver should be divided to attain, with as few crystallizations as possible, a market lead of a certain tenor in silver and an enriched lead. In practice the two methods by thirds and by eighths have become standards, especially the former. A variation of the method by thirds, the one with *intermediary crystals,* [*] aims to reduce the number of crystallizations. The contents of the kettle are divided into $\frac{2}{3}$ crystals and $\frac{1}{3}$ liquid lead; the liquid lead, however, undergoes directly a second crystal-

[1] *Berg- und Hüttenm. Ztg.,* 1863, pp. 64, 69, 77.
[2] Stetefeldt, *Berg- und Hüttenm. Ztg.* 1863, pp. 297, 381.

lization, so that intermediary crystals and final liquid lead will result. Thus the original lead is divided into $\frac{4}{8} = \frac{1}{2}$ normal crystals, assaying one-half as much as the original lead ; $\frac{3}{8}$ intermediary crystals, assaying the same as the original lead ; and $\frac{1}{8}$ liquid lead, four times as rich as the original lead. The method has, however, been abandoned where it was tried, as it complicated the process, and as considerable amounts of slightly enriched leads had to be kept on hand.

To carry out Pattinson's process successfully, the base bullion must not be very impure and a sufficient amount must be used. All the foreign metals contained in the lead interfere with the crystallization and the effectual separation of the liquid lead from the crystals. Ordinary lead can be sufficiently purified by poling (§ 109) and removing the dross that collects on the surface ; if tin, arsenic, and antimony are present to any extent the lead has to be softened (§ 97) at a bright-red heat before the crystallization can proceed. Of the metals' commonly found in base bullion, antimony, bismuth, and nickel are concentrated in the liquid lead ; arsenic in the crystals ; copper that has not been removed with the dross remains equally distributed in both products.

The kettles hold from 6 to 15 tons of lead, and the smallest permissible quantity is $2\frac{1}{2}$ tons.

The crystals are either taken out by a perforated ladle or the liquid lead is drawn off.

§ 92. **Description of Plant and Mode of Conducting the Process.**—The plant consists of a set of spherical kettles, from 8 to 15 (with the method by thirds), built closely together in a row. Each kettle has a separate fire-place, so constructed that the flame shall pass beneath and behind the kettle, thence into a flue encircling it, and finally into the chimney, which has a damper to regulate the draught. The details of the construction are the same as with the desilverizing-kettle of Parkes' process (§ 101).

The mode of operation in outline with the method by thirds is as follows. In the central kettle the base bullion is melted down, drossed, and poled, if necessary. The fire below is then withdrawn and transferred to a neighboring kettle. The cooling is promoted by sprinkling water on the surface from a rose. Crusts adhering to the sides of the kettle are pushed down into the lead, where they melt again. This is the work of one man, who also stirs the metal continuously until the smooth surface becomes rough with crystals.

[1] *Berg- und Hüttenm. Ztg.*, 1889, p. 116.

His partner now inserts at the rim of the kettle a long-handled skimmer that has been warmed, and works it across the bottom of the kettle to the opposite side, then back to the middle, where, after jerking it to remove the liquid lead, he discharges the dry crystals into the neighboring kettle, generally the one to the right ("down the house"). The operation is continued until two-thirds of the contents of the kettle has been removed in the form of crystals. The liquid lead is then ladled into the kettle on the left ("up the house"). To the kettle at the right, being $\frac{2}{3}$ full of crystals, $\frac{1}{3}$ of lead of the same tenor is added, and the kettle at the left, being $\frac{1}{3}$ full of liquid lead, is filled with a corresponding amount ($\frac{2}{3}$) of lead of its tenor. The kettles are heated, and the cooling, crystallizing, and ladling carried on in the same way as in the original bullion-kettle. This becomes again filled from the crystals of the kettle on the left and the liquid lead of that on the right. Thus the operations are continued, the lead of the kettles to the right decreasing in tenor till that of the last one, the market-pot, assays from 0.3 to 0.5 ounces silver per ton; that to the left increasing till the maximum of 650 ounces is reached.

From the foregoing it will be seen that before the whole plant can be in working order quite a number of crystallizations have to be carried on, so as to have on hand the necessary amounts of lead of different silver-contents required to fill the kettles.

Pattinson's process in its original form is still in use in England, Freiberg, and perhaps some other places. As it is improbable that it will be introduced anywhere in the United States, this general outline will suffice. Full details are given in the works of Percy,[1] Kerl,[2] Stölzel,[3] Roswag,[4] Grüner,[5] and the paper by Teichmann.[6]

In order to reduce the hard work necessary in withdrawing the crystals and ladling out the lead, as well as to insure a more regular crystallization and better separation of crystals and liquid lead, machinery was introduced into Pattinson's process; but the main modification of the original process, which has become the standard one of to-day, is that by Luce and Rozan, who stir by steam and draw the liquid lead off, leaving the crystals in the kettle.

[1] "Metallurgy of Lead," London, 1870, p. 121.
[2] "Grundriss der Metallhüttenkunde," Leipsic, 1881, p. 225.
[3] "Metallurgie," Brunswick, 1863–1866, p. 1122.
[4] "La désargentation de plomb," Paris, 1884, pp. 211 and 267.
[5] *Annales des mines*, 1868, xiii., p. 379.
[6] *Zeitschrift für Berg-, Hütten-, und Salinen- Wesen in Preussen*, xv., p. 40.

§ 93. Luce and Rozan's[1] Process (Steam-Pattinson Process).

—The advantages of the steam are, that it causes a regular crystallization and a good separation of the lead from the crystals, and that it poles the lead, which being much exposed to the influence of the air becomes purified. Thus moderately pure base bullion, containing from $\frac{1}{2}$ to $\frac{3}{4}$ per cent. of foreign metals, can be desilverized without previous softening. It is claimed that lead with a little antimony and copper is even preferable, as less dross forms than would be the case if it were free from these metals. Of course, lead containing appreciable amounts of arsenic and antimony has to be softened with this process, as with any other, before it can be satisfactorily desilverized.

The way in which the process is carried out at Přibram,[1] Bohemia, may serve as an example.

Figures 180, 181, and 183 show the general arrangement of the

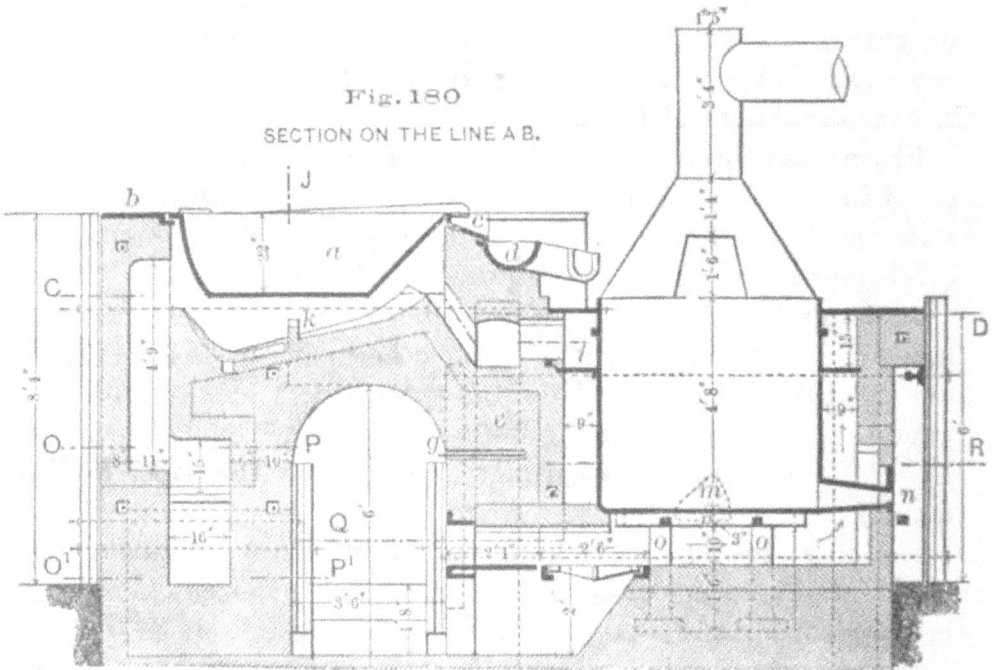

Fig. 180

SECTION ON THE LINE A B.

plant, consisting of two melting-pans *a*, one crystallizing-pot *l*, and two large conical moulds. The steam-crane is not shown; it is placed on the side of the crystallizing-pot, and serves to transfer the cakes of lead from the moulds to the storage-place, and thence to the melting-pans, and to tip the latter. The trough-shaped cast-

---

[1] Luce and Rozan, *Annales des mines*, 1878, iii., p. 160; Cookson, *Iron*, September 22, 1881; *Engineering and Mining Journal*, October 8, 1881.

[2] Zdráhal, *Oesterreichisches Jahrbuch*, xxxiv., p. 1; and Private notes, 1890.

iron melting-pans *a*, each holding 1,540 pounds of lead, are placed behind, and 2 feet 4 inches above, the top of the crystallizer *l*. They rest with their rims on the cast-iron frame *b*, and are emptied by tipping, by means of the crane, over the inclined plate *c*, which discharges the lead through the stationary cast-iron trough *d*, and

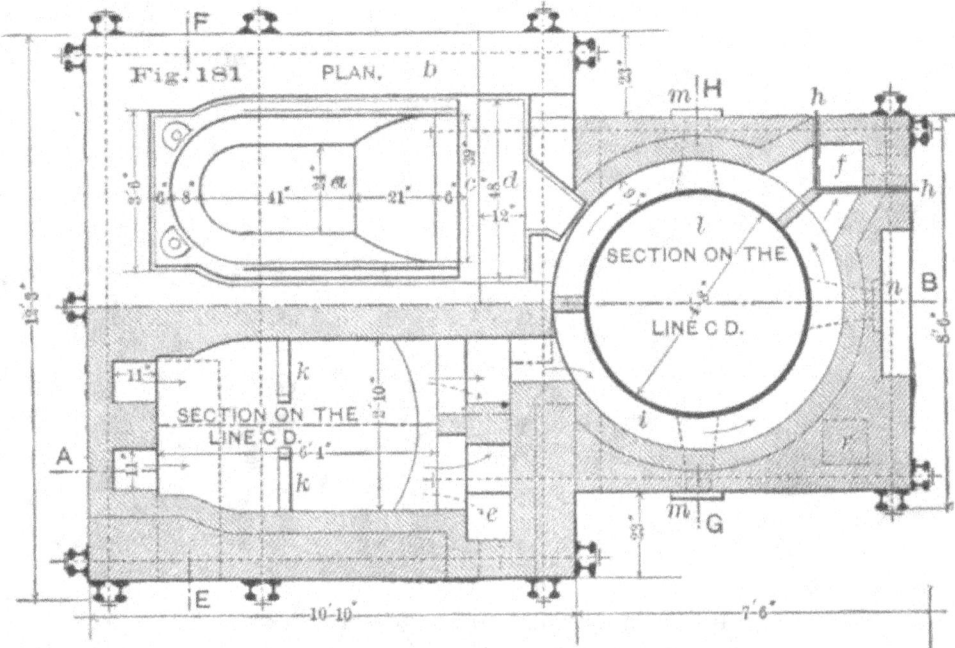

a movable sheet-iron trough (not shown), into the crystallizer *l*. Each pan has its separate fire-place (Figures 180 and 185) on the side, from which the gases, after passing upward (Figure 180) through a long flue, surround the bottom of the pan, and descend either directly through flue *e* (Figures 180, 181) to the chimney, or first encircle the upper part of the crystallizer (*l*, Figures 180,

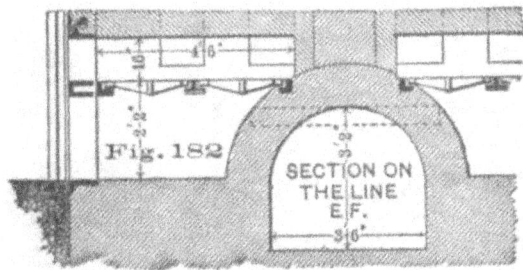

181), and then pass off through the flue *f* (Figure 181); the passage of the gases is regulated by the dampers *g* (Figure 180) and *h* (Figure 181). On the oval hearth (Figure 183) are built two small walls *k* (Figures 180, 181, 183), in order that the flame may pass close to the pan. Any lead coming from a leaking pan collects in the lowest part of the hearth (Figure 180), tamped with brasque, and is discharged outside of the brickwork.

The crystallizer *l* (Figures 180, 181) is a flat-bottomed cylindrical pot holding 44,100 pounds of lead, or nearly three times as much as one melting-pan. It has (Figure 187) near the bottom

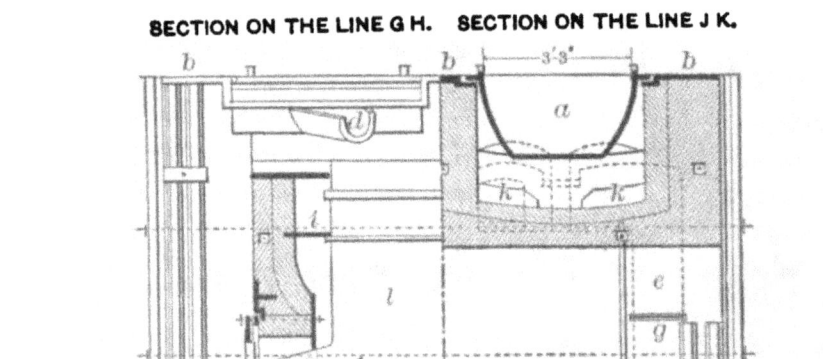

Fig. 183

SECTION ON THE LINE G H.   SECTION ON THE LINE J K.

two spouts, *A* and *B*, closed by slide-valves (Figures 191, 192, 193), for discharging the lead. At a right angle to the plane of these spouts is the steam-inlet *C* (Figures 186, 187). The pot rests (Figures 180, 185) on the cast-iron frame *p*, supported by four cast-iron pillars *o*. The top of the crystallizer (Figure 180) is covered with a conical hood, ending in a sheet-iron pipe, through which steam and dust are carried off to be condensed. The hood has three openings closed by doors—one at the front above the steam-

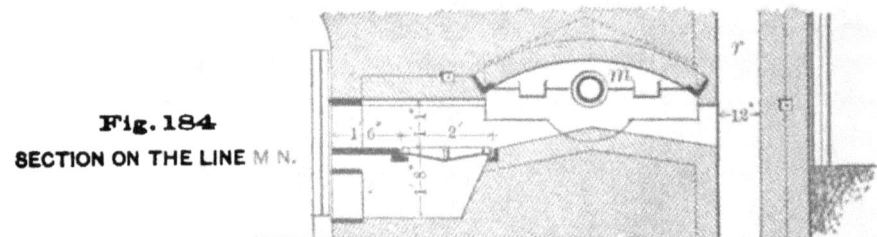

Fig. 184

SECTION ON THE LINE M N.

inlet, and two at the sides above the lead-spouts; and a small hole near the top, for the water inlet-pipe. The crystallizer is fired (Figures 180, 185) from the passage below the melting-pan *a ;* the flames pass along the bottom, turn to the left, and encircle the lower part of the pot ; they are checked by being forced to make their way through the narrow passage *q* before passing downward and off through the flue *r*. On either side of the large central fire-place is (Figures 184, 185) a smaller one, which serves to heat the discharge-spouts *m* before using them. Each of the discharge-spouts *A* and *B* (Figure 187) has a perforated cast-iron straining-plate to keep back the crystals when the liquid lead is being run off. These are held in place by wrought-iron arms *b* and the cast-iron frame *c* (Figure 187, 194), which is fastened by key-bolts to the baffle-plate *d*. The spouts are closed by a slide-valve (Figures 191, 192, 193). To the flange of the spout (Figure 189) is fastened, with countersunk screws and a red-lead cement, a plate

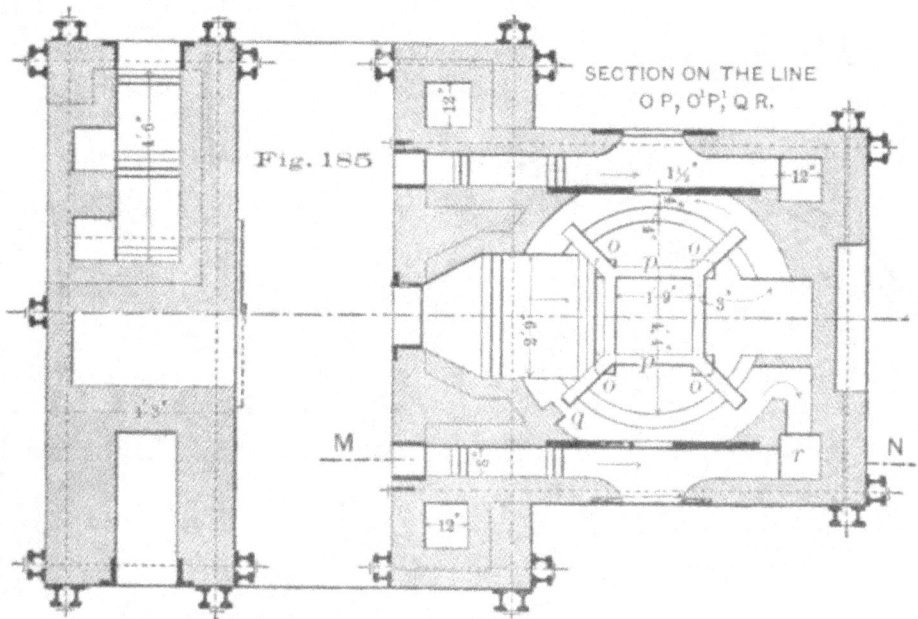

(Figure 190) of the same form, with one face planed smooth, having four openings to correspond to those of the flange; *i. e.*, the central lead-discharge and the three holes near the rounded corners. Through these the bolts *m'* and *m* are passed, *m'* serving as a pivot for the lever *o* and *m* to tighten the guide *n*. To the lever is fastened the plate *p*, also having one planed face. In Figure 191 is shown the position of the lever when the discharge is closed. In order to open it the nuts of the screw-bolts *m* are loosened, and the lever pushed into the second position, shown by the dotted lines. The lead from the crystallizer passes through the two lead-spouts into two tapering moulds (Figures 183, 195), each of which holds about 6,600 pounds of lead. The steam-inlet consists of the following parts (Figure 186). On the flat bottom of the crystallizer are four bosses *f* into which fit the screws *g*. On the collar of these is placed and keyed the cast-iron circular baffle-plate *d*, with its small opening *c* in the centre.   It servès to distribute the steam

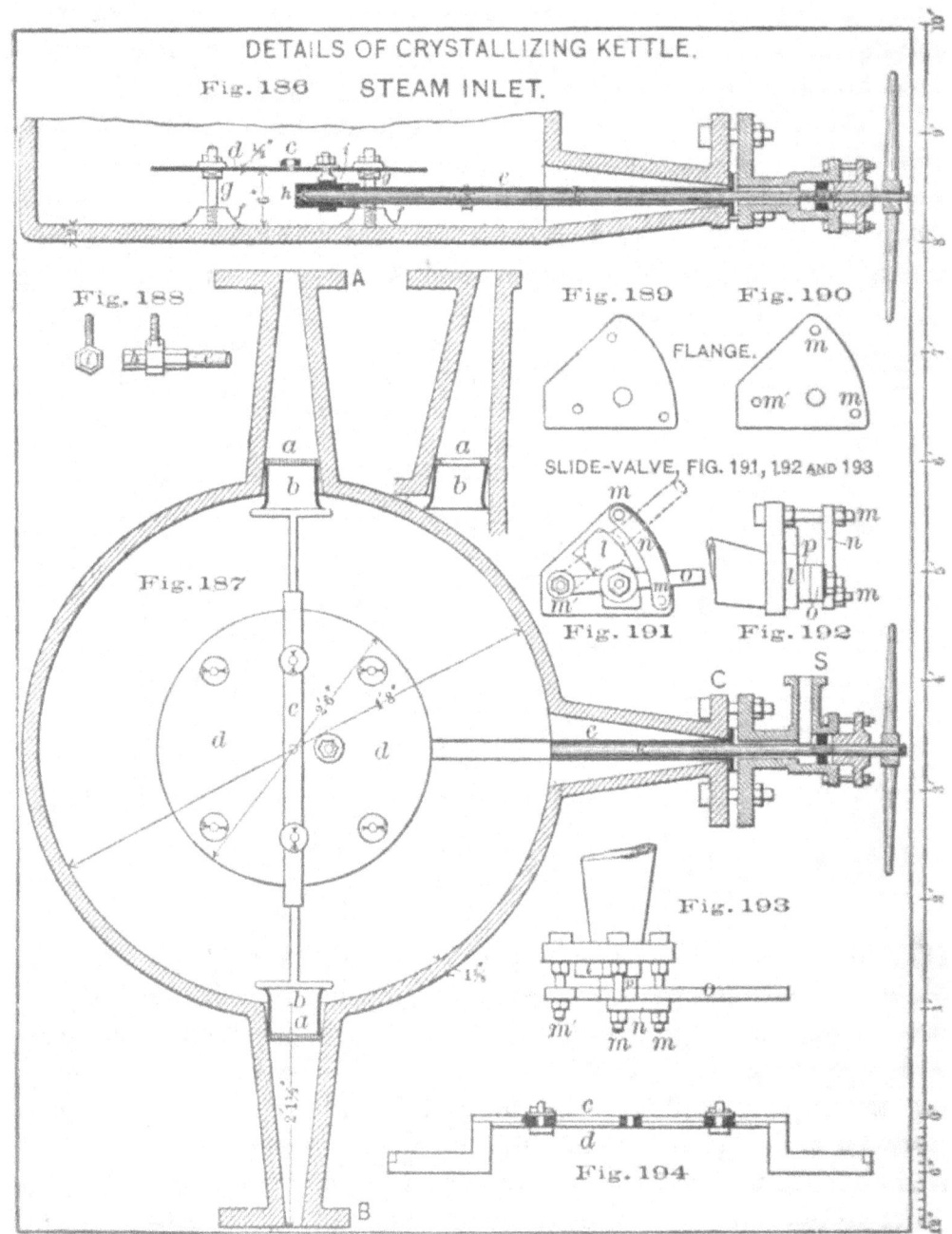

DETAILS OF CRYSTALLIZING KETTLE.

Fig. 186   STEAM INLET.

Fig. 188   Fig. 189   Fig. 190

FLANGE.

SLIDE-VALVE, FIG. 191, 192 AND 193

Fig. 191   Fig. 192

Fig. 187

Fig. 193

Fig. 194

evenly, and to make it rise regularly in the pot. From it is suspended by an eye-bolt *i* with hexagonal eye, the nozzle *h*, into which is screwed the steam-pipe *e*. Through it passes the rod *k*, moved to and fro at one end by the thread and cross-bar; the other end, which is conical, fits into the conical valve-seat of the nozzle *h*, and closes or opens the steam-outlet. The steam entering at *S* (Figure 187) passes through the small annular space between pipe and rod, and out at *h*, when the valve is open.

The mode of conducting the process is simple. Suppose the process to be going on and at the stage when the liquid lead has been drained off from the crystallizer; the valves have been again closed and the crystals liquefied. One pan will be full of liquid lead of the same tenor in silver as the melted crystals to be discharged into the crystallizing pot, while the other will contain two cakes of lead that are being melted down. They will have the same silver-contents as the crystals remaining in the crystallizer after the operation to be described has taken place. The melting down of two cakes takes about six hours.

The lead from the pan is run out by inserting two hooks, fas-

Fig. 195.

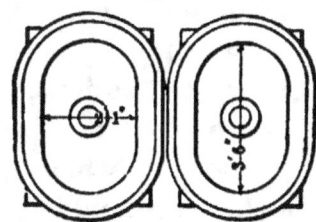

tened to the chain suspended from the pulley of the crane, into the ears of the pan, and raising it slowly. After the lead has been discharged into the crystallizer the doors of the hood are closed, steam is introduced for two minutes, and shut off again to remove the pulverulent dross which has collected on the surface. Now the crystallization proper begins. The fire below the crystallizer is withdrawn and divided between the two small fire-places on either side, from which the lead-spouts are warmed. Steam is turned on, and a small jet of water is allowed to play at short intervals upon the surface of the lead. Every time the water·is let on there are small explosions, and as soon as they become too violent the water is shut off again, while the steam enters continuously. The steam has forty-five pounds pressure to the square inch, and care must be taken to have it dry. About fifteen minutes after introducing the steam, the lead that has been splashed up on the upper edge of the pot, or on the hood, has to be removed. The steam is shut off, the doors in the hood are opened one after the other, and the solidified lead-crusts broken off with a chisel-pointed bar, and pushed back into the lead. This shutting off of steam to remove the lead is repeated at least twice. While the crystallization is going on, the two cakes of lead required to fill again the melting-pan, just emptied into the crystallizer, are hoisted from below with the crane, and deposited one on top of the other in the pan. The crystallization is finished when the normal amount of steam can no longer overcome the resistance offered by the crystals. The result is that the boiling ceases, and the surface of the crystalline mass of lead shows only a slow, wave-like motion. Two-thirds of the original lead have now been converted into crystals. Water and steam are shut off, the slide-valves are opened, and the liquid lead is discharged into the moulds, which takes from eight to ten minutes. In these have previously been placed iron hooks, by which the cakes of lead, when cold, may be lifted out. The whole process of crystallization lasts about one hour. While the lead is running off, the fire from the two small fire-places is returned to the grate below the crystallizer, and urged in preparation for the next operation. The melting and other work require three hours, so that one operation lasts four hours.

In starting a series of crystallizations, 41,876 pounds of base bullion are taken into operation, but the dross formed soon reduces this amount to 39,672 pounds, of which $\frac{1}{3}$, or 13,224 pounds, is obtained as liquid lead in the two moulds. Eleven crystallizations are necessary to obtain market-lead from liquated base bullion averaging 146.12 ounces silver to the ton. The following table shows the average assay-value in ounces per ton of the different leads produced during a whole year's work. The second column represents the results of the same process at Eureka,[1] Nev.

PŘIBRAM.	EUREKA.
Market Lead, 0.43	1.25, Market Lead.
0.87	2.5
1.75	5.0
3.21	9.0
6.41	18.0
10.21	30.0
18.96	50.0
29.16	75.0
40.83	100.0
55.41	150.0
93.33	460.0, Rich Lead.
142.91	
Rich Lead, 262.49	

[1] Curtis, "Silver-Lead Deposits of Eureka, Nev.," monograph vii., United States Geological Survey, 1884, p. 163.

Six charges are run in twenty-four hours ; two men working as partners attend to the crystallization, all the handling of the lead being done by the engineer and his helper.

The products of the process are rich lead, desilverized lead, dross, and flue dust. The rich lead is cupelled, the desilverized lead is refined in a reverberatory furnace and moulded into market lead, the dross and flue dust are worked with similar products from other parts of the works. The output of metal is shown by the following table :

Recovered in	From 100 Pounds Base Bullion Charged.	From 100 Ounces Silver Charged.	From 100 Pounds Lead Charged.
Rich Lead,	42.99	97.36	42.58
Desilverized Lead,	44.76	0.17	45.01
Scrap Lead,	3.02	0.61	3.03
Dross,	9.94	1.45	8.04
Flue dust,	0.46	0.07	0.35
Loss,	....	0.34	9.99
Total,	101.17	100.00	100.00

The material consumed for desilverizing 100 tons of base bullion excluding the refining of the lead, is :

Charcoal,....................25 bushels.
Bituminous Coal,..........26.60 tons (for melting and desilverizing).
    "   "   .... 8.23 " (for raising steam).

In comparing the processes of Luce-Rozan and of Pattinson, Cookson [1] comes to the conclusion that the former is to be preferred by far, as the softening of the lead is not so imperative, and the cost of labor only 20 per cent., and that of fuel 40 per cent., of the cost by Pattinson's process ; and, finally, as it produces only 33 per cent. the amount of drosses obtained by Pattinson. The disadvantages, greater outlay in capital, and expense of repair and renewal, are more than made up by the advantages.

[1] *Engineering and Mining Journal,* April 12, 1879.

CHAPTER X.

PARKES' PROCESS.

§ 94. **Introductory Remarks.**—Parkes' process is based on the fact that if from 1 to 2 per cent. zinc is stirred into melted base bullion, it will deprive the latter of its silver, and form an alloy, which, being less fusible than lead, and having a lower specific gravity, will become hard and float on the surface of the lead, whence it can be removed and treated separately; while the lead which has taken up some zinc is refined, and is then ready for the market. Karsten discovered in 1842 that argentiferous lead could be desilverized by the use of zinc, but his discovery could be applied in practice only when Parkes found the means (1850–1852) of working the zinc-silver-lead crust and refining the lead.

The theory on which the process is based has always been that silver has a greater affinity for zinc than for lead, and therefore combines with it when added to molten bullion. According to Roesing[1] the statement has to be modified by saying that, while silver has a greater affinity for zinc than for lead, it has less affinity for zinc-bearing lead than either for zinc or for pure lead, and that this is the main cause why argentiferous lead can be desilverized by means of zinc. According to Alder-Wright and Thompson,[2] zinc and silver form two definite alloys, $AgZn_5$ and Ag_4Zn_5.

The former is an unstable compound. It has the property of dissolving lead, and is itself dissolved by lead to a greater extent than either pure zinc or the alloy Ag_4Zn_5. If it be kept molten for some time, holding its maximum of lead in solution, it breaks up into Zn and Ag_4Zn_5, and releases some of the lead, which sinks to the bottom. Under the same conditions, if held in solution by lead, the homogeneous alloy will be divided into the above components, the lighter zinc rising to the surface.

The alloy Ag_4Zn_5 also dissolves lead, but to a smaller extent than a mixture of it with either $AgZn_5$ or free Ag would. It is, further, less soluble in lead than would be expected from the

---

[1] *Zeitschrift für Berg-, Hütten-, und Salinen-Wesen in Preussen,* xxxvii., p. 76.

[2] *Engineering and Mining Journal,* December 20, 1890.

amount of zinc it contains. When exposed to the air it assumes a coppery hue.

The low degree of solubility of Ag_4Zn_5 in lead which must retain some zinc explains perhaps more definitely what Roesing called the small affinity of silver for zinc-bearing lead.

Before zinc added to the base bullion can take up any quantity of silver, it combines with the gold and copper contained in the lead and saturates this, the amount taken up depending on the temperature of the lead (§ 11,²). By the use of zinc a market lead very low in copper is thus obtained, and by successive additions of zinc very small amounts of gold can be concentrated in a separate crust (the gold or copper crust) with some silver, and extracted at a profit. There is a difference of opinion[1] as to whether gold or copper combines first with the zinc. It would seem that it must be gold, as no desilverized lead is free from copper, but it never retains the least trace of gold.

In order to desilverize argentiferous lead with zinc successfully, it is necessary that the lead and zinc be reasonably pure. Tests made by Kirchhoff[2] on base bullion containing 4.5 per cent. foreign metals, such as copper, arsenic, antimony, bismuth, and zinc, showed that 2.87 per cent. zinc was required to desilverize the lead when the bullion had not been softened, while 1.75 per cent. was sufficient if softening had preceded the desilverization; the relative quantities of market lead produced were 43 and 72 per cent. of the bullion charged. The following table shows how the silver-contents decreased with each addition of zinc:

Number of additions.	Not Softened.		Softened.	
	Ounces silver per ton.	Pounds of zinc.	Ounces silver per ton.	Pounds of zinc.
After drossing...	85.60	...	94.90	...
1.	85.50	250	85.60	150
2.	85.30	250	47.60	150
3.	83.80	150	16.10	150
4.	83.50	100	1.70	150
5.	83.00	100	0.18	100
6.	48.20	100	.....	...
7.	8.20	100	.....	...
8.	0.80	70	.....	...
9.	0.15	30	.....	...

[1] Percy, "Metallurgy of Lead," p. 174.

[2] *Metallurgical Review*, vol. i., p. 224; or Dingler, *Polytechnisches Journal*, vol. ccxxviii., p. 265.

With the crude lead the first five zinc additions served only to remove the impurities to such a degree that the desilverization could begin. That the first addition of zinc to the softened lead took up so little silver shows that the lead must have been very coppery.

Of the three metals, copper, arsenic, and antimony, that principally interfere with desilverization, antimony is the least objectionable, as lead with as much as 0.7 per cent. of antimony, assaying 41 ounces silver to the ton, is desilverized without being softened, 1.3 per cent. of zinc being required, and 81.34 per cent. of market lead being produced.

Arsenic not only retards the desilverization greatly, but seems to prevent the zinc crust from separating satisfactorily from the lead. In skimming a kettle the usual beautiful, smooth, dark-blue surface is not seen, but it shows instead a rough, grayish-white surface ; and even if skimming be continued until the lead solidifies, the surface will hardly change in appearance.

That copper combines with zinc before silver has already been mentioned.

Of the other two metals contained in Kirchhoff's bullion, zinc has a favorable effect; bismuth is indifferent, as it remains for the most part alloyed with the lead.

The zinc used must be pure, if the desilverization is to proceed satisfactorily. In experimenting with cheap zinc containing iron, obtained from galvanizing works, the writer found that the process was so retarded, and the quantity of impure zinc required so great, that no saving at all was effected by the use of the inferior material. Jernegan [1] records a similar experience, and Föhr [2] mentions that he required four times the usual amount of zinc, which was found to be due to its impurity. It contained 2.75 per cent. lead, 0.61 iron, 0.077 copper, and traces of tin, arsenic, antimony, cadmium, sulphur, and carbon ; in good brands the iron appears only in the second decimal.

From the effect that foreign metals have on the result of the desilverization, it is clear that all argentiferous lead that contains them to any extent has to be softened. All or almost all American zinc-desilverizing works buy base bullion and ores in the open market to a greater or less extent. The bullion is therefore always liable to contain some arsenic or antimony, and every refinery softens the lead before it attempts to desilverize.

[1] *Trans. A. I. M. E.*, ii., p. 288.   [2] *Berg- und Hüttenm. Ztg.*, 1888, p. 38.

§ 95. **Outline of Plant and Process.**—The general plan of a ·desilverizing plant varies somewhat according to its location and the practice that prevails. All the arrangements, however, must be such as to require as little handling of lead and by-products as possible. In fact, the bulk of the lead, when charged into the softening-furnace, will in a modern plant never be handled again until it is ready for shipment. The result is that a section through a refinery will show the form of a terrace.

The general arrangement of a refining plant is given in plan and section in Figures 196 and 197. On the highest level are the softening-furnaces, which receive the base bullion and prepare it for the kettles. The latter are a stage lower, and there the softened bullion is desilverized. The apparatus for liquating zinc-crusts is also upon the same floor. In the drawing, each desilverizing-kettle has close to it only one liquating-kettle, with a small kettle for liquated lead, showing that no distinction is made between gold and silver crusts. The liquated crusts pass from this floor into an adjoining building, placed to the right or the left of the main building, the floor of which is on a level with the scale-floor. It contains two departments—the retort-room, where the crusts are distilled, and the cupelling-room, where the retort-bullion is turned into silver, or doré silver bars. In the plant shown by the figure only doré bars will result. Following the desilverized lead, the refining-furnaces are reached on the next level, in which the desilverized lead is dezincified. Thence it passes into the merchant-kettles, and from these into the moulds placed in the lead-pit. The market lead is loaded on trucks on the scale-floor, that are of the same construction as the bullion-receiving trucks; they are run on scales, and the weighed lead is transferred into the cars on the loading-track. The plant for working the by-products will be placed on the side of the main building, opposite to where the zinc-crusts are treated. The manner of dealing with these varies greatly in different refineries, and will be discussed later on.

Roesing [1] draws a comparison between the arrangement of a refinery on a horizontal plane and that on an inclined plane, which, on the whole, is not favorable to the latter.

According to the outline given, Parkes' process is best treated under the following heads :

[1] *Zeitschrift für Berg-, Hütten-, und Salinen-Wesen in Preussen*, xxxvi., p. 108; *Berg- und Hüttenm. Ztg.*, 1888, p. 337.

Receiving Base Bullion.
Softening Base Bullion.
Desilverizing Softened Bullion.
Refining Desilverized Lead.
Moulding Refined Lead.
Treatment of Zinc-Crusts.
Treatment of By-Products.
Table of Desilverization.
Conclusion.
Comparison between Pattinson's and Parkes' Processes.

§ 96. Receiving Base Bullion.

Receiving Base Bullion.—The base bullion arrives at the refinery in car-loads of from ten to fourteen tons. The track (receiving-track of Figure 196) is laid so low that the bottom of the car is on a level with the "upper platform" of the works. Along the whole length of this platform, and parallel with the railroad, runs a narrow-gauge track, from sixteen to twenty-two inches, which bears a number of strongly-built low bullion-trucks. They may be built as follows. A frame consisting of two pieces of channel-bar iron, three feet long, is fastened to the two axles of the wheels, and steadied by two iron bands running diagonally. The bullion is carried out from the car and loaded upon the truck standing before the car-door; when this is filled to a height convenient for lifting (about 3½ feet), it is moved on, and another takes its place. The trucks are run on scales placed at one or two points in the "bullion and scale-shed," the bullion is weighed, and is then sampled from one truck directly upon another, which then moves straight to the softening-furnaces, or to places near them, and no more handling is required before the bullion is charged into the furnaces. The bullion produced in the smelting department of the refinery is loaded at the blast-furnaces or reverberatory furnaces, on the same kind of truck, brought by an elevator to the "bullion and scale-shed," and then passes to the softening-furnaces.

§ 97. Softening Base Bullion.

Introductory Remarks.—The object of softening is to separate from the base bullion produced in the blast-furnace, impurities, such as copper, sulphur, tin, arsenic, antimony, etc., that would interfere with the desilverization. It comprises two processes, liquation and oxidation. By the former, metals and their compounds held in solution by the red-hot blast-furnace-lead are separated out again from the readily fusible lead by melting it down slowly at a low temperature. By the latter, metals alloyed with the lead, and more easily oxidized than the lead, are removed by heating it to a bright-red heat with access of air, with the result that these metals are converted into oxides, and, combining with the lead oxide, are drawn off.either as a powder or a slag from the surface of the metallic lead.

When base bullion is melted down slowly in a softening-furnace at a low heat, there rises to the surface a dark-colored, half-melted, pasty substance, the furnace-dross, consisting of a mixture of lead, copper, sulphur, arsenic, etc. The slower the melting down, the more effectual will be.the separation of the copper from the lead. The following analyses show the purification effected in the lead by liquating and the composition of the dross when it has been freed as much as possible from adhering lead :

	CLAUSTHAL.[1]		LAUTENTHAL.[2]		FREIBERG.[3]	
	Before Drossing.	After Drossing.	Before Drossing.	After Drossing.	Before Drossing.	Liquated Dross (5%).
Pb	98.92944	99.0239	98.96475	99.1883	96.667	62.40
Cu	0.1862	0.1096	0.2838	0.0907	0.940	17.97
Cd	trace	none	trace	none	....	....
Bi	0.0048	0.0050	0.0082	0.0088	0.066	none
Ag	0.1412	0.1420	0.1418	0.1440	0.544	0.17
As	0.0064	0.0053	0.0074	0.0082	0.449	2.32
Sb	0.7203	0.7066	0.5743	0.5554	0.820	0.98
Sn	none	none	none	none	0.210	0.04
Fe	0.0064	0.0042	0.0069	0.0048	0.027	0.48
Zn	0.0028	0.0017	0.0024	0.0015	0.022	0.07
Ni	0.0028	0.0017	0.0068	0.0088	} 0.055	1.09
Co	0.00016	trace	0.00035	trace		
S	....	....	....	....	0.200	4.00
O	....	....	....	....	....	1.87
Slag, ash, hearth-material						8.66

The analyses from Clausthal and Lautenthal demonstrate that the character of a comparatively pure lead is improved by melting down slowly and drossing. By comparing the two Freiberg analyses the degree will be seen to which the foreign matter of a very impure base bullion may be removed ; viz., nearly all the

[1] Hampe, *Zeitschrift für Berg-, Hütten-, und Salinen-Wesen in Preussen,* xviii., p. 208.
[2] *Ibid.*   [3] Schertel, *Berg- und Hüttenm. Zig.,* 1882, p. 293.

sulphur, 93 per cent. copper, 96 per cent. nickel and cobalt, 25 per cent. arsenic, and only 5.8 per cent. antimony, and 1.54 per cent. silver. Bismuth remained entirely in the liquated lead, and all the tin excepting 0.9 per cent.

On melting down this dross in a crucible, Schertel[1] obtained a product consisting of lead, speise, and matte, in well separated layers.

	Lead.	Speise.	Matte.
Ag...........................	0.34	....	....
Cu...........................	1.79	37.60	47.70
Pb...........................	96.50	25.68	32.80
Ni...........................	0.08	8.60	0.25
As...........................	0.75	27.00	1.15
S..........................	....	....	17.72

The absence of iron in either speise or matte proves that these impurities do not result from finely-divided blast-furnace speise or matte being dissolved by the lead, as has been often thought. It tends to show that, being held in solution by the metal, the impurities unite on liquating to form a compound that is not fusible at the temperature at which the lead was melted down, and that the concentration of copper in dross is due probably to the presence of sulphur and arsenic and not to the separation of an alloy of lead and copper.

If a sample be taken of the lead after drossing, poured into a small mould, and allowed to cool slowly, a crystalline, bright, pewter-white spot will appear on the slightly depressed, dull, grayish-white surface, which, in addition to the hardness of the lead, is characteristic for the presence of arsenic and antimony.

If the temperature be raised to a good red heat and the air permitted free access, the impurities contained in the lead will oxidize one after another : first tin, then arsenic and antimony. The surface at first will become quickly covered with dark-yellow skimmings, which vary from powdery to pasty, but are not fused on account of the tin ; these are called tin-skimmings. They consist mainly of antimoniate and stannate of lead and antimoniate of tin, and are worked by themselves (§ 120). As soon as the tin-skimmings have been drawn off from the surface of the lead, this begins to give off fumes of arsenic and antimony, and arseniate and antimoniate of lead begin to form ; the former is a light brown, the latter dark brown to black ; both are fused and are drawn off together as anti-

[1] *Loc. cit.*

mony-skimmings when the furnace has been sufficiently cooled down for them to solidify. Towards the end of the operation the antimony in the skimmings will be replaced by lead until the black color has changed to the greenish-yellow of litharge.

Samples are then taken to see how far the softening has progressed. Before the antimony has been removed, a sample of the bullion taken in a ladle will "work," *i.e.*, small particles of melted black skimmings will float on the surface of the lead, with a rotary motion which resembles that of particles of grease on hot water. As the softening approaches the finishing-point, the globules become less in number and smaller in size, a thin coating of yellow litharge forms more readily on the red-hot lead, and finally no more globules are seen and litharge forms quickly.

When a sample of the lead is poured into a mould, allowed to cool slowly, skimmed with a flat piece of wood, it will, when it has solidified, have lost the characteristics of arsenic and antimony, and the surface of the bar will have assumed a rich indigo-blue color ; the ease with which it can be scratched with the finger-nail, and the lustre on a freshly-made incision, show the change that has taken place. It can also be tested by cupelling a small sample ; if any incrustation is left on the cupel, the lead is not sufficiently softened.

§ 98. **Furnaces.**—The reverberatory is the furnace universally used in this country to soften base bullion. In some European works the liquation is carried on separately from the oxidizing smelting. The furnace is a real liquating-furnace with an inclined hearth, from which the lead runs off into an outside kettle, whence it is ladled out ready for the softening-furnace proper. The dross obtained is as free from lead as liquation can make it (see Freiberg analysis, *ante*). It is, however, more economical to make the entire softening process a continuous one, as it reduces the apparatus and the number of men necessary. Therefore in American refineries, both liquating and oxidizing smelting are carried on in the same kind of reverberatory furnace, and the drosses which contain considerable lead are all liquated in a single liquation-furnace, which is always kept running, and serves for the whole plant.

In some works the bullion is melted down in the desilverizing-kettle, drossed, heated to bright redness, and oxidized by introducing dry steam, which continually renews the surface of the lead, and thus hastens the elimination of arsenic and antimony.

This is the most expensive way. It consumes much fuel, is

lengthy, forms a very large amount of oxidized product, and ruins the kettle. At the level, where the antimoniate of lead is in contact with the iron, it is eaten out. Even a kettle made especially thick at this place lasts only a short time.

The reverberatory furnaces used for softening are generally large enough to hold from 8 to 10 per cent. more bullion than the kettle into which they discharge their contents. About ten years ago furnaces were constructed to hold from 15 to 20 tons of base bullion; their size has been gradually increased until most furnaces now hold from 33 to 35 tons of base bullion, furnishing 30 tons of softened bullion to the kettle. At some works, furnaces holding 50 tons are in successful operation, but they form the exception.

The construction of the different furnaces is the same in many points. The hearth is elliptical or rectangular in plan; the length being to the width as $1\frac{1}{2}$: 1, or often as 2 : 1. It is built of fire-brick, enclosed by an iron pan to prevent the leakage of lead. In section it is dish-shaped; it is shallow, its depth varying from 11 to 16 inches; in exceptional cases it reaches 22 inches. Its slope depends upon whether the tap-hole is located on the side or the flue end of the furnace, being from 2 to 5 inches.

As regards detail, there is considerable variety in construction. The pan which is to hold the hearth used to be made of cast iron; it rested on transverse rails supported by brick walls running longitudinally. Thus the bottom of the hearth was effectually cooled by the air circulating beneath the pan. In order to relieve it from strains, the pan was allowed to stand free, the skewbacks supporting the roof being separate heavy castings, held in place by buckstays and tie-rods. Notwithstanding all these precautions, a cast-iron pan generally cracked after it had been in use for a little while. To-day there may still be found a few old furnaces with cast-iron pans; a new furnace, however, will always be built into a wrought-iron pan. The softening of lead that contains a few per cent. of antimony often takes considerable time and requires a pretty high temperature. Thus the brick suffers greatly by the corroding action of the antimoniate of lead and the litharge forming on the surface of the lead. The best fire-brick soon begins to be eaten out if much hard lead comes to the furnace, and patching with a mixture of raw and burnt fire-clay, or raw clay and coke, after every few charges, has to be resorted to in order to preserve the side walls. As the furnace has to be somewhat cool to make this repairing effective, much time is wasted. An improvement was made by introducing a 2-inch pipe between the two courses of fire-brick forming the side-lining of the furnace, and allowing water to circulate through it. The inside course was eaten away by the litharge to a thickness of from 2 to 3 inches as quickly as before, but then the corrosion proceeded only very slowly, and the life of the side-walls was thus greatly prolonged. With this encouragement the water-cooling has been carried to the extreme of enclosing the wrought-iron pan holding the hearth in another one, leaving 3 or 4 inches space between the two, in which water circulates, thus cooling not only the sides but also the bottom of the furnace. The larger number of softening-furnaces in use to-day have two wrought-iron pans with water circulating between them. The outer pan is supported in the same manner as the former cast-iron pan; the inner pan rests also on rails which are laid in the same direction as the walls; stay-bolts connect the outer and inner pans. While without doubt this mode of cooling is very effective, there is unquestionably too much of it, considering the amount of fuel that is required to keep up the temperature necessary to soften the lead within the given time. Air-cooling alone has always been sufficient for the bottom; water-cooling is necessary for the sides only.

At several works the latest softening-furnaces have water jackets only at the sides, and one of these, given in Figures 198–204, is chosen to illustrate some of the details. The side elevation (Figure 198) shows the fire-place a and the hearth inclosed by the water jacket b and the pan c. Between fire-box and hearth is left an air space d (Figures 198, 199, 202). To counteract the bulging out of the pan and bending of the jacket, due to the expansion of the hearth, 3-inch I-beams are placed horizontally behind the jacket, and 7-inch I-beams behind the pan. They are not shown in the drawing. The discharge of the lead takes place at the flue end of the furnace through the spout e. The products of combustion pass off through the roof at f (Figure 202) into the horizontal flue g (Figure 200), which crosses the furnace, and then, passing down vertically close to it, leads into the main canal underground. The bullion is charged through two doors h on one side of the furnace (Figure 199); the drossing and skimming takes place on the opposite side, through three doors i (Figures 198, 199). The charging doors are on a level with the upper edge of the water jacket; the drossing and skimming doors are let in 3 inches. The pan of the hearth rests on 7-inch I-beams *j*, placed transversely, and these rest on two brick walls *k* running longitudinally. The hearth (Figure 201) is inclosed up to the level of the charging doors, *i.e.*, to the depth of 42 inches, by an iron pan made of ⅜-inch boiler iron. To the upper part of the pan is attached the water jacket, consisting of a boiler plate ⅜ of an inch thick and 21 inches wide, and two pieces of bar iron *m* 3 inches square, riveted to the pan. Stay-bolts, 9 inches apart, connect the two plates. The jacket has a 1¼-inch water inlet and outlet pipe. To insure the complete filling of the jacket, small pieces of pipe pass through the upper 3-inch bar. The jacket has further hand-holes to clean out at intervals the mud that settles out from the cooling-water. These details are not shown in the figures.

Another side jacket that is found at some furnaces has a somewhat different construction. It is open at the top and covers the entire side of the pan. Near the bottom it is slightly bent, and one row of rivets joins the bottom and side of the pan with the plate forming the jacket. Both kinds of jackets are in satisfactory use.

The manner followed in putting in the hearth (Figure 201) varies somewhat at different works. Usually a layer of brasque is first carefully tamped in, and then so cut out that the course of brick, laid endwise upon it, shall bring it into the desired shape and give it the necessary inclination toward the tap-hole. Assuming this to be at the flue end of the furnace, as is most common, the thickness of the brasque there will rarely exceed 2 inches, increasing to 5 inches at the centre of the fire-bridge. In the figure very little brasque *n* is observed in the centre ; it increases to a thickness of 7 inches at the sides. It is made so thin because there are two courses of brick instead of one, as is usual. The lower course *o* is an inferior grade of fire-brick, set dry, and grouted with a mixture of clay and cement, so as to stop up the joints and prevent any percolation of lead. The upper layer *p* is made of the best fire-brick available. In putting down the bottom the bricks have to be joined as tightly as possible. For this purpose they must first be carefully selected and fitted by rubbing them together until all roughness is removed. Each brick is dipped into water and then into a clay-mortar having the consistency of very thin gruel, then put in place and driven with the hammer against the brick it is to face. This makes the joint as close as possible and prevents the passage of lead. The sides of the furnace are built with the same care as the bottom. Commonly they rest on the curved working-bottom to prevent this from rising. In Figure 201 the sides *q* rest on the lower course, while the working-bottom *p* abuts upon them. Contrary to expectation, the bottom shows no tendency to rise. The reason for putting in the sides in this way is their convenience for repairing when eaten out. The corrosion and the lead do not, with a good bottom, reach the lower course of brick. When the side-wall has been torn out till sound bricks are reached, it can be built up again on a solid, smooth, horizontal layer, instead of having to be placed upon the protruding headers of the working-bottom. The roof is supported on either side by two rails *s* serving as skewbacks. Usually it has a slope from the bridge to the flue, and is not horizontal, as represented by Figure 202. The furnace is bound in the usual way with buck-stays (old rails) and tie-rods. The means of tapping the furnace deserves special mention. In Figure 200 the water jacket is seen to inclose the tapping-opening. Details are shown in Figures 203 and 204. The 2-inch tap-hole *t* is a conical opening in the cast-iron plate *u*, which fills the open space between the outer and inner plates of the jacket. Two bolts *v* pass through the jacket and plate, and then through the flanges of the spout placed on the outside, where they are tightened with nuts. Commonly the tap-hole is closed by simply ramming a clay plug into it. To make the breaking away of the clay impossible, the tap-hole in the figure is closed by an iron plug coated with clay. It is held in place by an iron wedge driven vertically between it and a horizontal piece of flat iron, held in place by the vertical ribs *w* on either side of the spout.

Finally, as to the firing of the furnace. The fuel commonly used is bituminous coal; at some works this has been replaced by crude oil. With a good grade of bituminous coal, natural draught is sufficient to soften the lead in the required time. Many works use slack coal, and then blast under the grate becomes necessary. At some works a series of small blast-pipes is introduced through the roof, above the fire-bridge, with very good effect. The writer recalls one instance where, with the use of blast both under the grate and in the roof, and an impure bituminous coal, Colorado base bullion of ordinary hardness was softened in 50-ton charges in 6 hours. This is probably the best work on record. The good effect of cold blast in the roof would suggest the admission of hot air, as is done in Lake Superior copper-refining furnaces,[1] where a flue as-

[1] Egleston, *Trans. A. I. M. E.*, ix., p. 690, plate ii.

cending in the side wall of the fire-place passes through the roof, and enters the furnace just above the bridge, delivering there the heated air to effect complete combustion. If necessary, additional heated air might enter the furnace through openings in the fire-bridge.

In the use of oil as fuel, air under a pressure of 8 or 9 ounces has begun to replace the different forms of Körting inspirators, although a larger number of burners is required to furnish the necessary amount of oil.

§ 99. **Mode of Conducting the Process.**—The mode of operating the softening-furnace is about as follows. The bullion is charged through the two charging-doors by means of a long-handled paddle, and melted down slowly.

The paddle is a rectangular iron bar about 8 feet long, made of 1⅛-inch iron; one end is flattened out for a distance of 2 feet 6 inches to the width of 3 inches, to receive a bar of bullion, while the other is rounded off and bent to a ring. A lug is often cast on either side of the door-frame to support a roller. It serves as bearing for the paddle instead of the door-frame, and thus facilitates the manipulation.

In some works the bullion is all charged at once; in others, where the charging, softening, desilverizing, refining, and moulding are given in contract to a crew of four men, the bullion is charged at intervals when the kettle and refining-furnace do not require their attention. The softening of the base bullion, as well as the refining of the desilverized lead, is regulated by the desilverization which takes the longest time (from 18 to 20 hours for a 30-ton kettle). When a kettle of bullion has been desilverized, the refining-furnace (average time required 12 hours) must be empty to receive the desilverized lead, and the softening-furnace (average time required 14 hours) ready to furnish softened bullion to the kettle. Upon this general scheme the whole of the refinery must be based to be properly continuous.

The bullion is melted an hour or more after it has been charged. It is stirred to detach some of the lead held in suspension by the dross; sometimes fine coal is spread over it and stirred in. This is very effective when the bullion is pure, so that little dross rises to the surface. With impure bullion considerable fuel is required to have any effect, and there is danger of the temperature becoming too high and the lead taking up again some of the impurities that had separated out. The dross is removed by a rabble or a rectangular skimmer.

The head of the rabble is made of $\frac{3}{8}$-inch iron, and is 3 by 12 inches, the handle of $\frac{3}{4}$-inch iron, and 10 feet long. The handle of the skimmer is of the same length and thickness as that of the paddle; the perforated part is made of $\frac{1}{4}$-inch iron, and is 10 by 12 inches, the perforations being $\frac{3}{8}$ inch in diameter.

With both tools about the same amount of lead is withdrawn with the dross from the furnace, so there is little choice between them; some prefer one, some the other. With either, the handle often rests in a hook suspended by a chain from the roof, thus facilitating the work. The operator removes from one side the dross, which his helper on the other side collects with a rabble, pushing it toward the door or upon the skimmer. The dross, while being removed from the furnace, is collected either in a slightly conical cast-iron mould running on wheels, *e.g.*, 2 by 3 feet at the base and 14 inches deep, made of $\frac{3}{8}$-inch iron, or preferably in an iron two-wheel barrow with perforated bottom, in order that some of the lead carried out with the dross may run off on the cast-iron plate in front of the skimming doors and be returned to the furnace.

The dross drawn off is weighed, and a sample is taken from the lead remaining in the furnace to be assayed. The weight and assay-value of the bullion charged being known, the weight of the dross and the assay of the residual hard lead give the data necessary to calculate the total silver contained in the dross, and with it its assay. The amount of dross formed is about 4 per cent. of the bullion charged, and assays about 80 per cent. of lead. It is freed from some of its lead in a liquating-furnace, and will then have a composition similar to that of the Freiberg analysis (see *ante*).

The tin-skimmings that form on raising the temperature after drossing are removed in the same way as the dross.

With the antimony-skimmings it is customary to cool the furnace by throwing open the doors, in order that the antimoniate of lead floating on the surface may harden, and be then taken off in the form of a thin crust. If the bullion is very hard, skimming once will not be sufficient to soften it. The furnace is therefore heated up again, and as soon as the surface of the lead is well covered, the cooling and skimming are repeated; ordinarily two operations are sufficient, but sometimes three are necessary. To hasten the cooling of the furnace, slacked lime is sometimes spread over the metal bath. Some refiners add lime to the furnace after drossing, with the idea that an antimoniate of lime is formed, and thus less lead is oxidized during the softening. This effect of lime still remains to be proved. Any addition of lime to the furnace has the great disadvantage of interfering with the subsequent liquation of the antimony-skimmings, and is therefore better dispensed with altogether.

If the bullion is very hard, the addition of litharge from the cupelling furnace greatly shortens the time required for softening.

Another method of hastening the softening is the introduction of dry steam to stir the lead, thus continually exposing a fresh surface to the oxidizing action of the air. This is done by introducing through each of the charging doors a 1-inch pipe, to the end of which are screwed, by means of a T, two pipes having a number of perforations on either side and closed at the ends. The main pipe is bent so that when it is introduced into the furnace and held in place by the closed furnace door, which has been weighted, the two pipes at its ends will be pressed down into the lead, and run parallel to the sides of the furnace. While the introduction of steam may shorten by one-third the time required for softening, it has the disadvantage that it forms a large amount of skimmings, and that the swash of the lead oxide and antimoniate strongly corrodes the sides of the furnace. It is, therefore, to be used only in extreme cases.

A third method to be mentioned is the one in use at Freiberg, where bullion rich in tin, arsenic, and antimony is softened. Blast is introduced on either side of the fire-bridge, and the skimmings are removed at the flue end of the furnace as fast as they form. The tool used is a long iron hook, to which is fastened a triangular piece of wood, say 8 inches long. With it the skimmings are drawn out of the furnace in a thin stream. In order to facilitate the work, and to enable the workmen to pass gently over the surface, and thus remove only skimmings, but no lead, the handle is supported by a hook suspended from the roof.

After the last skimmings have been removed, the doors are thrown open to cool the lead before it is tapped into the kettle. The skimmings are weighed and a sample of the softened bullion in the furnace is taken for assay; thus the silver contained in the skimmings can be calculated. The amount of skimmings found is about 5 per cent. of the weight of the bullion charged. The fuel consumed for the entire softening is about 156 pounds of soft coal per ton of bullion charged.

A sample of flue dust from the softening-furnace assayed 15.9 per cent. lead, 9.1 ounces silver, and 0.16 ounces gold.

A new furnace bottom absorbs a considerable amount of base bullion for which it is difficult to give any figure. One peculiarity still needs a satisfactory explanation, namely, that a larger proportion of gold collects than of silver, considering the average composition of the bullion treated.

§ 100. Desilverizing Softened Bullion.

Introductory Remarks.—From the softening-furnace the lead, when sufficiently cool, is tapped into the desilverizing-kettle, which has been whitewashed with lime water. and heated to the point where a splinter of dry wood thrown on the bottom will ignite readily. The whitewashing facilitates the removal of silver-crusts which adhere to the sides when the kettle is cooling. If the lead were tapped into a cold kettle, this would be liable to crack on the bottom, and the time for bringing the lead up to the required temperature would be unnecessarily prolonged. The lead runs into a trough of cast iron, $\frac{3}{8}$ inch thick, placed beneath the discharge-spout of the furnace. In order to decrease the amount of dross, the lead runs from the trough into a cast-iron pipe placed upright in the kettle.

The kettle-dross formed amounts to about one per cent. of the bullion charged. It is skimmed off and added to the next charge in the softening-furnace after the furnace-dross has been taken off. The kettle is now ready for the addition of zinc.

The quantity of zinc necessary varies according to the purity of the lead, and increases on the whole with the amount of silver present. Roswag's [1] new formula is

$$Z = 10.39 + 0.035 \, T,$$

where Z = kilogrammes zinc to be added to one metric ton of lead, and T = grammes silver in 100 kilogrammes lead.

This corresponds to

$$Z' = 23.32 + 0.223 \, T',$$

where Z = pounds zinc to be added to 2,000 pounds of lead and T' = ounces silver per ton.

Roswag also formulated the quantities for zinc Illing [2] found

---

[1] " La désargentation de plomb," Paris, 1884, p. 241.

[2] *Ztschrft. f. B. H. u. S. W. i. P.,* xvi., p. 51.

necessary to desilverize lead of varying tenor in silver. They are :

Z'' kilogrammes $= 11.66 + 0.0325\ T''$

(grammes silver in 100 kilogrammes lead)

for every metric ton of base bullion ; which corresponds to

Z''' pounds $= 20.78 + 0.24\ T'''$ (ounces silver per ton)

for every 2,000 pounds of base bullion.

These general formulæ probably give an approximate idea of the total amount required to desilverize base bullion running low in silver, say 30 ounces. For rich bullion the figures are too high. How the amount of zinc to be added increases with the silver-contents, irrespective of the zinc recovered later by distillation, is shown in the following figures given by Plattner.[1]

Assay of base bullion, ounces per ton.	Per cent. of zinc added.
28.09	1.34
111.56	1.84
148.16	1.96
245.00	2.45

In practice it has so far not been found possible to desilverize a rich base bullion with a single addition of zinc. The richer the bullion the less difference is there between the assay-values of the zinc-silver crust and the residual lead. From three to four zincings are therefore necessary. Low-grade bullion can be and is desilverized by two zincings. The aim in desilverizing must be to concentrate as much silver as possible into one zinc-crust, so as to utilize all the power of the zinc. This is best done by first adding sufficient zinc to remove the gold and the copper (as gold-crust) with as little silver as possible, and to saturate the lead ; then the bulk of the zinc is added, which takes up enough silver to form the rich zinc-silver-lead alloy (the first silver crust). One or two subsequent additions of zinc will completely desilverize the lead, but the zinc contained in these crusts will, under suitable conditions, combine with more silver ; hence the second and third silver-crusts are added as fresh zinc for the first desilverization, and two-thirds of the zinc contained in them is available as new zinc. The crust obtained from a kettle assaying 60 ounces or less per ton always

[1] *Berg- und. Hüttenm. Ztg.*, 1889, p. 117.

goes back to the kettles. The second silver-crust ranges ordinarily from 18 to 30 ounces, the third silver-crust from 0.3 to 0.5 ounces per ton, while the first never ought to run lower than 2,000 ounces. The following analyses show the composition of liquated first zinc-crusts obtained from low-grade bullion that has been drossed only and still retained the antimony.

	Altenau.[1]	Lauten-thal.[2]	Mechernich.[3]			Friedrichshütte.[4]	
	1880.	1880.	1875.[3]	1884.[4]	Average. 1884.[4]	1869.	1884.
Pb.....	75.675	77.820	48.80	49.70	45—48	34.66	81.2
Zn.....	11.78	12.11	89 00	34.00	35—45	20.19	12.15—10.15
Cn.....	1.12	0.82	5.83	6.00	2—3	....	....
Ag.....	1.855	2.420	1.22	1.75	1.5—1.7	1.21	1.075—1.202
As.....	trace	trace	....	....	....	....	....
Sb.....	....	....	0.86	....	....	0.88	....
Cd.....	trace	trace	....	....	....	....	....
Ni.....	trace	trace	....	....	....	....	....
Fe.....	....	....	1.28	....	....	0.96	....
PbO ...	4.75	4.00	...	....	....	....	....
ZnO....	0.60	0.44	....	....	....	....	....
Bi_2O_3 ..	1.72	0.87	....	....	....	....	...
H_2O_3...	0.63	0.98	....	....	....	....	....
Fe_2O_3 ..	1.87	1.04	....	....	....	....	....

All the crusts run much lower in silver than any from American refineries. Those from Altenau and Lautenthal retain more lead than any of the others, as they are to be melted down again in a kettle and decomposed by steam, which could not be satisfactorily done if the liquation had been carried any farther.

It is usually stated that the total zinc required is added in three separate portions : first $\frac{2}{3}$, then $\frac{1}{4}$, and finally the remaining $\frac{1}{12}$. While this corresponds approximately to actual practice, it cannot be implicitly followed, as the amount of the successive additions must depend on the assay of the bullion. The rate at which the silver in the lead decreases with the additions of zinc is influenced by a number of circumstances that cannot be determined in advance. Each refinery has its own table to show the amount of zinc that shall be added if the bullion assays a certain number of

[1] *Ztschrft. f. B. H. u. S. W. i. P.*, xxviii., p. 262.

[2] *Ibid.*

[3] *Berg- und Hüttenm. Ztg.*, 1875, p. 129. *Engineering and Mining Journal,* March 17, 1877.

[4] *Ztschrft. f. B. H. u. S. W. i. P.*, xxxiv., p. 92.

ounces of gold and silver to the ton. Some of these tables are very complicated. The following is one of the simplest. By it, the gold and silver from a 30-ton kettle are extracted separately with three zinc-additions.

TABLE OF ZINC-ADDITIONS FOR GOLD.

Up to 0.10 ounces gold per ton, 250 pounds of zinc.
0.10 — 0.80   "   "   "   "   300   "   "   "
0.30 — 0.50   "   "   "   "   350   "   "   "
0.50 — 0.70   "   "   "   "   400   "   "   "
0.70 — 0.90   "   "   "   "   450   "   "   "
etc.   etc.

It is to be noted that gold and copper are extracted from the lead without saturating this with zinc. For instance, 30 tons of lead, taking up 0.6 per cent. of zinc, require 360 pounds of zinc, while 0.30 ounces gold per ton are extracted by an addition of 300 pounds.

When the gold-crust has been skimmed off the kettle, 500 pounds of zinc are added for bullion assaying from 150 to 250 ounces silver. With bullion running as high as 300 or 400 ounces, 550 pounds are given. After removing the first silver-crust the kettle will assay from 10 to 50 ounces silver per ton, generally from 30 to 40 ounces; in exceptional cases it may run as high as 70 ounces. The second and final addition of silver-zinc, varying from 400 to 600 pounds, will reduce the silver-contents of the kettle down to a trace, even if it has been as high as 70 ounces. Bullion running low in gold and silver, say from 0.05 to 0.10 ounces gold and from 50 to 125 ounces silver to the ton, receives one gold-zinc and one silver-zinc, the resulting lead running less than 0.2 ounces silver to the ton and generally a trace.

Lately Edelmann and Rössler[1] have carried on an interesting series of experiments with the object of concentrating the silver in a richer alloy than usual. The ordinary silver-crust, they say, consists of a small quantity of zinc-silver alloy distributed through zinc-bearing lead which is partly oxidized. As the oxidized part obstructs a satisfactory separation of the zinc-bearing lead from the zinc-silver alloy, the improvements must consist in preventing the formation of oxides; then the direct output of desilverized lead would be increased, an alloy rich in silver produced, and the consumption of zinc reduced. They suggest heating the lead, freed from copper by means of zinc, up to incipient redness (500° C.),

[1] *Berg- und Hüttenm. Ztg.*, 1890, pp. 245, 429; 1891, p. 123. *Engineering and Mining Journal*, November 15, 1890; April 4, May 16, 1891.

introducing the zinc, heated to a higher temperature than the lead, at the bottom of the kettle, which does away with stirring and skimming the little oxidized crusts from the surface, keeping the first rich crusts separate from the following ones running lower in silver, and utilizing these with fresh zinc. The addition of aluminum to the zinc, in some of their experiments, had a favorable effect in preventing oxidation, and thus assisted in obtaining crusts very rich in silver. At a Belgian refinery [1] crusts running 20 per cent. silver have been obtained by adding aluminum with the zinc, and experiments of electrolyzing these rich crusts are under way. In this connection Rösing's electrolytic treatment of silver-crusts [2] is of special interest. Other suggestions regarding improvements in desilverization have been made by Rösing,[3] Honold,[4] and Schlapp.[5]

§ 101. **Desilverizing-Kettles.**—The kettles (Figure 231) used for desilverization are spherical in form. They are from 3 feet to 3 feet 4 inches deep, and their diameter varies according to the required capacity. Most kettles used to-day hold 30 tons; the dimensions of such a kettle are shown in the figure. A circular kettle of a greater capacity would present difficulties in working. In the few instances where kettles holding over 30 tons are used, they are made oblong and have rounded ends. Such a kettle, *e.g.*, holding 55 tons of lead when actually filled, and 45 tons when filled to 5 inches from the top, is 12 feet long, 7 feet wide, and 3 feet 3 inches deep.

The kettles are all of cast iron. The iron used should be dense and strong, but not hard. Lake Superior No. 2 iron, which is neutral, strong, and tough, mixed with car-wheel iron, giving the necessary density, furnishes a material filling every requirement. Kettles are best cast bottom down, as it gives greater density. The other way is easier for the foundry. In any case they should be cast in brick-work if a smooth surface is to be obtained, which is necessary for the scraping of the kettle; any blisters, besides weakening the kettle, are liable to assist the corroding effect of

---

[1] Private communication, December, 1891.

[2] *Berg- und Hüttenm. Ztg.*, 1886, pp. 468, 478, 488; abstract in *Engineering and Mining Journal*, May 22, December 25, 1886.

[3] *Chemiker Zeitung*, 1889, p. 1059.

[4] *Berg- und Hüttenm. Ztg.*, 1890, p. 187; 1891, p. 342. *Stahl und Eisen*, 1891, p. 152.

[5] American patent, No. 380,524, April 3, 1888.

the zinc, and to retain particles of zinc-crust which may enrich already desilverized lead. Kettles have often been made 2¼ inches thick at the bottom, tapering to 1½ or 2 inches at the rim. At present they are made of uniform thickness throughout, and rarely over 1½ inches in thickness. Such a kettle lasts from one to one and a half years, being in continual use. Steel kettles are not found in American desilverizing works. They have come into use in European works, where the lead is refined by means of steam in the same kettle in which it was desilverized. A kettle is usually suspended by its rim, which rests on a circular cast-iron ring covering the top and sides of a brick wall. Figures 225, 229, 230 show three supporting rings; see Figure 225 for the desilverizing kettle, Figure 229 for the liquating-kettle, and Figure 230 for the liquated-lead kettle. The casting (Figure 225) consists of four separate pieces, shown in section by Figure 226. They are fastened together by bolts passing through flanges, as seen in Figure 227. The casting rests on the working-platform of the kettle, as drawn in the front elevation (Figure 205). It is made 9¼ inches wide, and encloses the 9-inch wall, which rises 18 inches above the main brick-work. With many kettles the support of the rim consists of a circular iron ring, covering only the top of the brick-work. In such cases this must be thicker than 9 inches, if it is not to give way to the pressure of the weight of the kettle filled with lead. The side-wall thus often reaches a thickness of 18 inches, the iron support-ring either entirely covering it or leaving 2 or 3 inches exposed. Müller[1] gives it as his experience that a kettle lasts longer if suspended from a rib cast on the kettle at half its depth. This rib would then form the partition-wall between the fire-place and the encircling flue, which simplifies the construction of the brick-work. The place from which the kettles are fired is seen in the front elevation (Figure 205). The horizontal section (Figure 208) shows the plan of the brick-work with the ash-pits of the three kettles.

Figures 206, 207, 209 give more detail of the brick-work, and show the road of the products of combustion from the grate to the flue leading to the chimney. In the desilverizing-kettle the flame goes from the fire-place *d* (Figures 206, 209), first back and upward; it then passes around the kettle to the right, in a circular flue (as indicated by the arrows), and leaves this at *e*, entering a vertical flue leading to the main chimney. In the liquating-kettle the

[1] *Berg- und Hüttemn. Ztg.*, 1889, p. 218.

products of combustion go from the fire-place *f* (Figures 207, 209), after passing under the kettle, straight into the flue *g*. The gases from beneath the liquated-lead kettle *h* (Figure 209) go to the left, and join those of the liquating-kettle.

Desilverizing-kettles were formerly emptied by a discharge-pipe cast in the bottom of the kettle and running out through the brick-work. It was closed either by a slide-valve on the outside (similar to the lead discharge of the Luce-Rozan crystallizer, Figures 191–193), or by a clamp and thumb-screw on the inside. At present the Steitz siphon is in common use, and is preferable to any other means for emptying a kettle. A common form of it is shown in Figure 235. It consists of a piece of gas-pipe *a* from 2 to 2½ inches in diameter, bent so as to reach from the rim of the kettle to the bottom. Here it has an elbow *b* screwed to it to prevent the lead column from breaking. To the other is attached, also by an elbow *c*, the vertical section-arm *d*, having a cast-iron stop-cock *e* near the lower end. The siphon discharges the lead into a cast-iron trough of ⅜-inch iron, which carries it into the refining-furnace.

§ 102. **Liquating Apparatus.**—In connection with the desilverizing-kettle must be discussed the apparatus required for liquating the zinc-crusts. Two sorts are in use. The first is shown in Figures 232 and 233. It consists of a shallow kettle placed on the same level with the desilverizing-kettle and close to it. The bottom is convex in the centre, in order that the discharging-spout may be shorter than would be possible if the kettle had the usual spherical form. Formerly a perforated cast-iron disk (an old skimmer) was placed inside the kettle, over the opening into the spout, to prevent particles of crust from passing off with the liquid lead; this has been given up at some works, as the perforations easily become clogged, and when open do not prevent fine particles from being carried off by the liquated lead. This runs from the spout into a small spherical kettle, whence it is bailed out after it has been skimmed.

The drawings given show for every desilverizing-kettle one liquating-kettle with its liquated-lead kettle. This presupposes that no distinction is made between gold-crusts and silver-crusts, with the result that all the silver produced contains gold, and has to be parted. Where the crusts are kept separate the desilverizing-kettle will have a liquating-kettle on either side, one for the gold-crust, the other for the silver-crust, the liquated crusts as well as the liquated lead from the two small spherical kettles being kept separate. Only the silver resulting from the gold-crust will then have to be parted. With a plant where the two crusts are kept separate, the distance between the centres of the two desilverizing-kettles of 32 feet, as shown in the general plan (Figure 196), will be too small; it will have to be enlarged sufficiently to leave a passage-way between the two small liquated-lead kettles.

The second apparatus is a reverberatory-furnace placed on the floor of the desilverizing-kettles. It may be built as follows. The hearth is a cast-iron plate, 10 by 5 feet, slightly trough-shaped, and having a rim 4 inches high along the sides. To the lower end a spout is attached, which discharges the liquated lead outside of the furnace into a small liquated-lead kettle of the same form as the one described, the rim of which is on a level with the working-floor. The plate has an inclination of 3 inches. It lies on a bed of brasque, which is tamped into a wrought-iron pan supported by transverse rails resting on two longitudinal walls. The furnace has two working doors on one side. The writer prefers the reverberatory furnace to the liquating-kettle, as he obtained from it in a shorter time than from the kettle a dryer crust, and thus a richer bullion, without driving some of the silver from the crust back into the liquated lead, as often happens with the kettle. The reason for the better result in the reverberatory furnace is probably that the lead can be gradually eliminated at a slowly increasing temperature in a reducing atmosphere, which prevents the oxidation that always takes place in a liquating-kettle, even if it be covered with a sheet-iron plate, on raising the temperature to the required degree. Supposing, however, the liquation to be equally good with the two apparatus, there remains this advantage for the reverberatory, that the liquation of the silver-crusts of several systems can be performed continuously in one furnace, which, being separated from the desilverizing-kettles, does not disturb the work there, and which collects all the rich crusts into one place, whence they are delivered through a chute into the bins of the retort-room. With a small plant a liquating reverberatory furnace of any reasonable dimensions would not have sufficient work to do to pay for the extra labor. Here the liquating-kettle must be used, even if it obstructs to some extent the work in the desilverizing-kettle. The preference for the reverberatory furnace is not general, as several important refining works adhere to the liquating-kettle.

The practice of liquating zinc-crusts in a spherical kettle without a lead-discharge and removing the liquated crust floating on the liquid lead with a skimmer is antiquated, if the crust is to be distilled, and justly so, as it is impossible to obtain in this way a dry crust that is satisfactory. If the temperature of the lead be raised sufficiently high to obtain a dry crust, a considerable quantity of it will be redissolved by the lead. It will rise again to the surface when the lead cools, but it will be rich in lead, and at the same time low in silver. It must, therefore, be returned to the next charge that is to be liquated, instead of going directly to the retort, as it would with either of the two apparatus just discussed.

§ 103. **Mode of Conducting the Desilverization.**—This is as follows. To the lead in the kettle, which represents, after the kettle-dross has been removed, about 90 per cent. of the bullion charged into the softening-furnace, is added lead obtained by liquating the gold-crust from a previous charge. It is then heated above the melting-point of zinc (412° C.) and receives the first zinc to remove the gold and the copper. The heating takes a very short time; in half an hour from the time when the kettle is drossed the zinc will usually be melted down.

It is customary to place the slabs of zinc on top of the lead and to heat this until the zinc has become thoroughly melted, when the stirring-in can begin. Objections have been and are raised to this method of adding the zinc, which, floating on the lead, is exposed for some time to the oxidizing action of the air. To avoid this, the zinc has been enclosed in a perforated iron box, which is forced down to the bottom of the kettle and held there by an upright iron rod fitting into the centre of a piece of flat iron which reaches across the kettle, and is fastened to it by set-screws. There can be no question that the zinc will melt more quickly at the bottom of the kettle, will not be exposed to the air, and rising upward in thin little streams will combine more readily with the lead and the silver, and will require less stirring-in than if melted while floating on the lead. French refineries use this method frequently. According to Edelmann and Rössler,[1] it is advantageous to melt the zinc in a separate vessel and to pour it into the lead, as it is then taken up quickly by the lead. It seems, however, doubtful whether less dross is formed than if the zinc is melted down floating on the lead. Figures comparing the results of these three methods of incorporating the zinc would be of interest.

In order to bring the melted zinc into intimate contact with the

[1] *Loc. cit.*

argentiferous lead it is stirred in for from one-half to three-quarters of an hour. When this is done by hand, a paddle is used, consisting of a perforated disk 12 inches in diameter, riveted to a handle 6 feet long, having a cross-piece as hand-hold. Two men standing opposite each other do the stirring. They insert the paddles vertically at the rim of the kettle, push them downward towards the centre, raise them, using the rim of the kettle as a fulcrum, and draw them, with the disks gliding on the lead, from the centre towards the periphery, giving the lead a rotary motion which they reverse every five minutes, thus insuring an intimate mixing of zinc and lead. This stirring-in has always been hard for the workmen, and various mechanical stirrers have been devised, but stirring-in by hand remained the universal method until steam-stirring came into use. The effect of steam when introduced into lead containing zinc varies greatly according to the temperature of the lead.

1. If the temperature of the lead be below the melting-point of zinc, *i.e.*, the temperature when the kettle is skimmed, the steam will bring to the surface a zinc-crust, and with it some of the silver contained in the lead.

2. If the temperature be slightly above the melting-point of zinc (stirring-in time), the steam will cause a thorough mixing of zinc and lead.

3. If the temperature be between a dark-red and an incipient cherry-red, the steam will cause a scum to rise, containing about 3 per cent. of zinc, which does not, however, take any silver away from the lead.[1]

4. If it be a clear cherry-red, the zinc will decompose the steam ; the resulting zinc oxide (mixed with lead oxide) collects as a powder on the surface of the lead.

The steam must be absolutely dry if violent explosions are to be avoided. The condensed water is separated by a steam-trap placed beneath the working-platform. To the vertical pipe coming through the platform is fastened by means of a coupling, a small piece of pipe, to which is joined by two elbows (with a nipple intervening) the horizontal pipe which reaches to the centre of the kettle. An elbow connects it with the vertical pipe, that will reach 24 inches into the kettle when in place. Before the steam-valve is opened, the vertical pipe is turned up, in order that the steam may first pass out into the air to warm the pipe and to insure the expulsion of any condensed water. The pipe is then turned

[1] Rösing, *Z. f. B. H. u. S. W. i. P.*, xxxvii., pp. 76 and 77.

down and weighted with a bar of lead to keep it in place. When the steam is turned on, the waves of lead caused by the first ascending bubbles will drive the zinc floating on the lead toward the rim of the kettle, and hardly any of it would become incorporated with the lead if it were not pushed towards the centre of the kettle to be drawn into the lead by the downward current close to the steam-pipe. The zinc-crusts that rise to the surface soon after stirring-in has begun are also pushed toward the centre, that they may take up more silver. Thus the zinc and then the crusts pass down at the centre, and come up again nearer the circumference of the kettle, whence they are again pushed towards the centre.

The tool used for this purpose is a wooden hoe, consisting of an inch board, 12 by 18 inches, into the centre of which is inserted an inch lath from 8 to 10 feet long.

Towards the end of the stirring-in, which lasts from one-half to three-quarters of an hour, the fire beneath the kettle is removed or damped with wet slack coal, the fire-doors and ash-pit doors and the damper in the flue are thrown wide open, and the kettle is allowed to cool. At St. Louis, A. Meyer[1] tried to hasten the cooling by means of water-cooled pipes bent to the shape of the kettle, but his apparatus has found no favor, as the pipes did not cool the lead enough to pay for all the trouble and inconvenience connected with their use. At St. Louis they were given up many years ago. A better means of hastening the cooling, although not much used, as it injures the brick, is to sprinkle water from a hose against the inside walls. If the zinc is stirred in by steam, its use will be found effective afterward in cooling.

After from two to three hours the lead has cooled down so far that the crust begins to adhere to the sides of the kettle. It is then removed with a skimmer, the disk of which, made of $\frac{1}{8}$-inch iron, is from 14 to 18 inches in diameter; and the handle, of $1\frac{1}{4}$-inch iron gaspipe of the same weight, is 7 feet long, having a cross-piece as hand-hold. The work is facilitated by suspending the skimmer by a hook, which acts as a lever. Two men work together. One man pushes with a wooden hoe the crust towards his partner, who takes it up with the skimmer. Before discharging it into the liquating kettle or the mould, it is important that the skimmer be well jerked several times, in order that as much lead as possible may be drained off into the kettle and the crust obtained become

[1] *Mining and Scientific Press*, 1882, vol. xliv., No. 5 ; *Berg- und Hüttenm. Ztg.*, 1882, p. 391.

dry. Toward the end of the operation both men have to work very slowly and carefully to avoid pushing the crusts back again into the lead, which would retard the work very much. When the crusts have been removed from the surface, the alloy adhering to the sides of the kettle has to be brought to the surface, which is done by scraping them first with a chisel-pointed bar and then with a wooden lath. The blade of the bar is of steel, 2 by 4 inches, and the handle of 1-inch round iron. Skimming and scraping are generally repeated twice, after which no more crusts will rise to the surface. It takes about an hour to perform this operation. When finished, the fire under the kettle is again started, and this heated to melt down and stir in the next zinc. The time required for heating up varies from one to two hours, according to the amount of lead (obtained from liquating zinc-crusts) that is added to the kettle.

A sample is taken from the lead to see whether all the gold has been extracted and how much silver. It is well to cupel eight samples of an assay-ton each, and dissolve the silver buttons together to ascertain if all the gold has been taken out.

The gold-crust obtained by the first skimming is collected either in one of the liquating kettles or, if the reverberatory furnace is in use, in flat, slightly conical moulds that have been whitewashed.

While the kettle is being heated up, liquated lead from the first silver-crust of the previous charge is added, and when the lead is sufficiently hot, the zinc for the extraction of the bulk of the silver. The poor second silver-crust obtained from the previous charge is used with third crust, if there was any, and fresh zinc. The operations are the same as for the gold-crust.

The second and third crusts are also sometimes collected in moulds, but oftener the skimmer is discharged on a whitewashed iron plate. There the crusts remain until needed, when one end of the plate is raised by block and tackle, and the two crusts are slid into the kettle together. Some works discharge only the second silver-crust on the plate, while the third goes into moulds.

After each skimming, samples are taken, and $\frac{1}{4}$ assay-ton is assayed for silver to check the progress of the desilverization. After the last silver-crust has been removed, the assay should show 0.2 ounces silver per ton, or less if corroding lead is being made. Should it prove to be slightly higher, say 0.4 or 0.6 ounces, the introduction of steam at the low temperature of the kettle after the last crust has been taken off will be effective in causing more silver-bearing zinc-crust to be given off by the lead, as stated above. If the steam is used from one-half to three-quarters of an hour, the silver-contents of the kettle will be reduced, and thereby generally an entire zincing saved.

In a 30-ton kettle five hours are allowed for each zincing if the silver is extracted by four additions of zinc. The first five hours include the time the softened bullion is being run into the kettle, the last five hours the time during which the desilverized lead is being syphoned into the refining-furnace. When the lead comes from the softening-furnace it is usually hot enough to melt the first zinc quickly, and the melting down of the last zinc and skimming the final crust require but little time, so that the hour necessary to tap the softening-furnace and empty the kettle can be included in the standard five hours. With a skilled crew it is possible to desilverize a 30-ton kettle with four zincings in eighteen hours, but every moment has to be utilized to accomplish it. Some works, therefore, aim to extract the precious metals with three additions of zinc, and give six hours for each addition, keeping the gold and silver crust separate. When gold and silver are extracted together, a 45-ton kettle can be desilverized with three additions of zinc in the same length of time. The reduction of time may be still further increased by the substitution of oil or gas for coal as fuel.

The weight of the single crusts varies considerably, and also the amount of silver which they take up. As an illustration, 1,500 tons of softened bullion, containing 170 ounces silver and 0.5 ounces gold to the ton, gave 5 per cent. of liquated gold-crust, required 0.3 per cent. zinc, and reduced the silver-contents 15 ounces silver per ton; the liquated silver-crust was 8 per cent., required 1.5 per cent. zinc, and reduced the silver-contents to 30 ounces per ton. The 1.5 per cent. represents the entire zinc added to extract the silver. The second silver-crust reduces the assay of the kettle to about 3.0 ounces per ton, and the third to 0.2 ounces or less. Where only three zincs are added, the silver-contents are reduced by the second silver-zinc from 30 ounces silver to the ton to 0.2 or less ounces. From the zinc added to the kettle part is recovered by distilling the zinc-crusts and makes the amount actually consumed lower. This was one per cent. for a year's run for softening bullion in the kettle, which averaged 150 ounces silver and 0.5 ounces gold to the ton, with the usual amount of copper. The coal required for desilverization, including liquation, is about 54 pounds for a ton of unsoftened base bullion.

In connection with desilverization may be mentioned an excellent custom [1] of breaking up unwieldy old kettles that have been set aside on account of leakage or corrosion. This is done by filling the kettle with water and suspending a dynamite cartridge from a floating board so that it nearly touches the bottom. When this is exploded the water will be thrown up into the air and the kettle broken into five or more pieces, which can be easily handled and shipped to the foundry.

§ 104. **Method of Liquating.**—The liquation of the gold and silver crusts takes place, as stated in § 102, either in a liquating-kettle or a reverberatory furnace. It is essential that the temperature be raised gradually and that small quantities only be liquated at one time. If in the liquating-kettle the temperature is raised quickly, some of the zinc-crust will be taken up by the melted lead and carried off into the small liquated-lead kettle, whence it is again skimmed off as a crust rich in lead, to be treated with the next batch of crust. The kettle must therefore be large enough, as shown in Figures 232 and 233, to hold the silver-crust in not too thick a layer. With the reverberatory-furnace a charge is introduced through the door near the flue, and gradually moved toward the fire-bridge, whence it is drawn out through the second door upon an iron plate let into the floor. By moving the crust from the coolest part of the furnace to the hottest, a gradual elimination of the lead takes place, and there is little danger of dissolving the crust, which grows less and less fusible, in the lead, which runs off as soon as melted.

The zinc-crust, when still hot, can be easily broken up into small pieces. When removed from the kettle or the furnace it is spread on an iron plate and reduced to nut size, and smaller, by flattening with the back of a shovel and working with a rake. In this form it is readily charged through the narrow neck of the retort. Another way to break up the liquated crust is to transfer it from the liquating apparatus into a cast-iron box (24 by 18 inches, with a rim 3 inches high), the bottom of which has a series of slits (18 by 2 inches), alternating with ribs of the same width, the casting being $\frac{1}{2}$ inch thick. With a liquating kettle this box requires a frame as support ; with the reverberatory furnace it is attached to the door frame.

The lead recovered by liquating is from 40 to 60 per cent. of the weight of the crust charged. In desilverizing 250-ounce bullion,

[1] Lautenthal, Smelting and Refining Works. Private notes, 1890.

the lead from the gold-crust assays from 100 to 200 ounces silver per ton; that of the first silver-crust from 30 to 40 ounces. The lead liquated from the gold-crust is added to the desilverizing-kettle before the gold-zinc is given, that from the silver-crust before the first silver-zinc.

§ 105. Refining Desilverized Lead.

Introductory Remarks.—The desilverized lead retains, after the last crust has been removed, from 0.6 to 0.7 per cent. zinc, according to the temperature that prevailed at the last skimming. To remove this, and also small quantities of arsenic and antimony that were either not entirely taken away during the softening, or that were introduced with the zinc used in desilverizing, the lead must undergo a refining process.

From the desilverizing-kettle the lead is siphoned off into the apparatus used for refining, which is in most American refining works a reverberatory furnace, in a few instances only a spherical kettle.

The siphon (§ 101) is heated and filled by immersing it in the kettle, the stop-cock being open. When it has attained the temperature of the lead and is filled entirely with it, the stop-cock is closed with a key, the longer arm taken out and suspended, and the shorter one held down in the lead. The stop-cock is then opened, and the lead runs out into a cast-iron trough, which discharges into the refining-apparatus. To keep the siphon in place it is weighted by a couple of bars of lead. Should the lead-column break when the kettle has, for example, been half emptied, and it not be possible then to fill the siphon again in the usual way, it will be necessary to invert and fill it by ladling; for this purpose an iron funnel may be used to avoid delay. But the breaking of the lead-column is a very rare accident.

§ 106. Refining in the Reverberatory Furnace.—The general construction of the reverberatory furnace used for refining is the same as that for softening. Formerly it was customary to make the refining-furnace smaller than the softening-furnace, in order that it might correspond to the smaller amount of lead it has to treat. For instance, the hearth of a softening-furnace of a 25-ton kettle was made 13 by 9 feet and 13 inches deep, having 13-inch side-walls; that of the refining-furnace 12 by 8 feet, and of the same depth, with the same thickness of walls. At present it is commoner to give the refining-furnace the same dimensions as the softening-furnace, only making the hearth slightly shallower, thus

simplifying the iron parts. If in the softening-furnace, shown in Figures 198–204, the depth of the hearth is reduced 3 inches, it will have the capacity necessary to receive the desilverized lead from the kettle, which will fill it to just below the skimming doors. In some refining-furnaces the lowest point of the hearth is placed beneath the flue, as in the softening-furnace described; in others, below the central skimming door. By referring to the general plan (Figure 196), it will be seen that the lead is discharged from below the door next to the flue. The arrangement for tapping is usually the same in both furnaces. If, however, the refined lead is to be conveyed directly from the furnace into the moulds, as is still found in a few instances, instead of being moulded from the " Merchant-kettle," the tap will be slightly different from that of the softening-furnace (see § 108).

The mode of operating is similar to that in the softening-furnace. When the furnace is filled, the fire is urged, as a high heat is required to burn off the zinc. This is partly volatilized and carried off with the fumes, and partly oxidized and scorified by the litharge which forms at the same time. Some refiners add lime to the charge, as in softening. After heating about four hours the surface of the lead will be covered with a heavy litharge-like skimming. The doors are thrown open, the skimming is removed, and the second heat given, after which cooling and skimming are repeated. A third heat is often necessary to slag the last traces of zinc and antimony. When these are completely removed, the litharge drawn from the surface of the lead by means of a rabble should be in large, thin plates. It should, while hot, have a bright-yellow color when seen in bulk, and a greenish-yellow when held up to the light, but not one brown spot (antimony) should be visible. If these large flakes of litharge should become dark or show spots after having been exposed for some time to the air, the lead is not sufficiently refined to satisfy the requirements for corroding-lead. In firing the refining-furnace it is essential for the temperature to be raised quickly to the necessary intensity, and kept there. If it is allowed to fall even a little, the burning-off of the zinc will be greatly retarded, and with it the dependent operations.

As in softening the base bullion, steam was introduced through pipes to expose more surface to the oxidizing action of the air, so in refining, steam is also used, though not so often now as formerly. In addition to the mere mechanical effect of stirring, it acts chemically by being decomposed by the zinc (§ 103), and thus hastens a great deal the elimination of zinc from the lead; but, as in softening, the side-walls of the furnace are more easily attacked, and the percentage of skimmings increased.

The time required for refining with a 30-ton plant is about fourteen hours; the refining skimmings amount to from 4 to 5 per cent. of the bullion charged, and contain about 90 per cent. of lead. The coal consumed in refining is about 120 pounds per ton of unsoftened bullion.

§ 107. **Refining in the Kettle.**—The second method of refining desilverized lead is the one invented by Cordurié, who introduced steam into the lead heated up to cherry-redness, the oxidized zinc collecting in the form of a powder on the lead. As the air cannot be excluded from the surface, and as it is also carried in by the steam, some of the lead is oxidized, and the pulverulent yellow mass floating on the surface consists of a mixture of lead and zinc oxide and finely divided shots of lead. The composition of these oxides when taken from the kettles at the Lautenthal[1] Smelting and Refining Works (Prussia) is

$$Sb_2O_5, \quad 1.893$$
$$Fe_2O_3, \quad 0.986$$
$$ZnO, \quad 23.775$$
$$PbO, \quad 37.933$$
$$Pb, \quad 34.236$$

The base bullion at Lautenthal is desilverized without previous softening; hence the high percentage of antimony.

The larger shots of lead of the oxides are separated by screening, the finer ones by washing over an inclined plane. The impalpable powder, forming 15.44 per cent of the whole, floats off and is settled in vats, dried in a reverberatory furnace, and forms a reddish-yellow paint of good covering power. It consists of

$$67-60 \text{ per cent. } ZnO,$$
$$33-40 \text{ per cent. } PbO.$$

The residual shots of lead are smelted at intervals for a second-class lead, as they contain some antimony.

At Lautenthal[2] the cast-iron kettles (5 feet $4\frac{1}{8}$ inches in diameter, and 2 feet $9\frac{1}{2}$ inches deep, holding $12\frac{1}{4}$ tons of lead) are heated, after the desilverization is finished, in four hours, to a

---

[1] *Z. f. B. H. u. S. W. i. P.*, xxxviii. p. 272.
[2] Private notes, 1890.

cherry-red ; superheated steam having a pressure of from 29 to 36 pounds per square inch is then introduced through a cast-iron pipe bent to the form of the kettle, so that the steam enters at the bottom. After two hours all the zinc has been oxidized. In order to decrease the loss of heat by radiation, to keep off the air, and to prevent the oxides from being lost, the kettle is covered by a movable sheet-iron cylinder, 7 feet 8 inches high, which has near the lower rim two opposite doors (4¾ inches square) and the opening for the steam-pipe. It ends in a conical hood which leads the vapor and dust through a sheet-iron pipe (1 foot 1¾ inches in diameter) into the main flue, terminating in a dust-chamber. The cylinder with its hood and pipe is suspended by a running differential pulley.

For every 100 pounds of unsoftened bullion 4.67 pounds of paint are produced, which is higher than the percentage of skimmings in the reverberatory furnace, as the paint forms but a small part of the total lead taken out of the kettle.

The great drawback of this method is the wear and tear in kettles. According to Schmieder,[1] at Tarnowitz, where the lead is very free from antimony, cast-iron kettles hold out only for twenty charges, while steel kettles are good for ninety charges. The life of cast-iron refining kettles varies greatly ; extreme figures are 30 days and 120 days. Of kettles from the same foundry, cast under apparently the same conditions, one will last only a short time, while another, in the writer's experience, has lasted over a year.

As by the use of steam a considerable amount of air is carried into the lead and some of this oxidized, Rössler[2] tried to replace it by different gases. With carbonic acid, the lead being heated to 700° C., in a short time all the zinc was converted to white oxide, and could be skimmed off from the lead. As carbonic acid cannot be easily obtained pure to be used for such a purpose, he tried a mixture of carbonic acid and nitrogen, obtained by pressing air through a cylinder filled with glowing coal. The result was a gray powder, in which some of the zinc was present as metal in a finely-divided state. By the use of a mixture of carbonic oxide and nitrogen, drawn from a gas-producer, a powder of a darker gray was obtained, containing up to 75 per cent. of its zinc as metal in a finely-divided state, the rest being oxidized by the carbonic acid present. The refined lead was entirely free from zinc if the tem-

[1] *Berg- und. Hüttenm. Ztg.*, 1887, p. 377.
[2] *Berg- und Hüttenm. Ztg.*, 1890, p. 248.

perature was kept above 700° C., otherwise a zinc-crust formed on top of the lead.

§ 108. **Moulding from the Refining Furnace.** (Figure 236.)— The moulding of lead was formerly universally done by ladling it from a kettle. This has, however, become antiquated. In some works the lead is moulded directly from the reverberatory furnace in which it has been refined. At the lowest part of the furnace, on the side away from the desilverizing-kettle, a piece of 2-inch gas-pipe is screwed into the pan if of cast-iron; if of wrought-iron, into the flanges placed on either side and fastened together with bolts. If the furnace has a wrought-iron jacket, the pipe *b* (Figure 236) is screwed into the cast-iron plate placed between the

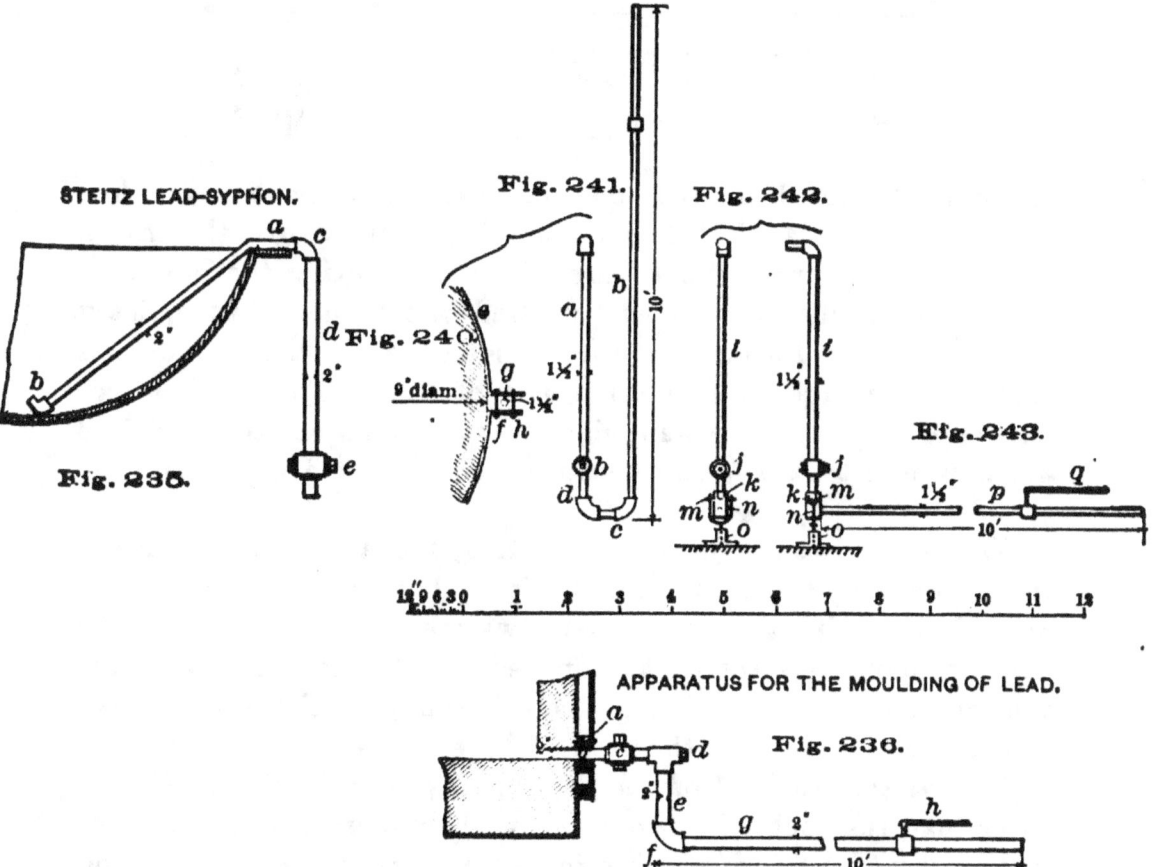

STEITZ LEAD-SYPHON.

Fig. 241. Fig. 242.

Fig. 240.

Fig. 243.

Fig. 235.

APPARATUS FOR THE MOULDING OF LEAD.

Fig. 236.

two sides of the jacket at the tap-hole. The pipe *b* is joined by a cast-iron stop-cock *c* to a T, whose horizontal arm is closed with a plug *d*. To the vertical end of the T is attached a nipple *e*, with an elbow *f* at the lower end, into which is screwed a long pipe *g* (from 7 to 10 feet), which can be moved horizontally by the arm *h*, while it discharges the lead into moulds placed in a semi-circle, the centre of which lies beneath the nipple. The moulds commonly used now (Figures 237–239), differ from the ordinary

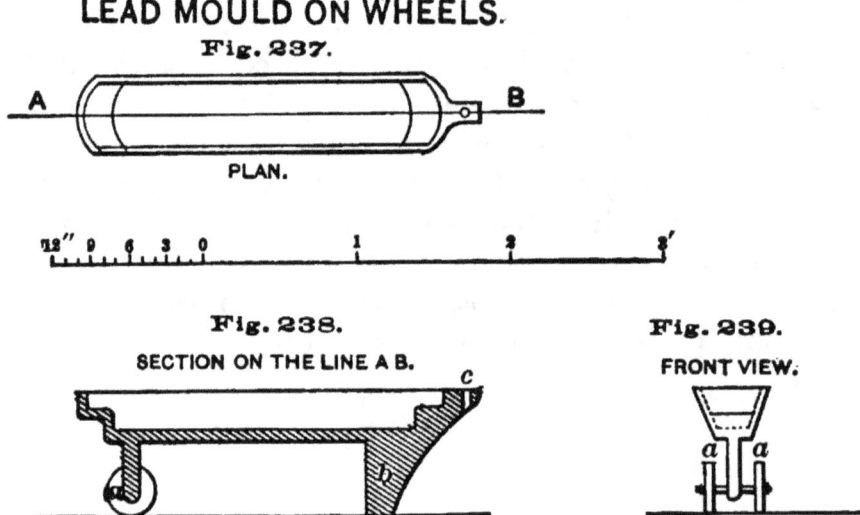

LEAD MOULD ON WHEELS.

Fig. 237.

A   B

PLAN.

Fig. 238.

SECTION ON THE LINE A B.

Fig. 239.

FRONT VIEW.

blast-furnace moulds in that one end rests on two wheels *a*, while the other has a leg *b*. The lip of the mould above this has a hole *c*. By passing a hook through it and tilting the mould, it is run away, the bar tipped out, and the mould then quickly returned to its former place. At some works the moulds are made large enough to hold three bars of lead. The lip then has, instead of the hole *c*, a rectangular socket running horizontally, into which is inserted a slightly bent iron handle to move and tilt the mould.

When the furnace is to be emptied, a charcoal fire is started under the stop-cock, and the horizontal pipe immersed in the lead of the refining furnace to be warmed. It is then screwed into the elbow, the stop-cock is opened, and the lead run into the first mould of the semi-circle. This warming of the pipe is, however, not necessary. If the stop-cock be opened entirely, the first lead arriving at the end of the horizontal pipe will still be liquid. After that, the cock will have to be slightly closed, as the moulds would otherwise fill too quickly for the man, who has also to attend to the skimming of the surface of the bars. This he does with two thin pieces of board the width of the mould, collecting the dross between them and dropping it on the floor. Another method is to rake off the dross on to the floor with a bent piece of hoop-iron. The former method gives a cleaner bar.

One mould after another is thus filled. When the lead in the first three or four moulds has solidified, it is chilled with water, the pigs are trimmed with a sharp, chisel-pointed bar, and the moulds run off to the wall of the lead-pit (Figure 196), where the lead is to be dumped before weighing, and brought back empty to their places. In this way from forty to sixty moulds that form the semi-circle are filled one after the other.

This method of moulding has the advantage that the vertical distance between the shipping level of refined lead and the receiving level of base bullion can be less than with the methods to be discussed ; then the filling of the moulds need not be continuous, as it must with the siphon. With a kettle holding thirty tons of lead, the moulding can thus be given in contract to the four men in addition to the charging, softening, desilverizing, and refining. They do it in two separate operations, moulding part of the lead between the third and fourth zincings, and the rest while the last crust is rising, the total time required for moulding being six hours. The method has the disadvantage that the furnace has to be cooled down considerably for the lead to attain the right temperature for moulding. Thus after every moulding, two hours or more are required to heat up the furnace again for the next charge. This cooling and heating-up of the furnace with every charge cannot be good for the lining, therefore the moulding of the lead directly from the refining furnace has not found so much favor as might be expected.

§ 109. **Moulding from the Merchant-Kettle.**—It has become more common to tap the refined lead from the reverberatory furnace into a kettle—the merchant-kettle—(Figure 196) heated from below, and to let it cool there till it has attained the correct temperature for moulding. In this case the tapping of the furnace is done in the same way as from the softening furnace.

If the desilverized lead is refined by means of steam in a kettle, the moulding is done either from the refining-kettle or the lead is siphoned off into a merchant-kettle below, to store the lead until it is time for moulding, and thus have the refining-kettle ready for another charge of desilverized lead.

Some refining works pole the lead in the merchant-kettle at a low temperature, under the impression that an especially fine grade of corroding-lead is thus obtained. Poling is, however, not necessary, as all the impurities not only can be but ought to be completely removed in one operation, be it in the reverberatory furnace or in a refining-kettle.

As this poling at a low temperature is the best method of purifying the otherwise pure leads obtained from smelting the clean, non-argentiferous ores of the Mississippi Valley, a few remarks are in place. The gases and vapors from the wood stir up the lead and expose continually new surfaces to the oxidizing action of the air. Thus the small amounts of arsenic, antimony, copper, zinc, and iron are slowly oxidized and collect on the surface as a dross. A crutch serves to keep the stick of wood horizontally depressed in the molten lead. It consists of a piece of flat iron long enough to reach about 1½ feet over the rim of the kettle, upon which it is placed, and weighted with a couple of bars of lead on either side. To it are riveted two arms, say 2 feet 6 inches long and 2 feet apart, forked at the ends which, reaching into the lead, receive the wood ; they are connected half-way down by a cross-piece of flat iron. If the lead is to be poled with a billet of green wood, this and the crutch are best put in place before the kettle is filled, as thus the tedious work of depressing a green stick of wood into hot lead is avoided. It is much simpler and just as effective to introduce steam near the bottom of the kettle by means of a one-inch pipe bent to the form of the kettle and passing downward through the lead.

The moulding of the lead from the merchant-kettle in American refining-works is almost always done with the Steitz siphon.

In some of the older European works, where the height necessary for the siphon cannot be had without rebuilding the entire plant, the Rösing lead-pump* has come into universal use.

Two moulding apparatus are shown in Figures 240–243. In Figures 240, 241, *a* represents the longer arm of the siphon, made of 1½-inch pipe, with the cast-iron stop-cock *b*. At the lower end it is joined by two elbows, with a nipple intervening, to the swinging pipe *b*. This can be turned down around the centre *c*, and when in that position it can be moved in a horizontal circle, having its centre at *d*. The siphon is filled just like the one shown in Figure 235, with the exception that the handling is done with the swinging pipe *b*, instead of with a pair of tongs. As in moulding, the swinging pipe has to describe nearly a semi-circle, the ordinary way of keeping the siphon in place by weighting with a couple of bars of lead is not sufficient. For this purpose two iron hoops about 2 feet 3 inches apart pass around the brick-work of the

---

· * *Engineering and Mining Journal*, November 28, 1885, and *Berg- und Hüttenm. Ztg.*, 1889, p. 262.

kettle. The two ends of a hoop are bent, as shown in Figure 240, and tied by a bolt. The vertical arm of the siphon *g* is held in place between the two hoop-ends, the fixed bolt *f* and the movable one *h*.

Another arrangement for moulding is shown in Figures 242–243. *i* represents the longer arm of the siphon with the stop-cock *j*; it discharges the lead into a 3-inch pipe *k*, closed at the bottom. This has two trunnions *m*, which swing in the bearings *n*. They are joined to a pivot rotating in the socket *o*. The pipe *k* is connected with the swinging arm *p*, which is moved with the handle *q* over the moulds. At some works the bottom of pipe *k* is closed with a cap having a socket, into which the pivot, fastened to the floor, fits loosely. Thus a number of slight variations in detail are found at different works.

§ 110. **Labor, Fuel, Output of Lead.**—In many refineries the working of the softening furnace, desilverizing-kettle, and refining furnace is given in contract as a whole, instead of having separate men working by the day at each furnace. Three men with a good head man can, if the work is well systematized, do everything that is necessary with a 30-ton plant. If the lead is moulded straight from the refining furnace, it is possible for them to attend to it also, but this is almost to overwork them. Therefore it has become the general custom to store the lead in the merchant-kettle and to mould from it. A separate contract is made for the moulding and loading into cars. This is often taken by a set of men who unload, weigh, and sample the base bullion, and deliver it at the softening furnaces. They also move the base bullion and lead produced at the blast-furnaces or the different reverberatory furnaces. By having a good head man for every desilverizing-plant and one or two contractors for handling the raw material, the by-products and the market-lead, the labor in the refinery becomes very much simplified and cheapened.

The fuel consumed in softening, desilverizing, liquating, refining, and moulding is about 330 pounds of soft coal per ton of base bullion.

The amount of lead recovered in the form of market-lead varies somewhat according to the purity of the base bullion. It is about 80 per cent. of the bullion charged, or 88 per cent. of the softened lead in the kettle.

§ 111. **Treatment of Zinc-Crusts.**—The working of the zinc-crust has been and still is the weak point of Parkes' process. Many methods[1] have been tried, but only few survive. They are all based on the volatility of the zinc and the readiness with which it is oxidized. Two only will be discussed.

§ 112. **Flach's Process.**—This consists in smelting the zinc-crust in a blast furnace with a large percentage of slag, some matte and fluxes, the slag aimed at being ferruginous and low in silica, and the pressure of the blast not exceeding $\frac{1}{2}$ inch mercury. The zinc is partly taken up by the slag and matte; a large part of it passes off into the dust-chambers; the resulting rich bullion retains but little. If the crust contains any copper, it is taken up by the matte. In smelting zinc-crust with 120 per cent. of slag, 30 per cent. of matte, 11 per cent. of fuel, and puddle-cinder as iron flux, in a small 36-inch circular blast furnace with an Arent's siphon-tap, the writer found that after two days' running, a mushy substance collected on top of the lead, that refused to be taken up either by slag or matte. In order to keep up a good communication with the lead in the crucible, the soft mush had to be repeatedly removed from the front of the furnace. It seems, therefore, that if in exceptional cases it should be necessary to smelt zinc-crust, it would be advisable to fill up part of the crucible with brasque and to tap the lead and matte, when the mush will be carried out of the furnace.

While there is no doubt that smelting the zinc-crust in the blast-furnace furnishes quickly a rich bullion to be cupelled, the process has been abandoned as a regular method of treating the crusts, as none of the zinc is recovered, and the great losses of silver and lead are made up only in a very small degree by resmelting the zinc-bearing by-products.

§ 113. **Distillation of Zinc-Crusts.**[2]—This process, first used by Parkes, has undergone many improvements and has become the one universally used in the United States since Balbach first used graphite retorts. The method, therefore, often bears his name.

§ 114. **Furnaces.**—A furnace, to be suited for the process, must permit the raising and sustaining of a high temperature, and at the same time be of such a construction that a broken retort can be readily exchanged, and the rich bullion that has run out quickly

---

[1] Kerl, "Grundriss der Metallhüttenkunde," 1881, p. 314. Roswag "Désargentation de plomb," 1884, p. 296.

[2] Eilers, *Trans. A. I. M. E.*, iii., p. 314. Rösing, *Zeitschrift für Berg-, Hütten-, und Salinen-Wesen.*, xxxiv., p. 91; *Berg- und Hüttenm. Ztg.*, 1886, p. 421.

and completely collected. Of the different forms three may be cited.

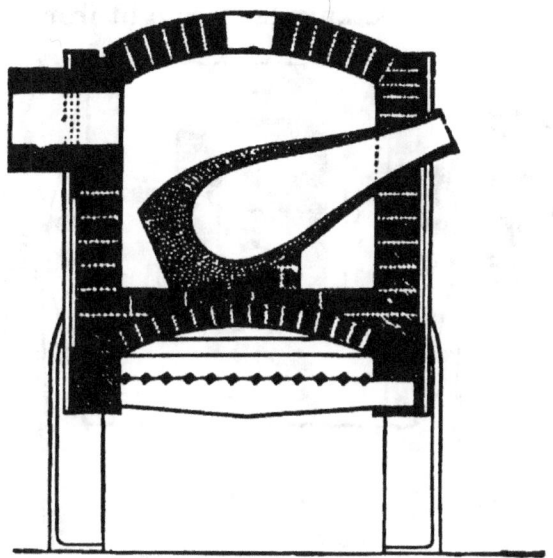

Fig. 244.

1. *The Faber du Faur Furnace* (Figures 244-248).—This is a crucible furnace of cubical form, built into cast-iron frame-work that swings on trunnions, in order that the furnace may be turned over and the contents of the retort emptied.   The furnace is closed

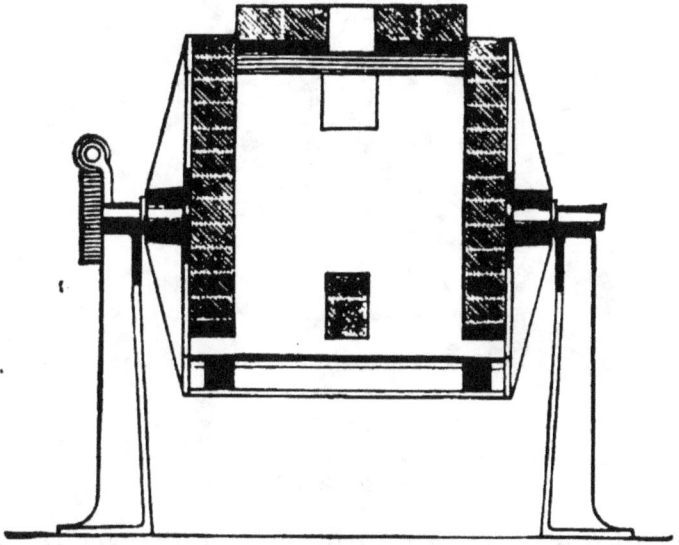

Fig. 245.

at the top by an arched roof, which usually has one opening, the charging opening for the coke ; the products of combustion pass off through a flue, which is generally placed at the back, as in the drawing, but sometimes at the side and occasionally in the roof. At the front is an opening for the neck of the retort. The bottom is formed by two sets of rectangular wrought-iron grate-bars placed

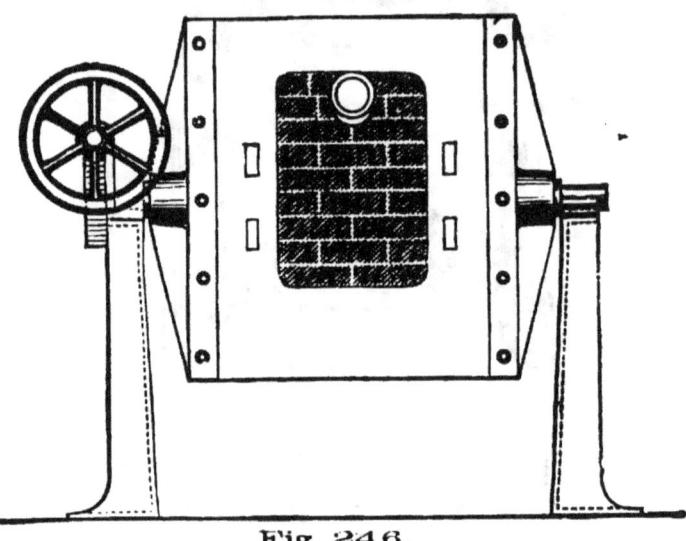

Fig. 246.

on edge. The retort rests on a small brick pillar, which is supported either by a brick arch (as in the figure) or by an iron plate running from front to back, and protected from the heat by two

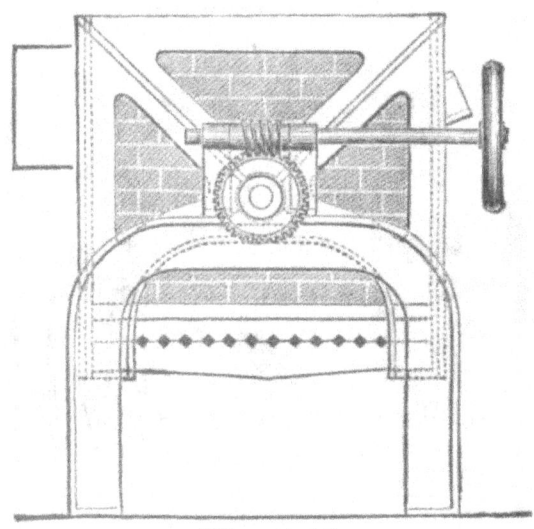

Fig. 247.

courses of fire-brick. The rotation of the furnace is effected by a worm-gear, sometimes simply by means of a lever. The furnace in its original form, as represented by the drawing, was about 4½ feet cube, and was intended for a pear-shaped retort holding 250 pounds of liquated crust. It was lined with a full course of brick, excepting at the front, where the brickwork was $4\frac{1}{2}$ inches thick. At present the retorts, while retaining their original form, are made larger and thinner; they hold as much as 1,000 pounds of crust; the furnace has retained its original size, but is lined on all sides .with a half course of brick. The old retorts were made of raw and burnt clay mixed with about 25 per cent. of graphite to protect the clay from the corrosive action of the lead, and were very

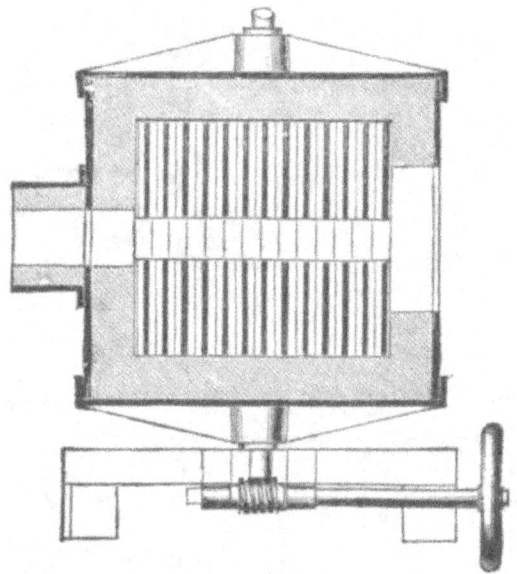

Fig. 248.

thick at the bottom; at present they are made of graphite to which sufficient clay (perhaps 50 per cent.) has been added to give strength and stability to the retort. A 1,000-pound retort is 36 inches high, 8 wide at the neck, 18 at the belly, and 13 at the bottom. It is $1\frac{1}{4}$ inches thick at the neck, and increases to 2 inches at the bottom.

The furnace has replaced most other furnaces since the patents of the inventor expired, proving the general favor it has won. A set of retorting-furnaces is arranged in two ways. They are placed on either side of a horizontal flue in such a way that the openings into it shall not be opposite each other, or they are built around a central stack, say eight in number, each flue extending into the stack and then continuing upward for a few feet. Both arrangements are so chosen as to avoid any obstruction of the draught.

2. *The Tatham Furnace* (Figures 249–251).—This is a stationary crucible furnace. The retort, which holds 500 pounds of zinc-

crust, has the usual inclined position.   It rests on a support at the
back and protrudes at the front through a small arched opening.
The top of the furnace is covered with a clay tile; the products of
combustion pass off through three small openings on one side,
leading into a flue that is common to two furnaces, and terminates.
in the main flue leading to the stack.   Each flue can be shut off by

Fig. 249.

SECTION ON THE LINE A B.

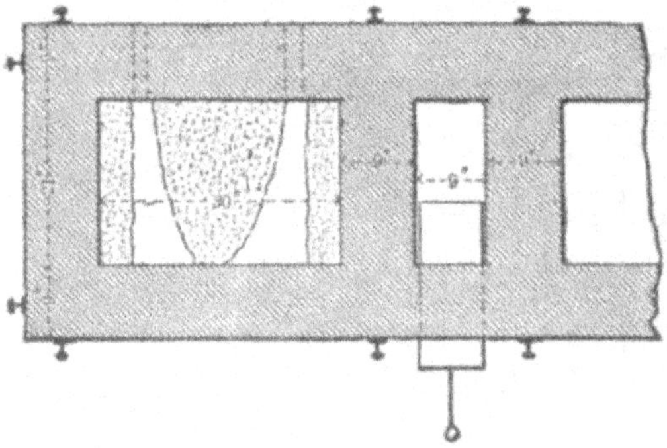

TATHAM
RETORTING-FURNACE.

Fig. 251.

SECTION ON THE LINE C D.

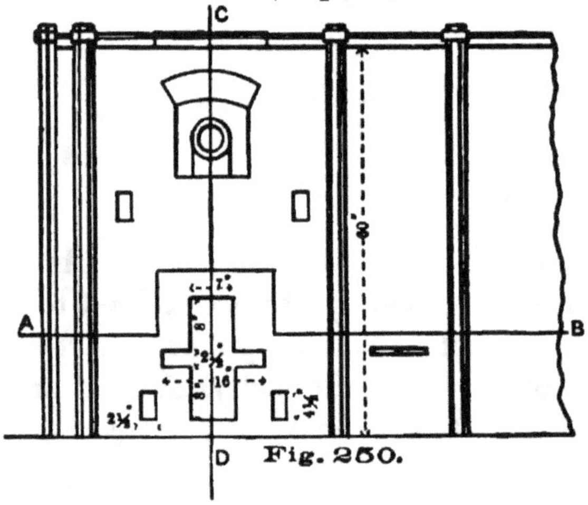

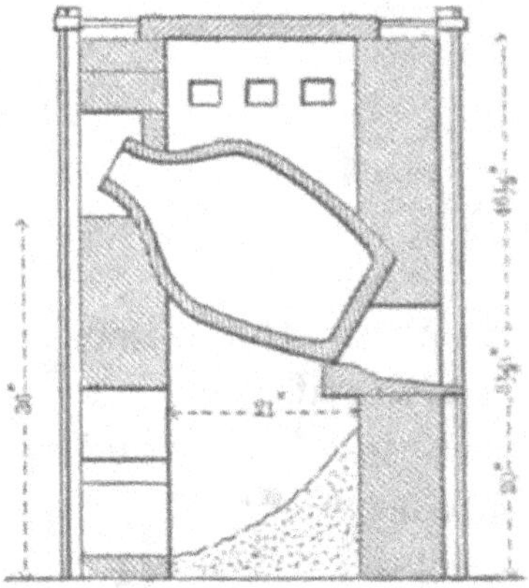

FRONT ELEVATION.

Fig. 250.

a damper.   The bottom of the furnace, inclining from the back to
the front, is made of brasque.   It has a ridge in the centre and is
elevated at the sides in order that any lead may run directly
through the two gutters towards the front and out of the furnace.
The coke is fed at the top, and the ashes and clinkers are removed
at the bottom through the large opening in the front.   Both front
and back have stoking-holes.   The admission of air is regulated
by openings which are closed by bricks.   The rich lead is tapped
from the bottom of the retort and the residue raked out through
the neck.   The tap-hole is bored $1\frac{1}{2}$ inches away from the side-

wall. The entire contents of the retort can, however, be removed through the neck, as is done with most stationary retorts. The ladle used for this purpose is a 6-inch piece of 3-inch gas-pipe closed at one end, and having an iron rod 5 feet long riveted to the other. It is good for six charges. Steitz[1] several years ago constructed a siphon to discharge the contents of a stationary retort. Sometimes it worked well and again it did not. As no reliance could be placed on it, all attempts at using it have been abandoned.

This furnace has replaced the Brodie furnace[2] at the Delaware Lead Works, which had two retorts, one above the other. The results obtained with the Tatham furnace by the writer have been very satisfactory.

3. *Other Furnaces.*—In addition to the furnaces described, which are heated with coke, a few furnaces are in use that are constructed like a reverberatory furnace for the use of bituminous coal. The fireplace may be built on the side of the retort, at the back, or at the front, the aim always being to expose the retort as much as possible to the full action of the flame. The writer has worked a retort fired from the side with good results; retorts fired from the back appear to be satisfactory, while firing from the front has not proved so effective. The reason that retorts heated with bituminous coal have in so many cases given way to those using coke is because they require so much care to keep the temperature uniformly at a white heat, which is absolutely necessary if the retorting is to be finished in the required time. Further, the facility with which the hot contents of a tilting-retort can be discharged has probably influenced the discarding of stationary retorts. The use of gaseous fuel has not been successful here as it has in Germany;[3] coal-oil is being experimented with at present, and promises very satisfactory results.

§ 115. **Condensers.**—The condensers used for collecting the zinc differ very much in form and material. Some are simply old retorts; others are plumbago crucibles (diameter at bottom, 7 inches; at top, 11 inches; height at front, 18 inches; at back, 22 inches; thickness, 1 inch). The former are supported by a specially constructed buggy; the latter rest on tripods, which also hold the receivers for the distilled zinc. Again, they are made of cast iron having the form of a truncated cone. One condenser of this

[1] Egleston, "Silver, Gold, and Mercury," New York, 1887, vol. i., p. 102.

[2] *Trans. A. I. M. E.*, iii., p. 324.   [3] Rösing, *Loc. cit.*

class is 2 feet long, and has·handles on either side, by which it is suspended on two hooks from the iron frame of the furnace. A conical condenser is also made of clay, being about 3 feet long and supported by a tripod. Another form, finally, is that of a sheet-iron cylinder lined with specially moulded fire-bricks. At the base it has, in addition to the tapping-hole, two small pivots around which a thin chain passes and is hooked to the frame of the furnace, thus supporting the condenser.

The condensed zinc is rarely allowed to run off continuously. Usually there is a tap-hole closed by a clay plug or a conical piece of coke, and the condensed zinc is discharged into a suitable mould only three or four hours after the distillation begins. It has been found that more zinc will collect in the condenser if it contains some liquid zinc. Most condensers have a second opening on the upper side for fumes to pass through while the distillation is going on. They go into a sheet-iron pipe leading to the main flue. It has always been considered essential to have this second opening if the distillation is to proceed in a satisfactory way, but at some works using an old retort as condenser the second opening has been dispensed with, the condenser being put in place only when the distillation begins, and not, as is customary, when the retort is well filled.

§ 116. **The Method of Working.**—The method of working is about the same whichever furnace is used. When a new retort has been put in place, it is carefully warmed and brought up to a dull red heat. It is then ready to be charged. The zinc-crust, mixed with from 1 to 3 per cent. of charcoal, is brought in an iron wheel-barrow in front of the retort and charged with a trough-shaped scoop, filled on the wheel-barrow with a small shovel. The temperature is then quickly raised to a white heat, the crust softens and sinks in the retort after half an hour, is pushed down, more crust is added, and this is repeated till the retort is completely filled. The larger rim of the condenser receives a heavy lute of clay, is passed over the neck of the retort, and is made to adhere to the front wall of the furnace ; the lower end rests on its support. About an hour after charging, blue powder and then metallic zinc begin to collect in the condenser. The distillation is finished after from six to eight hours, according to the size of the charge, the percentage of zinc in the crust, and the draught of the furnace. Too much emphasis cannot be laid upon the draught. The slightest obstruction means a failure in distilling off the zinc in the required time, and many consequent disturbances, inconveniences, and losses. The main points to be looked after during the operation are the quick raising of the temperature, and the keeping it high. The higher the temperature, the more rapid will be the distillation, the better the output of metallic zinc, and the lower the percentage of zinc remaining in the rich lead. If the temperature is lowered, blue powder forms, and some of the crust floating on the lead may harden and be suddenly broken by the zinc-vapors developed beneath the crust, when the heat is raised again, and thus cause an explosion. The effect of this is to loosen the condenser from the neck of the retort. At most works it is the practice to introduce at certain intervals, through the upper opening in the condenser, an iron rod, free the neck of the retort with it from oxidized zinc, and then stir up the crust floating on the lead. The saying that the condenser ought to smoke well if the distillation is to go on satisfactorily, is a mistaken one when a condenser with only one opening—the tap-hole, which is kept closed—is giving excellent results.

When the zinc ceases to collect in the condenser the distillation is finished. The last zinc is then removed and the condenser taken off and scraped clean. In the meantime the vapors in the retort pass off into the air. It is assisted by throwing a few chips of wood into the retort. It has now to be emptied. With a tilting-furnace, a slag-pot, lined with brick to prevent the hot metal from cracking the cold bottom, is wheeled in front of the furnace and the lead emptied into it. While the lead is running out the sample for assay is taken. It is then allowed to cool and, after the removal of the dross from the surface, is ladled into small moulds, so as to obtain bars of suitable size for the subsequent treatment in the English cupelling-furnace. The residue in the retort, consisting of slag and charcoal, is removed with an iron scraper. It is essential for the life of the retort that it be well cleaned after each distillation. The retort is now raised, some fine charcoal is thrown into it to prevent the oxidation of globules of lead adhering to the sides, it is then turned back to its normal position, and is ready to be charged again. The grate-bars are cleaned, clinkers adhering to the furnace-walls or sides of the retort are removed, fresh coke is added, and all is then ready for the next operation. The time required for discharging and refilling is about twenty minutes. In scraping, care must be taken to do it gently, so as not to wear off the lower side of the retort. To avoid this, as well as to change the line of contact with the surface of the lead, the position of the retort is sometimes changed, after it has worked about twenty charges, by turning it 180 degrees.

With the Tatham furnace the operations are similar, except that the lead is tapped from the bottom, and only the residue raked out through the neck.

§ 117. **Tools.**—The tools required by one man in retorting are: 2 scrapers (6 feet long, ⅝-inch round iron flattened out at one end to the width of 3 inches and bent up 3 inches) to stir the contents of the retort after distilling and to rake out the residue after tilting); 2 pokers (5 feet long, of ¾-inch iron) to remove the clinkers from the grate ; 1 bar (8 feet long, ¾-inch steel) to break off clinkers from the walls ; 1 scoop to feed the coke ; 1 ladle (6 inches in diameter and 3 inches deep, with a 4-foot handle) ; 2 slag-pots lined with brick ; a wheel-barrow to receive the zinc-crust ; 1 scoop and shovel to charge the retort ; 2 moulds for every retort to receive the zinc ; 10 bullion-moulds.

§ 118. **Results.**—The weight of the charge is from 800 to 1000 pounds of zinc-crust, and it takes from six to eight hours for one operation ; less time is required in winter than in summer. The crust yields from 70 to 80 per cent. of lead containing from 4 to 10 per cent. of silver and from 0.75 to 1.50 per cent. of zinc. The zinc recovered in the form of metal is about 10 per cent. of the weight of the crust, and that in the form of blue powder about 1 per cent. Of all the zinc required for desilverizing, over 60 per cent. is recovered to be used again in the kettles. The quantity of dross varies a great deal ; from 5 to 8 per cent. of the weight of the crust is a fair figure. A retort lasts now forty charges ; formerly twenty-five charges was considered a good average. For every ton of crust 1,100 pounds of coke are required. To use a good quality of coke of uniform size (egg-coke), although more expensive at first, is cheaper in the end, when the life of the retort and the better results obtained are considered. One man attends to from two to four furnaces in a twelve-hour shift.

§ 119. **Roesing's Suggested Improvements.**—Roesing[1] proposes to facilitate the distilling of zinc-crust in a very novel way. He heats the retort, having a basic lining in a tilting-furnace. When the crust has become soft, he introduces some fuel to drive out all the air, and pours pig-iron, heated above its melting point, into the retort, which volatilizes the zinc to be caught in a con-

[1] *Berg-und Hüttenm. Ztg.*, 1890, p. 369.

denser. By tilting the retort, first the iron is poured off to be used over again and then the rich lead.

§ 120. **Comparison of the Two Processes.**—In comparing the smelting of the zinc-crust in the blast furnace with the distillation, it is clear that if the cost of distilling is made up by the amount of zinc recovered, it is preferable to smelting, on account of the greater loss of silver incurred in the latter method. General practice has shown that there is even a margin of profit in distilling the silver-crust, and hence the smelting has been almost entirely abolished. There is one case when it may be doubtful, which is more profitable. The gold-crust is very heavy, compared with its low contents in silver. In retorting, very little metallic zinc is recovered, most of it being obtained in the form of blue powder ; the retort-bullion assays from 400 to 800 ounces silver per ton and is rich in copper. Some works, therefore, prefer to smelt this crust with the addition of matte, obtaining thus a bullion which is free from copper and is easily cupelled.

§ 121. **Treatment of By-Products.**—This is a very important part of refining. It requires to be carried on simultaneously with the main operations, so that the by-products may be disposed of at once and not allowed to accumulate.

§ 122. **Softening Furnace Dross and Skimmings.**—These are usually first liquated in a reverberatory furnace, in order to extract as much lead and silver as possible. One furnace of reasonable dimensions, if kept constantly in use, is able to liquate both prod-ucts of a plant treating 100 tons base bullion a day. While a furnace for liquating the dross ought properly to have a some-what different construction from the one used to treat the anti-mony skimmings, it is found more economical to have one furnace for the two operations. The reverberatory (for example, one with a hearth 8 by 12 feet and 10 inches deep, and having two working-doors on one side only) is built just like a softening furnace. It also has a water jacket to counteract the corroding influence of the antimony skimmings when they are being melted down. The two tap-holes of such a furnace are placed in one of the sides near the centre.

In liquating *dross*, the furnace is charged and heated until the lead that has liquated forms a bath on which the dry dross floats, the operation being continued until sufficient dry dross has accumulated on the lead, that it may be raked out through the furnace-door. The dross directly in contact with the lead is not touched till towards the end. When the level of the lead comes near the furnace-door, enough is tapped into an outside kettle to leave a bath of lead for the charge next following to float on; charging, heating, and drossing are continued until the batch of dross is worked up. The temperature is always kept low, in order that no dross may be taken up again by the lead. The liquated dross is smelted in the blast furnace with a sulphur-bearing ore or by-products (galena, matte low in copper) to form a matte, or to increase the percentage of the matte that has added to the charge. The lead goes to the softening furnace.

In liquating the *softening furnace skimmings*, two objects are aimed at, the elimination of metallic lead and the desilverization of the skimmings. Both are accomplished at once by melting down the skimmings with a reducing (*i.e.*, a smoky) flame in the reverberatory furnace. All the lead carried off from the softening furnace will collect on the bottom of the hearth, and, from the melted skimmings floating on top, part of the lead oxide, converted into metal by the reducing flame, will, while descending through the skimmings, desilverize these and carry the silver with it into the metal-bath below. If necessary, some fine coal is spread over the charge. When the furnace is filled and everything well melted, the liquated skimmings are tapped from the upper tap-hole into slag-pots, and the lead from the lower tap-hole into a kettle, whence it is ladled into bars to go back to the softening furnace.

Another and better way is to have only one tap-hole, and to tap the entire contents of the furnace into a kettle that has an overflow spout. While the lead is flowing into the kettle some can be ladled out, if necessary. As soon as the liquated skimmings appear, they will rise in the kettle and overflow through the spout into slag-pots placed beneath. When the flow ceases, the skimmings still floating on the lead will soon harden and are then removed; the lead in the kettle is moulded by hand into bars and goes to the softening furnace.

When the cake of liquated skimmings is rolled out from the slag-pot and broken, there will be found two products: liquated skimmings assaying three ounces silver or less to the ton, forming the bulk of the cone, and at the bottom a small cake of a silvery-white antimonial speise of lead and copper, which contains as much as 40 per cent. copper, and with 250-ounce bullion assays often as high as 300 ounces silver to the ton. Formerly it was considered preferable to tap the skimmings or let them overflow on an iron plate, as they chilled quickly and were easily broken to pieces, but they ran higher in silver.

The liquated skimmings are smelted in a small (36-inch) blast furnace, with about 100 per cent. of slag and 11 per cent of fuel, care being taken to avoid a slag that is very ferruginous, as speise might otherwise form. In order to reduce the loss of antimony by volatilization (which is considerable), non-argentiferous galena is added to the charge. The sulphur acts as a reducing agent, doing thus to some extent the work of carbon or carbonic oxide. The amount of galena added is so regulated that no matte shall form ; it is from 13 to 28 per cent. of the weight of the skimmings. While this addition reduces the percentage of antimony in the resulting hard lead, this is of no consequence, as, if hard lead assays from 14 to 20 per cent. antimony (a common figure), this is paid for at the same rate as when the lead contains from 7 to 10 per cent.

The resulting slag is always liable to be rich in lead and to contain considerable antimony. Smelting the slag all again in the ore blast-furnace would bring the antimony back to the base bullion, and with every smelting of hard lead, fresh ore-slag would be required. Therefore, a certain amount of slag is kept apart and run over and over again, and only the small excess, produced in every run from coke-ash and fluxes given back to the ore-charge ; the rest is saved for the next antimony run.

Another way to treat the softening skimmings is to smelt them, without liquating, in the blast furnace and to make a hard bullion. This is charged into the softening furnace with soft lead poor in silver (say from the Mississippi Valley), in the proportion of one hard lead to four soft lead. The amount of silver in the original hard bullion is reduced, the softening facilitated, and the resulting skimmings are low enough in silver to be smelted directly into marketable hard lead, which, however, ordinarily runs as high as 6 ounces silver per ton.

The grade and grain of the hard lead are much improved by poling it in a kettle for a few hours or liquating it in a reverberatory furnace at a low temperature. The reason for this is that the copper remaining in the softening skimmings enters the hard lead on being smelted. The dross from liquating the crude hard lead has been found to contain as much as 40 per cent. copper, but generally contains about 10 per cent. The hard lead is usually moulded by ladling from the kettle where it has been poled, or from the one into which it ran from the liquating furnace. The surface of the refined hard lead, when liquid, soon becomes covered with dross, and in order to obtain clean bars without being forced to waste much lead by skimming, it is advisable to place a wrought-iron ring on the lead just large enough for the ladle to pass through and to dip the lead from the ring. Thus only a very small surface will have to be kept bright.

Another way of treating the dross and skimmings has been introduced at some works within a few years, by which the copper is eliminated from the skimmings before they are smelted in the blast furnace. It is to smelt dross and skimmings together in a reverberatory furnace with some galena free from silver. The result is base bullion, a copper-lead matte free from iron, and refined skimmings. The skimmings are discharged through the upper, matte and lead through the lower tap-hole. The skimmings are low enough in silver to be smelted for hard lead, and being free from copper and other impurities can be ladled directly from the lead-well into marketable bars. Thus the common operation of liquating or poling hard lead from the blast-furnace is made unnecessary. The bullion goes back to the softening furnace. The matte is converted in a cupelling furnace with some siliceous material into copper bottoms, 60 per cent. copper-matte and slag, the bottoms collecting the gold. The slag runs off into an iron pot and, when the test is full of metal and matte, is tapped into moulds, placed on a truck beneath the test, by drilling with an augur a hole into the breast. The tap-hole is again closed from the inside by inserting a clay stopper at the back through the tuyere-hole; the furnace is then ready for another charge. When the moulds are cold, the bottoms are separated from the matte, and this is converted in another cupelling furnace into metallic copper to be used in precipitating silver from silver sulphate solution obtained in parting doré silver by means of sulphuric acid. The advantage for refining works in concentrating the copper obtained from base bullion in this way lies in the fact that copper refiners pay for only 93 per cent. of the silver contained in the matte, making no allowance for the lead, and charge two and a half times as much for desilverizing matte as they do for metallic copper, or $40 per ton as against $100.

ANALYSES OF HARD LEAD AND HARD LEAD DROSS.

| | LAUTENTHAL. | | CLAUSTHAL. | PRIBRAM. |
	Before Poling.	After Poling.	Dross from liquat-ing Hard Lead.	
Sb............	12.753	15.390	38.763	24.270
As...........	......	......	......	0.109
Cu...........	1.861	0.152	37.648	0.160
Pb..........	85.291	84.650	22.962	74.886
Ag...........	0.0035	0.0030	......	0.009
Ni............	......	......	(S 0.240)	0.015
Fe..........	trace	trace	0.139	0.018
Zn...........	trace	trace	0.282	0.009
Sn..........	......	......	......	0.524
Reference.......	1	2	3	4

§ 123. **Tin Skimmings.**—Tin ore occurs so rarely with lead-silver ores that it is only an exceptional case when tin skimmings are formed in softening base bullion. At Freiberg (Saxony) the tin of the ores becomes concentrated to some extent in the base bullion produced. The skimmings that rise to the surface in the softening furnace after the dross has been removed have the following composition : [5]

$$PbO, \quad 70.35$$
$$SnO_2, \quad 12.53$$
$$Sb_2O_3, \quad 12.50$$
$$As_2O_3, \quad 4.73$$
$$CuO, \quad 0.61$$

and contain 72.9 ounces silver to the ton. Plattner has introduced a method of desilverizing these skimmings and concentrating the tin in a marketable alloy which contains

$$Sn, \quad 33$$
$$Sb, \quad 14$$
$$As, \quad 1$$

the rest being lead. Details with analyses of intermediary products are given in the reference.

[1] Private notes. [2] *Ibid.*
[3] *Berg- und Hüttenm. Ztg.*, 1870, p. 169.
[4] *Oesterreichisches Jahrbuch,* xxxix., p. 64.
[5] Plattner, *Berg- und Hüttenm. Ztg.*, 1883, p. 417.

§ 124. **Kettle Dross.**—The impurities contained in this dross are very few. It consists principally of lead oxides mixed with metallic lead, and is usually put back into the softening furnace with the next charge after the furnace dross has been drawn off. This is the best way of disposing of it.

§ 125. **Refining Skimmings and Polings.**—These are best treated in a reverberatory furnace that is used for no other purpose, except perhaps for liquating hard lead, in which case the lead resulting from the refining skimmings is sold as second-class lead on account of the copper it has taken up. If the reverberatory is used only for reducing refining skimmings, the resulting lead can be worked in with the regular refining charges and corroding-lead produced. The hearth of a furnace suited for this purpose may be 10 by 5 feet and 9 inches deep, built of fire-brick enclosed in a wrought-iron pan. It will have an inclination of 3 inches from the bridge to the flue, where the main working-door is placed ; a second door is in the middle of one of the sides to introduce and distribute the charge. The tap-hole below the flue discharges the lead into a small spherical kettle having a fire-place beneath. The English (§ 43) or Silesian (§ 44) reverberatory also serves the purpose very well.

The mode of operation is as follows. The hearth of the furnace, dark-red from a previous charge, is covered with a 2-inch layer of fine coal to protect it from the corroding effect of the skimmings. It is then filled with skimmings mixed with 10 per cent. of fine coal, leaving room for the gases to pass. Sometimes the charge reaches only to the working-doors, being renewed from time to time as it shrinks. As the charge heats, lead will flow into the sump, whence it is tapped at intervals. The fire is gradually urged ; when no more lead flows the charge is rabbled, and, when exhausted, drawn.

Four tons of skimmings, yielding about 60 per cent., are worked in twelve hours, two men attending the furnace.

The residue, which contains lead, zinc, antimony, perhaps some arsenic, and coal-ashes, is added to the smelting charge of liquated softening skimmings.

The refining skimmings and polings are sometimes, though rarely, smelted in the ore blast-furnace with the original softening skimmings to reduce the silver-contents of the latter. They are also sometimes charged into the softening furnace after the dross has been removed, in order to oxidize the arsenic and antimony of the bullion, but some zinc enters the hard lead when the skimmings come to be smelted in the blast-furnace, making it unfit for some of its uses, as for example in acid works.

Owing to the scarcity of lead ores now prevailing, nearly all refiners use up their refining skimmings for lead-flux in smelting dry argentiferous ores.

§ 126. **Rich Lead and Metallic Zinc.**—The former goes to the cupelling furnace (§ 133) ; the latter is used in desilverizing. It always contains a few ounces of silver.

§ 127. **Retort Dross and Blue Powder.**—The retort dross is very rich in silver, which must be extracted quickly. The dross, if formed in small quantities, is worked off in the regular cupelling a little at a time, or at some works it is charged back into the retort. If there is too much for this, it is allowed to accumulate, and is scorified on the bath of lead free from or low in silver, with which a new test is usually charged, in the cupelling furnace. Sometimes the retort dross is added to the softening furnace, after its own dross has been removed, that the silver may be taken up by the lead, and the impurities oxidized and taken up by the skimmings. It is not often that the dross is added to the charge of the bullion blast-furnace.

The blue powder, consisting of a mixture of finely divided metallic zinc and zinc oxide, always contains some silver, say from 4 to 5 ounces per ton. It is not readily disposed of. If distilled by itself with the addition of charcoal, it yields from 33 to 55 per cent. of zinc. At some works it is fed back to the retort with the following charge of zinc-crust ; at others it is added with the first zinc to the desilverizing-kettle, that the metallic zinc may be taken up. If this is done, 50 pounds of blue powder are charged at a time into the kettle before the lead is tapped into it from the softening furnace. It is stirred into the bullion while this is rising in the kettle. It does not remove much silver from the kettle, but serves rather to saturate the lead with zinc and to assist in removing the gold and copper. Sometimes the blue powder is screened to remove all coarse particles and is then sold to zinc-works.

§ 128. **Litharge** is reduced in a reverberatory furnace or goes to the bullion blast furnace ; sometimes it is added to the softening furnace after drossing. While the reduction of litharge in the reverberatory furnace is preferable from a purely metallurgical point of view, the blast-furnace is in more common use, as refining works are thus often enabled to smelt dry silver ores at their own price, as the Western smelters do not compete for them on account of the usual scarcity of lead ores. It is important for a refinery to smelt some ore in treating by-products in order to make new slag, and not be forced to smelt old slag over and over, which is expensive and causes losses in lead and silver, as these slags become gradually rich in zinc.

§ 129. **Old Retorts, Cupel-Bottoms, etc.**—These are added to the charges of the bullion blast-furnace.

§ 130. **Table of Desilverization.**—The following table gives a summary of the work done in an American desilverizing works, as shown in detail in the preceding pages :

1. Base Bullion : softened in reverberatory furnace yields soft bullion ', softening dross ', softening skimmings '.

2. Soft Bullion : desilverized in kettle yields kettle-dross (back to 1), gold-crust ', silver-crust ", desilverized lead ".

3. Gold-Crust : liquated yields liquated gold-crust ', and gold-bearing base bullion '.

4. Liquated Gold-Crust : retorted yields retort dross ", blue powder ", zinc ', retort (doré) bullion '.

5. Retort (Doré) Bullion : cupelled yields doré silver ', litharge ", cupel-bottom ".

6. Doré Silver : sold, or parted by sulphuric acid or electrolysis, yields gold, silver, blue or green vitriol, or precipitated copper.

7. Softening Dross : liquated in reverberatory I. yields base bullion ', and liquated dross ".

8. Softening Skimmings : liquated in reverberatory I. yields base bullion ', and liquated skimmings '.

9. Liquated Skimmings : smelted in 17 yield hard lead " and slag ".

10. Hard Lead : poled in kettle or liquated in reverberatory II. yields market hard lead and dross ".

11. Silver-Crust : goes to 3, 4, and 5 in separate apparatus, yielding fine silver and by-products treated like 4 and 5.

12. Desilverized Lead : refined in reverberatory yields corroding-lead and refining skimmings ".

13. Refining Skimmings : reduced in reverberatory II. yield lead (either to 12 or sold) and dross ", or smelted with ore in 16.

14. Retort Dross : either fed on cupel ', goes back to retort ', softening furnace ', or bullion blast-furnace ".

15. Blue Powder : goes either to 2 (as first zinc), or back to retort ', or screened and sold.

16. Bullion Blast-Furnace.

17. Hard-Lead Blast-Furnace.

§ 131. **General Remarks.**—A few general remarks regarding Parkes' process are in place.

The output of silver with average base bullion as it comes from Western smelting works (running from 150 to 250 ounces silver, and 0.5 ounces and less gold per ton, being pretty free from dross, but containing some arsenic and antimony) is never less than 99.5 per cent., and generally there is a slight surplus of silver; with gold it is from 98 to 100 per cent. (a surplus is rare); with lead from 99 to 99.5 per cent.

A refinery makes its deduction for loss in treatment and for cost of working, the base bullion being delivered f. o. b. refinery. Some refining works sum up both items and make a general charge by the ounce of silver; *e. g.*, one cent for every ounce of silver, the lead being paid on New York quotations and the freight to New York being deducted. Others make a working charge by the ton and a deduction for loss in silver; *e. g.*, the working charge varies from $10 to $14 per ton, and the reduction amounts to 2 or 3 ounces silver per ton, the lead being paid as above. A number of other ways of settling between smelter and refiner are in use, but the two quoted are the most common ones.

No detailed statement can be made about the cost of refining that would be generally applicable, as the single items vary greatly in different refineries. In a general way it may be said that the actual operating expenses, including interest and taxes, vary from $5 to $6 per ton of base bullion. If the general expenses, such as salaries, marketing, etc., be added, this figure rises to $8 and $9. If, finally, the loss in metal and incidentals is added to the last figure, giving the total final cost, this will be found to vary from $10 to $12 per ton.

In a refinery where so many by-products result, it is essential to know how much silver, gold, and lead is contained in each of them. These quantities vary a great deal with the different kinds of bullion treated. It is therefore essential to be able to follow up the metals from the time they enter the refinery until they leave it.

Below is reproduced the scheme of the Kettle-Book, the most important one of all the books. It shows the various products and the amount of fuel consumed :

KETTLE-BOOK.*

189— Date.		Lot.	Name of Bullion.	Charge.	Gross Weight	Assay: Ozs. per ton.		Contents: Total Ounces.	
Month.	Day.	No.		No.	Lbs.	Ag.	Au.	Ag.	Au.

Softening Dross.	Softening Skimmings.	Kettle Dross.	Net Weight in Kettle.	Assay: Ozs. per ton.		Contents: Total Ounces.	
Lbs.	Lbs.	Lbs.	Lbs.	Ag.	Au.	Ag.	Au.

Zinc Added: Lbs. for.		Gold-Crust.	Assay: Ozs. per ton.	Contents: Total Ounces.
Ag.	Au.	Lbs.	Ag.	Ag.

Liquated Gold-Crust.	Liquated Silver-Crust.	Refining Skimmings.	Refined Lead.	Coal.	REMARKS.
Lbs.	Lbs.	Lbs.	Lbs.	Tons.	

It is to be noted that the "net weight in kettle," before adding the first zinc, is found by deducting the sum of weights of softening dross, skimmings, and kettle dross from the base bullion charged into the softening furnace. The "net weight in kettle"

* Owing to the form of this volume, it has been necessary to put the headings in four rows, one beneath the other, while in the actual kettle-book, they simply run across the two opposite pages in one line.

after adding the first zinc, is found in a similar way, by adding the pounds of zinc used for extracting the gold to the net weight before the first zinc addition was made and deducting the weight of the gold-crust from this sum.

The principal other books kept in a refinery are : the retort-book, the cupellation-book, and the books for reverberatory furnace I. (liquating softening dross and skimmings) and II. (reducing refining skimmings and liquating hard lead), and special assay-books for the softening furnace, desilverizing-kettle, retorts, and cupelling furnaces.

§ 132. **Relative Advantages of Parkes' and Pattinson's Processes.**—The many advantages[1] (§ 90) that Parkes' process has over Pattinson's have made it *the* desilverizing process used in the United States, there being only one refinery (at Eureka, Nevada), that desilverizes with the Luce-Rozan process. There is, however, one instance in which Pattinson is to be preferred to Parkes. It is when the base bullion is rich in bismuth. During the crystallization bismuth follows the liquid lead (§ 91). It is thus concentrated in the rich lead, and can be recovered when this is being cupelled. The bismuth-contents of the market-lead thus become very much lowered, although not quite removed. In Parkes' process the bismuth enters only to a small extent (§ 94) into the zinc-crust, with the result that the refined lead may become richer in bismuth than the original bullion. This difference is shown by the two following analyses by Hampe[2] from Lautenthal market-lead where Parkes' process replaced that of Pattinson.

Lautenthal.	Pb.	Cu.	Sb.	As.	Bi.	Ag.	Fe.	Zn.	Ni.
Pattinson........	99.9662	0.015	0.010	none	0.0006	0.0022	0.004	0.001	0.001
Parkes..........	99.982139	0.001413	0.005698	none	0.005487	0.000460	0.002289	0.000634	0.000680

An interesting combination of the two processes is found at Freiberg,[3] where the original base bullion, containing from 116 to 233 ounces silver per ton, and from 0.02 to 0.06, rarely 0.16 per cent. bismuth, is concentrated by crystallization to a liquid lead with 0.17 per cent. bismuth, assaying 583 ounces silver per ton,

[1] Phillips, *Engineering and Mining Journal*, May 27, 1887.

[2] *Zeitschrift für Berg-, Hütten-, und Salinen-Wesen in Preussen*, xviii., p. 195.

[3] *Engineering and Mining Journal*, December 4, 1886 ; *Berg- und Hüttenm. Ztg.*, 1887, pp. 45, 192.

which is cupelled, while the crystals averaging 30 ounces silver per ton are desilverized with zinc, the bismuth in the market-lead not exceeding 0.02 per cent.

Large shipments of base bullion with considerable quantities of bismuth are of rare occurrence in refining works. Occasional ones are worked in with other bullion that is free from, or at least low in, bismuth.

CHAPTER XI.

CUPELLATION.

§ 133. **Introductory Remarks.**—The process of cupellation has for its object the final separation of lead and silver. It consists in melting and heating in a reverberatory furnace argentiferous lead with access of air to the temperature at which litharge forms on its surface. This is run off and is in part absorbed by the hearth, while the silver, having scarcely any affinity for oxygen, remains behind in the metallic state. The oxidation of the lead is principally effected through the action of the blast playing over its surface, but is also assisted by the litharge when formed, as this absorbs oxygen and gives it off again to the lead and its impurities. The most important of these are copper, arsenic, antimony, bismuth, silver, and gold.

The bulk of the copper is removed with the dross, all the arsenic and antimony with the skimmings, just as in softening base bullion (§ 97). The copper remaining with the lead after drossing, is taken up only very gradually by the litharge. As it has less affinity for oxygen than lead, its oxidation must be caused not so much by the action of the air as by a large quantity of litharge acting on a small amount of copper. With reversed conditions cuprous oxide[1] oxidizes lead. The oxidizing action of cuprous oxide contained in the litharge seems to be the cause of the fact that cupriferous lead is cupelled quicker and with less loss in lead and silver than if the lead were free from copper. Kerl[2] states that in cupelling thirty tons argentiferous lead containing from $\frac{3}{4}$ to 1 per cent. copper, the operation lasted twenty-four hours less than when copper was absent, and that the losses in metal were as 2:7.

Bismuth is concentrated with the silver in the lead until towards the end of the process, and then greatly retards the progress of the work. It is finally oxidized and enters the litharge, giving this a greenish color, and is absorbed by the hearth-material, while the

---

[1] Berthier, "Traité des essais," Liége, 1847, vol. ii., p. 572.
[2] "Grundriss der Metallhüttenkunde," 1881, p. 270.

24

silver also retains some bismuth with great tenacity. If the bismuth is to be recovered, the silver is concentrated only to a certain degree (say to 50 or 60 per cent.) in the lead, which is then cupelled in a separate furnace ; the saturated part of the hearth of this furnace and the litharge form the raw material for the extraction of bismuth in the wet way.

Silver is always present in the litharge, probably in both forms, as oxide and as metal. Rose[1] states that silver oxide begins to lose its oxygen at 250° C. ; according to Sainte Claire-Deville and Debray,[2] as well as Troost and Haute-Feuille[3], the oxide appears to be able to exist at a very elevated temperature. Finally Wait[4] dissolved from litharge containing 2.94 per cent. silver, by means of acetic acid, 18.67 and 19.25 per cent. of the silver present. As metallic silver is insoluble in acetic acid, the silver dissolved must have been present as oxide. If silver-bearing litharge remains in contact any length of time with metallic lead having little silver, it loses its silver. Thus at the beginning of the cupellation little silver is contained in the litharge. During the progress of the operation the lead becomes richer, more silver is liable to be oxidized, and less of it again reduced by the enriched metallic lead. Fine particles of argentiferous lead cannot be prevented from being carried away by the litharge. From both causes, therefore, the tenor in silver of the litharge will increase with the progress of the cupellation.

Gold, finally, follows the silver in the cupellation, but none of it, or perhaps a trace, are found in the litharge.

According to the general construction of the furnace, and the consequent mode of operating, cupellation is generally discussed under two heads :

A. German Cupellation.[5]
B. English Cupellation.

[1] "Handbuch der analytischen Chemie," Leipsic, 1867, i., vol. 339.

[2] Graham-Otto-Michaelis, "Anorganische Chemie," Brunswick, 1884, iii., p. 985.

[3] *Ibid.*

[4] *Trans. A. I. M. E.*, xv., p. 463.

[5] The writer is fully aware that the German cupelling furnace is not much found in this country ; in fact, he knows of only one refinery that uses it. If, nevertheless, it is treated here in more detail than may seem necessary, the reason is that operations which it has in common with the English furnace can be more easily dealt with here than later.

A. German Cupellation.

§ 134. **Characteristics.**—The characteristics of this method are : a large reverberatory furnace with a fixed bed and a movable roof; the fact that the bullion to be cupelled is all charged at once, and the silver not refined in the same furnace where the cupellation was carried on.

§ 135. **The Furnace.**—The furnace selected for illustration is the one in operation at Přibram [1] (Figures 252–255). It differs somewhat from the generally accepted circular form, [2] and is an improvement on it. Figure 252 shows the fire-place *a* at the right, and the flue *b* at the left, of the hearth. Figure 253 represents a horizontal section of one furnace and the fire-place *a'* of a second furnace, the furnaces being built in pairs. The products of combustion pass downward through four separate flues *b*, which unite in one main flue *c*, leading to the dust-chambers. In Figures 254 and 255 are seen two vertical sections on the same line EF, Figure 254 representing the furnace before, Figure 255 after, tamping in the hearth. The furnace is built of common red brick, with the exception of the parts that are exposed to the flame, as indicated by the cross-hatching. In the upper part of the foundation and in the side-walls small channels *d* and *e* reaching outward are left open. These serve as drains for the moisture. At the back of the furnace are three openings *f*, through which the tuyere-pipes are introduced. At the front is the litharge-channel *g*, which can be closed by a sliding-door ; a cast-iron breast-plate *h* serves as support for the upper hearth *i*. The movable arched roof *k* rests on an L ring *l*, and is removed with a differential pulley suspended from a traveller. On the foundation is built a brick bottom *m*, the brick being set dry. Beneath its lowest point, just below the cavity *n*, is a cast-iron plate *o*, to prevent any leakage of metal through the drain *d*, should the working-hearth crack or be injured in any other way. The flues leading from the fire-place *a* to the hearth are shown in *j*. The fire-brick part of the furnace is encased in cast-iron plates that have openings corresponding to the upper drains. The whole furnace is well bound together by buckstays and tie-rods. The fuel used is a mixture of bituminous coal and lignite ; the ash-pit is closed as the blast is introduced beneath the grate-bars.

[1] *Oesterreichisches Jahrbuch*, xxxix., p. 46. Private notes, 1890.
[2] *Zeitschrift für Berg-, Hütten-, und Salinen-Wesen in Preussen*, xxii., p. 89, and *Berg- und Hüttenm. Ztg.*, 1872, p. 415.

§ 136. **Plattner's Cupelling Furnace.**—In this connection may be mentioned the Plattner[1] modified German cupelling furnace. It has the form of a reverberatory furnace; the hearth is rectangular in plan (13 by 8½ feet), and receives its blast from two pipes on either side of the fire-bridge; the litharge-channel is at the opposite end, beneath the flue, which carries off the products of combustion as well as the lead-fumes. It would seem as if this modification might, with much advantage, be applied to English cupelling furnaces. Less fuel is required and less metal volatilized, because the litharge need not be heated to such a degree to remain liquid; and no fumes enter the cupelling-room, as they are all carried off with the fuel-gases.

§ 137. **Mode of Conducting the Process.**—The operations necessary to work a charge are six in number: preparing the working-bottom, charging and firing the furnace, softening the bullion, cupelling the softened bullion, removing the crude silver, and refining the crude silver.

1. *Preparing the Working-Bottom.*—To be suitable for preparing the hearth, the material must be one that will not be attacked by the litharge, nor crack, must be sufficiently porous to absorb some litharge, and free from any reducing agent (organic matter, metallic sulphide). The material most used is a marl. The composition of that used in the Hartz Mountains varies according to Kerl and Wimmer[2] within the following limits.

$$SiO_2 \quad 21.22 - 22.24$$
$$Al_2O_3 \quad 5.39 - 6.76$$
$$Fe_2O_3 \quad 3.54 - 5.39$$
$$CaO \quad 65.65 - 66.41$$
$$MgO \quad 1.05 - 2.22$$

Usually the hearth-material is a mixture of dolomite or limestone with fire-clay. For instance, at Tarnowitz[3] a dolomite of the composition,

$$SiO_2 \quad 6.00$$
$$Al_2O_3 \quad 7.00$$
$$Fe_2O_3 \quad 4.10$$
$$CaCo_3 \quad 49.86$$
$$MgCo_3 \quad 32.82$$

[1] Drawings in Arche, "Die Gewinnung der Metalle," Heft i., Plate i., Leipsic, 1888. Results in *Berg- und Hüttenm. Ztg.*, 1866, p. 211.

[2] *Berg- und Hüttenm. Ztg.*, 1858, p. 241.

[3] *Zeitschrift für Berg-, Hütten- und Salinen-Wesen in Prussen*, xxxii., p. 107.

is mixed with 25 per cent. of clay. At Přibram three parts of limestone are ground together with one part of clay so as to pass a 5-mesh sieve; at other places, an 8-mesh sieve. The hearth-material has to be moistened before it is beaten down in the furnace. For this purpose it is spread on the floor, sprinkled with water from a rose, and turned over and over with a shovel, that the moisture may be equally disseminated through the powder. If left over night, it must be covered with wet cloths and worked again the next day. The material is of the right consistency if, when pressed in the hand, it coheres to a lump, but has not sufficient moisture to adhere to the hand. To obtain a uniform material, it is sifted, just before using, through a coarse hand-sieve, and any lumps that may have formed are broken up or thrown aside. Sometimes it is introduced all at once, sometimes in two separate layers. The latter is the way at Přibram, as Figure 255 shows, *i* being the upper and *i'* the lower bottom. Before the filling is put in, the brick bottom is sprinkled with water, that it may not take up any moisture from the hearth-material. At Přibram the lower bottom *i'* is first tamped down to the form shown in the drawing. The tool required, the tamping-iron, is a cast-iron disk of about six inches in diameter and one inch in thickness, with a socket into which fits a wooden handle about four feet long. The tamping is begun at the centre, proceeds in the form of a spiral to the side-walls, and returns in the same manner, care being taken that the circular indentations should overlap in part those made in working outward. By giving attention to this point, the surface will be evenly beaten down, which is essential. Before putting down the upper bottom, the surface of the lower one is roughened by scratching it with the point of a chisel. This is done that the bottom to be put down may adhere to the one already in place. The tamping of the upper bottom is done in the same way as the lower one, only the surface, when finished, must be perfectly smooth. It is of prime importance that the hearth should have just the right degree of hardness. This is easily indicated to the ear and hand after a little experience. If too hard, it will crack and not be sufficiently porous ; if not hard enough it will absorb too much lead. If the material was too dry, the hearth will peel when heated ; if too wet, it cannot be beaten to the desired hardness. That it shall not adhere to the tamping-irons, these are slightly warmed. The thickness of the hearth at the bottom and sides varies somewhat ; the least is perhaps six inches at the bottom and eight inches at the sides. The general rule for the curvature of the hearth is that the more concave the bed is, the easier will be the cupelling and the harder the finishing ; the flatter the bed, the harder the cupellation and the easier the finishing.   When the hearth is completed, a cavity *n* (one inch deep and thirty-four inches in diameter) is cut in the lowest point to receive the silver.   This is located a little to one side of the medial line, towards the fire-bridge, that the silver may be kept easily molten at the end of the operation.

2. *Charging and Firing the Furnace.*—The furnace now receives its charge of 25 tons of lead enriched by Luce and Rozan's process.   Sometimes the bottom is covered with straw before charging, to prevent its being damaged during the operation.   The bars are placed in such a way as to leave an open space reaching from the tuyeres to the litharge channel.   The hood is then lowered on a clay lute placed on top of the furnace.   The litharge-channel is closed by lowering the door, the fire kindled on the grate, and soon the blast below let on.

3. *Softening the Bullion.*—The lead melts down slowly.   The dross rises to the surface and is drawn off through the litharge-channel.   The temperature is raised and the blast put on through three tuyeres, the skimmings form and are drawn off, and finally pure litharge takes their place.

4. *Cupelling the Softened Bullion.*—The temperature and the blast are now lowered, and are kept low during the larger part of the operation.   They are raised only towards the end, when the enriched silver-lead alloy requires a higher temperature to give up the last parts of lead.   When the skimmings have been removed and the cupellation has somewhat progressed, the convex surface of the lead will be exposed to the action of the blast, while the lead near the periphery will be covered with litharge.   The width of this rim depends on the rate at which the litharge is allowed to run off through the litharge-channel.

As litharge melts only at 954° C. (§ 5), a temperature of about 980° C. has to be maintained, if it is to remain liquid.   Lead, melting at 325° C. (§4), would be volatilized to a considerable extent, if fully exposed to the action of the blast at this high temperature. The litharge is therefore allowed to run off only to such an extent as to give the rim a width varying from 12 to 15 inches at the beginning, and of 5 inches toward the end, of the operation.   The blast playing on the surface of the lead forms small waves and drives the litharge towards the channel.   The pressure is about 8 ounces per square inch, and about 300 cubic feet of air are delivered per minute through the three tuyeres. In order to remove the litharge, a gutter is cut into the channel by means of a scraper. This is an iron rod, 8 feet long and ⅜ inches in diameter, flattened out at both ends. One of these is bent to encircle a wooden handle, while the other, only slightly flattened, is sharpened and bent to the form of a hook. In cutting the gutter, the entire edge of the tool must be used, and not one of the corners. If larger pieces of the breast are to be cut out, it is done with a chisel-pointed bar, say ¾ of an inch in diameter. The rate at which the litharge runs off depends on the depth of the gutter and the strength of the blast. The depth is correct when the litharge runs off in a thin stream which stops as soon as the blast is lowered. If it runs too fast, the rim of litharge in the furnace decreases and lead is volatilized, while the litharge is not sufficiently desilverized from too short a contact with the lead beneath it. If it runs too slowly, the rim of litharge becomes too broad and the cupellation is retarded ; there is again loss in silver by the higher temperature that is required to keep the larger amount of litharge liquid, and if the temperature be not raised sufficiently, lead will be carried out mechanically by the litharge. The litharge gutter is first cut into the breast on the side farthest away from the bridge and gradually moved towards the opposite side, that the final litharge may be drawn off as near the fire-bridge as possible. The current of the litharge, when flowing out of the furnace, is directed in such a way as to form a large cake in front of the furnace. In some instances a U-shaped piece of sheet-iron is placed upright in front of the channel, that the litharge may collect in a rectangular block. Before removing this, the litharge in the centre that is still liquid is tapped from near the bottom of the cake. The litharge resulting from the cupellation is graded according to the silver-contents and the percentage of impurities into marketable litharge and into a by-product to be treated by a separate process. As the cupellation progresses towards the end, the temperature is raised and the blast increased, the side tuyere-openings are closed, and two tuyere-pipes introduced through the central opening ; finally, the last film of litharge disappears from the surface of the lead with a characteristic phenomenon, the so-called brightening,[1] which every reader has watched while making a silver assay.

[1] Van Riemsdijk. *Berg- und Hüttenm. Ztg.*, 1880, pp. 247, 275, and Bock, *Op. cit.*, 1880, p. 409, have made interesting investigations on this subject.

The products of the Přibram cupelling furnace have, according to Dietrich, the following compositions :

	Dross.	Tin Skimmings.	Antimony Skimmings.	Red Lithare.	Green Litharge.	Cupel-Bottom.	Flue Dust.
Pb......	30.75	13.40	.....	......	......	......	.....
PbO.....	55 27	64.97	77.95	98.370	98.140	68.860	64.41
PbO combined with (AsSb)₂O₆.	.....	.....	11.87	......	......	......	.....
CuO.....	1.99	0.29	0.28	0.069	0.080	0.070	.....
Bi₂O₃.....	.....	.....	.....	......	......	trace	trace
MoO₄.....	.....	.....	.....	......	......	trace	.....
As₂O₆.....	1.42	1.87	0.92	0.010	0.009	......	⎮trace
Sb₂O₆.....	1.83	6.76	5.85	0.074	0.067	0.530	11.40
SnO₂	0.72	10.31	trace	......	......	.....	.....
Ag........	.....	.....	0.004	......	......	.....	.....
Ag₂O.....	0.307	0.189	.....	0.004	0.0048	0.170	0.013
Al₂O₃.....	⎫	⎫	0.32	0.072	0.056	2.120	.....
Fe₂O₃.....	⎬ 0.54	⎬ 0.23	0.14	0.010	0.014	0.300	.....
ZnO......	0.13	0.05	trace	0.009	0.012	trace	0.50
Ni........	0.09	trace	.....	......	......	......	.....
NiO......	.....	.....	0.04	0.005	0.005	......	trace
CaO......	0.45	0.43	0.95	0.256	0.362	......	trace
CaCO₃....	.....	.....	.....	......	......	24.100	.....
MgO......	.....	.....	.....	......	......	trace	.....
CO₂......	.....	.....	.....	0.383	0.432	......	.....
SiO₂......	1.75	0.94	0.37	0.320	0.350	2.970	4.35 ⎱
							+ ash ⎰
S.........	2.30	0.65	0.09	......	......	......	.....
SO₃	4.12	1.37	.....	0.034	0.027	0.040	16.65
C........	.....	.....	trace	......	......	......	trace

CRUDE SILVER.

	Přibram.[1]	Freiberg.[2]	Wyandotte, Mich.[3]	
Ag.....................	95	92.180	98.691	99.593
Pb.....................	5	4.210	1.090	0.260
Cu.....................	..	2.104	0.117	0.106
NiCo..................	..	0.600	0.004	0.008
Fe.....................	..	......	0.090	0.031
Bi.....................	trace	......	0.0058	......
Au.....................	trace	......	0.0023	0.0015

[1] *Loc. cit.*
[2] Stölzel, "Metallurgie," Brunswick, 1868–1886, p. 1182.
[3] *Trans. A. I. M. E.,* ii., p. 97.

The time required to cupel the twenty-five tons of rich lead is eighty hours.   It is divided as follows :

	Hours.
Preparing the hearth-material, making the hearth, and charging the lead	8
Melting down and wheeling the necessary coal	16
Drossing	6
Drawing the tin skimmings	6
Drawing the antimony skimmings	3
Running off market-litharge	23
Running off rich litharge	18
	80

The cupellation is in charge of three men, each with a helper, working in eight-hour shifts.   For every 100 tons of base bullion are consumed 19.63 tons of coal and 23 bushels of hearth-material (which includes the refining of the silver).   The loss in silver is 0.83 per cent.; that in lead, 4.33 per cent.   The figures do not include the loss endured in re-treating some of the by-products. Thirty-six per cent. of the litharge is low enough in silver to be sold in the market, the bullion from which it is made averaging 167 ounces silver to the ton.

5. *Removing the Crude Silver.*—After the silver has brightened, the blast is shut off, the tuyere-pipes are removed, and the litharge gutter is closed with a ball of clay.   Two knife-shaped pieces of wrought iron are introduced through the litharge-channel and pressed into the cake of crude silver.   First warm, then cold, water is allowed to run into the furnace, and the silver then removed through the central opening at the back.   It is cleaned, weighed, etc.   The furnace is left to cool till the next day, when the hearth is examined for small particles of silver that adhered to it.   The upper hearth is then removed with a pick.   Part of it is soaked with litharge to a depth of 2 or 3 inches.   This is screened off from the unsoaked part, which is mixed in with the hearth-material for the next charge, while the lead-soaked part goes to the blast furnace.

6. *Refining the Crude Silver.*—The refining of the crude silver has for its object the removing of impurities, which vary from 2 to 10 per cent.   Formerly this was done exclusively in a small oval reverberatory furnace, having a working door at one side or at the flue-end, and a working bottom of similar composition to the cupelling bottom, the reason being that the loss in silver and the consumption of fuel were considered very much smaller than if

the firing was done in the big cupelling furnace. Ohl[1] and Foehr[2] have since proved this not to correspond to the facts, and the refining in a separate furnace has received a check. Since the discovery by Roessler[3] how to refine crude silver in a plumbago crucible by means of silver sulphate, the refining in a separate furnace has been abolished entirely at some works.

Refining in a crucible is carried on at small works by melting down the silver, uncovering the crucible that the air may oxidize the impurities. These are stiffened by sprinkling bone-ash or hearth-material on the silver and then removed with a skimmer, the operation being repeated till no more impurities rise to the surface. A slag obtained by Curtis[4] at Wyandotte, Mich., sand being used in refining, contained, in addition to silicate of lead, the following metals :

$$
\begin{array}{ll}
(NiCo)O & 0.550 \\
CuO & 0.203 \\
Bi_2O_3 & 0.026 \\
Ag & 1.837 \\
Sb_2O_4 & 0.639 \\
As_2O_3 & 0.005 \\
\hline
& 3.260
\end{array}
$$

Roessler found that if silver sulphate is added to melted silver in a crucible, first the lead and then the bismuth are converted to sulphates, the silver being at the same time set free. Copper is not removed by silver sulphate. By keeping separate the different slags he concentrates the bismuth in a comparatively small amount of slag, to be treated separately, while the first slag contains most of the lead. To prevent the crucible from being attacked, he introduces a layer of quartz-sand on top of the silver, and then stirs in the silver sulphate in the middle. The sand serves at the same time to stiffen the slag, which is then removed with a skimmer. The process, as seen by the writer at the Lautenthal Smelting and Refining Works in 1890, differs slightly from the manner indicated above. It is as follows. The silver sulphate is produced by dissolving silver in sulphuric acid of 66° B. in a small cast-iron kettle. The solution is allowed to cool, is then diluted to

---

[1] *Berg- und Hüttenm. Ztg.*, 1879, p. 274.
[2] *Ibid*, 1885, p. 381.
[3] *Berg- und Hüttenm. Ztg.*, 1889, p. 387.
[4] *Trans. A. I. M. E.*, ii., p. 98.

60° B., when nearly all the silver sulphate will fall out as a slightly yellow cheesy mass. The supernatant liquor is drawn off as much as possible, and the remaining dilute acid driven off by heating. Special arrangements are required to cool the vapors, as they carry finely-divided silver sulphate along with them. The temperature is raised to redness in order to fuse the silver sulphate, which when liquid is cast into moulds and is ready for use. The color of the melted sulphate is grayish-green ; it is hygroscopic, and is therefore kept in a lead-lined wooden box ; 1,000 parts contain 650 parts of silver.

Crude silver of a fineness varying from 950 to 980 thousandths is melted down in a plumbago crucible holding 700 pounds. The crucible is heated with coke in a small cylindrical furnace having in the lower part two 1¼-inch openings for the blast-pipes. On trying to stir in the silver sulphate as advised by Roessler, it was found that sometimes it got beneath the layer of sand, spread over the silver, and corroded the crucible. To prevent this, a wrought-iron ring (⅛ inch thick, 10½ inches in diameter, and 7 inches high) is coated on either side with a 3-inch layer of clay, and placed on the silver. Into the centre are introduced with a ladle from 6 to 8 pounds of sulphate (the size of a hen's egg) that has been warmed. As soon as it comes in contact with the silver, this begins to boil. When the effect decreases, the silver is stirred with an iron rod to assist the action of the sulphate. From twenty-five to thirty minutes after the sulphate has been added, this is completely decomposed, a slag has collected on the surface of the silver, and quartz is added to stiffen it, that it may be removed with a skimmer. A second, a third, and, if necessary, a fourth addition of silver sulphate is given to make the silver fine. The test made for fineness is to dissolve some silver into nitric acid and to supersaturate with ammonia. No precipitate must form even after standing.

The amount of silver sulphate required to fine the silver is about 1½ times the total quantity base metal present. Thus 700 pounds of crude silver, being 970 thousandths fine, contain 21 pounds of base metal, which would require 31 pounds of silver sulphate to be added in three portions. If the test with ammonia should prove this not to be sufficient, an extra addition is made. In 1890, 107,031 pounds of crude silver with an average fineness of 970 thousandths required 6,009 pounds of silver sulphate, which corresponds to about 2 parts of sulphate to 1 part of base metal. All the silver of the sulphate is not taken up by the silver in the crucible; part of it enters the slag, as shown by an analysis made by Hampe.[1]

SiO_2	40.7
P_2O_5	0.64
SO_3	0.61
S	0.15
FeO	13.47
Al_2O_3	0.43
Bi_2O_3	6.01
PbO	33.50
Ag_2O	2.05 $(=1.88$ Ag.$)$
Cu	0.45
Sb	0.02
CaO	1.73
MgO	0.25
K_2O	0.64
Na_2O	0.26

The main advantage of Roessler's method of refining is to be found in the larger direct output of silver that is obtained and the concentration of the bismuth in a comparatively small amount of slag that is more easily worked than cupel-bottom and litharge obtained in the reverberatory used for refining the silver.

B. English Cupellation.

§ 138. **Characteristics.**—The characteristics of this method are a small reverberatory furnace with a movable bed and a fixed roof, and the fact that the bullion to be cupelled is charged gradually and the silver is refined in the same furnace where the cupellation was carried on.

§ 139. **The Furnace.**—This has undergone many changes from the original English furnace as described by Percy.[2] Figures 256–260 represent a form of cupelling furnace that is commonly used in American refining works. The vertical section (Figure 257) shows the general construction of the furnace, with the fire-place *a*, the vertical flue *b*, to the right, and the space *c* between the fire-bridge-wall *d* and the flue-wall *e*. This is closed at the top by the compass-ring *f* and the test when it has been put in place. The upper part of the furnace is encased in cast-iron plates; the side-castings

[1] *Berg- und Hüttenm. Ztg.*, 1891, p. 187; *Engineering and Mining Journal,* November 14, 1891.

[2] "Metallurgy of Lead," p. 178.

have a strengthening rib to resist the thrust of the roof. In addition, the front *g* of the ash-pit *h*, the flue-end of the furnace, as well as the inner sides of *i* and *j*, are protected by castings. The usual buckstays and tie-rods have been left out in the drawings. To be noted are the large grate area (4 feet 6 inches by 2

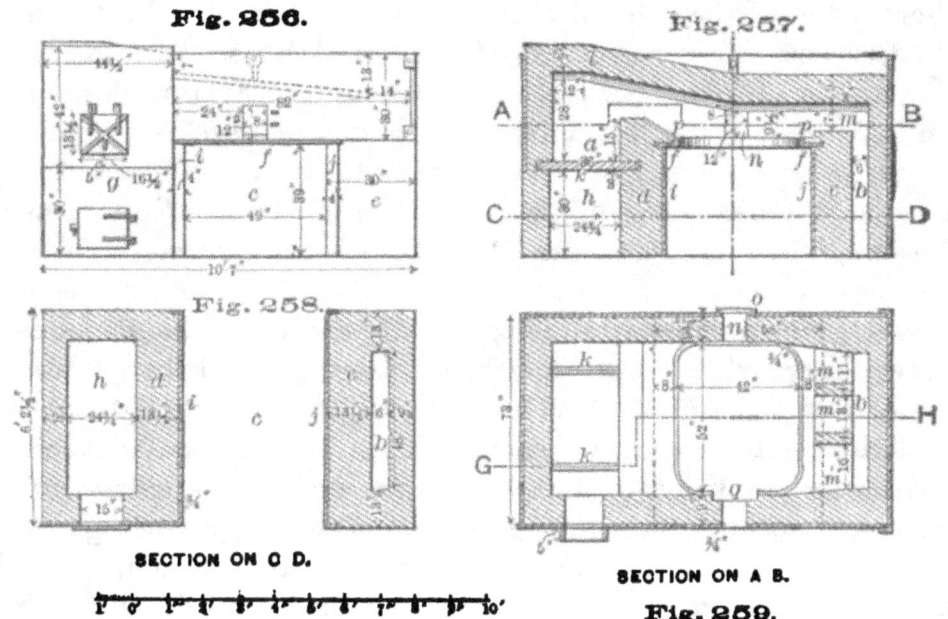

Fig. 256.

Fig. 257.

Fig. 258.

SECTION ON C D.

SECTION ON A B.

Fig. 259.

feet ¾ inch) in comparison with the hearth area (4 feet 4 inches by 3 feet 6 inches), the height from the grate bearer *k* to the roof *l* (2 feet 4 inches) and the short distance (9½ inches) between the roof and the top of the compass-ring, as they are all essential for good working. The flame from the grate-bars is directed downward by the pitch of the roof, and, being forced to pass through the small

Fig. 260.

SECTION ON E F.

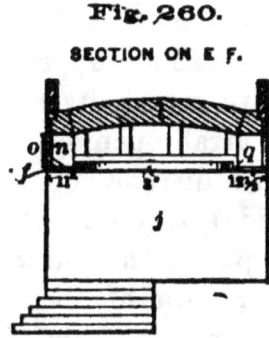

space between the shallow roof and the hearth, exerts all its heating force on the lead. By paying special attention to this part of the construction, it is possible to obtain a sufficiently high temperature to refine silver without being forced to use special kinds of bituminous coal. In fact, with forced blast under the grate, slack coal is good enough for cupelling, and nut coal is required only for fining, the coal being of ordinary grade. The grate in the drawing is so arranged that firing, as well as stoking, is done from the front. Another plan is to place the grate-bars parallel with the short sides of the fire-place, and to stoke from the short side of the furnace. If in such a case forced blast is used, the casting at the stoking side has two oblong openings, each to be closed by a cast-iron door, say 21 inches long and 6 inches high, having its two hinges on the lower side. The horizontal flue m (Figure 257) is seen in Figures 259 and 260 to be divided into three smaller flues m', increasing in width (11, 13, 16 inches) from the back of the furnace toward the front, the object being to prevent the flame from taking the shortest line toward the centre of the flue, and to draw it somewhat toward the front, whereby the litharge floating on the front part of the lead is kept hot. At some works this flue is divided into five or six smaller flues. If in the furnace described the conditions of the draught are such that the flame rushes too much toward the centre, this is remedied by placing one or more fire-bricks in the flue, which will correct the evil. At the back of the furnace there is only one door n, through which passes the blast-pipe and through which is fed one pig of lead at a time. Many furnaces have three small openings, a central one for the blast-pipe and two lateral ones, through each of which a pig of lead is gradually pushed forward and melted away. Having only one opening, to be closed by a sliding door running in a vertical frame o, simplifies the casting. The compass-ring, whose inner contour must, of course, correspond to the form of the test that is to pass through it, is intended for a rectangular cast-iron test having rounded corners. The upper rim p serves to hold in place the brickwork with which its surface is covered. It extends all around the central opening, with the exception of the front, where it stops for a distance of 16 inches, leaving room for the slot g (4 inches wide), through which the litharge is to run down into the litharge-pot. The outer contour of the compass-ring has the rectangular form of the furnace, and reaches from front to back. It thus forms at the front the support for the working tools, and at the back the support for the prop which gives the blast-pipe the desired pitch. At the front the compass-ring is often left open the entire length of the litharge-slot. In such a case the support for the tools is a special cross-bar (skimming-bar) held in place by screws, or by being let into the brick-work, or fastened in some other convenient manner.

§ 140. **Test-Rings.**—The test (4 feet by 2 feet 6 inches), which originally consisted of an oval frame of wrought-iron (4½ inches

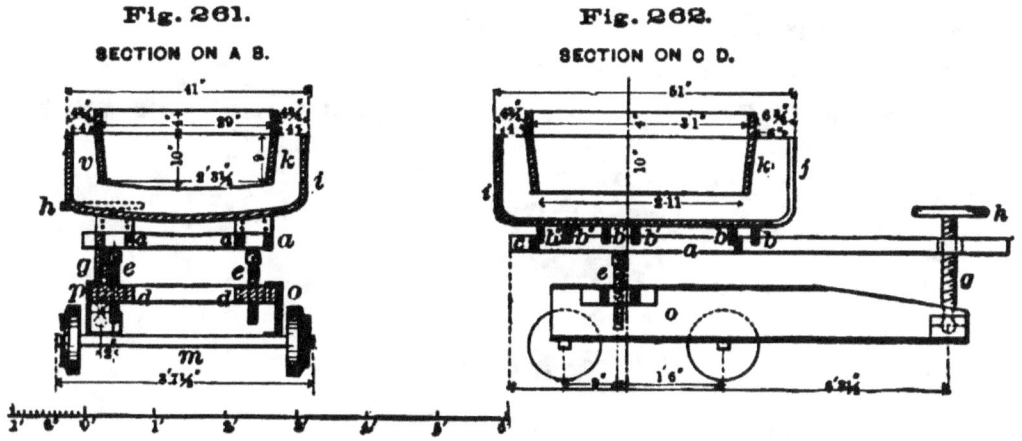

Fig. 261.   Fig. 262.

SECTION ON A B.   SECTION ON C D.

wide, ½ inch thick) filled with bone-ash, has undergone many changes in construction, manner of support, and filling material. A few of the tests in use at present are represented in Figures 261–269. Figures 261 and 262 show a cast-iron test *i* resting on a test-

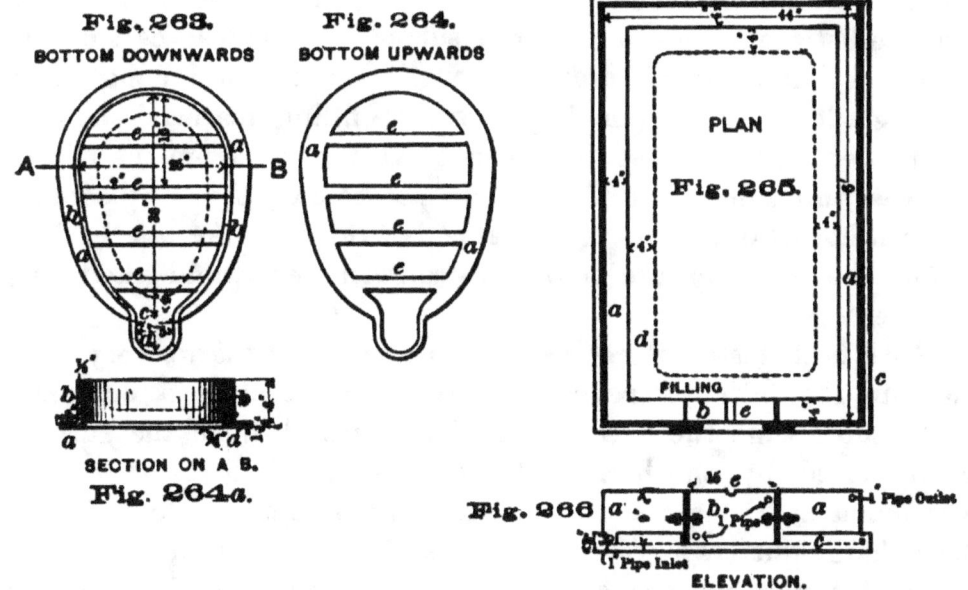

Fig. 263.   Fig. 264.

BOTTOM DOWNWARDS   BOTTOM UPWARDS

SECTION ON A B.

Fig. 264a.

PLAN

Fig. 265.

FILLING

Fig. 266   ELEVATION.

carriage. It has a concave bottom and a cast-iron pattern of the cavity on the inside, both of which will be discussed farther on. The test-ring is rectangular in plan and has rounded corners. It fits into the cupelling furnace shown in Figures 256–260. When in place, its upper surface will be on a line with the upper rim of the compass-ring p (Figure 257). In front it has a 3-inch slot j which is closed when the filling material forming the hearth is being rammed in. The rectangular form of test offers a large surface for oxidation, and therefore more lead can be cupelled on it than on an oval hearth.

Figures 263–264a represent an oval cast-iron test-ring, having a horizontal flange a. When in place, the upper side of this will be

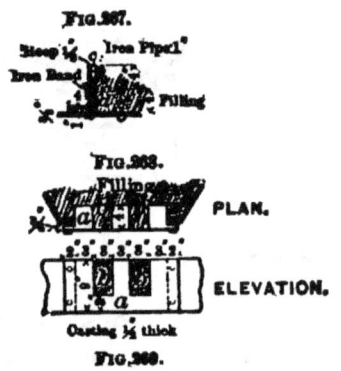

close to or in contact with the lower side of the compass-ring f (Figure 257), and the upper edge of the test-ring itself will be in line with the upper edge of the compass-ring, thus making the distance between the roof and the surface of the lead as small as possible. The test-ring protrudes over the horizontal flange at c, forming a loop d. In tamping down the filling material the loop is not filled, in order that the litharge overflowing from the hearth may pass through it into the litharge-pot below; thus any contact between hot litharge and iron frame is avoided. Across the bottom of the test-ring are four cast-iron arms e to hold the filling in place.

With both tests the corrosive action of the litharge very soon eats out the filling, especially at the front, and in a comparatively short time the test has to be removed from the furnace and replaced by another. To counteract the corrosion of the front, and at the same time permit the raising or lowering of the litharge-gutter, a water jacket, as shown in Figures 268 and 269, is fastened by means of bolts to the test-ring, be this cast iron or wrought iron. The jacket has the same depth as the test-ring and forms the breast. The litharge runs off through gutters cut into the filling b (Figure 269). As this is cooled by the water circulating in the jacket, it is eaten out only very slowly. The jacket itself does not come in contact with the hot litharge in the furnace, as it is protected from it by a 3-inch rim of filling. This wears out somewhat, but never or rarely so far as to bring the casting into direct contact with the hot litharge.

By this arrangement only the front is protected. A device that protects the sides alone is shown in Figure 267. Here *a* is a cast-iron test-ring, resting on a bed-plate *b*. It is surrounded by an iron hoop tied by an iron band. On top of the test-ring are placed two 1-inch pipes in which water circulates. The filling *d* is rammed down in the usual way. The lead, while the furnace is running, being always kept about at the same height, the litharge can show its bad effect only on the level of the water-pipes, and these effectively counteract to a great extent the corrosive action.

A combination of the arrangement shown in Figure 267 with those in Figures 268 and 269 might protect both sides and front of the test-ring.

This is effectively done by the Steitz water-jacket test, as represented in Figures 265 and 266. Here *a* is a rectangular water jacket made of boiler-iron. The open space in front is closed by a cast-iron water jacket having a litharge-gutter *e*; it is fastened with bolts to the wrought-iron jacket. The jackets are placed on a cast-iron test-plate *c*, which supports the filling *d*. Here both jackets are protected from the hot litharge forming on the surface of the lead in the furnace. The gutter *e* alone is attacked by the litharge, and is eaten out after some time; the breast jacket then has to be exchanged for another. This is quickly done, while the furnace is being used, without any difficulty whatever. The effect of jacketing is that the filling lasts longer than with the other test-rings. The cast-iron front jacket, in addition to preserving the breast better than any of the other arrangements, has for concentrating another advantage, that the depth of the litharge-gutter does not have to be regulated by the cupeller, but always remains the same. This renders it, however, unavailable for bringing very rich, say 70 per cent., bullion up to fine silver, as the uniform level of the gutter prolongs indefinitely the removal of the last lead contained in the silver.

§ 141. **Test-Ring Supports.**—The manner of bringing the tests into position and fixing them there has of late years undergone various changes. The old method consisted in driving four iron wedges between the bottom of the test-ring and two transverse bars, the ends of which were inserted 4 inches below the test-

25

frame into the fire-bridge wall *d* and the flue wall *e* (Figure 258). It is not much used now.

Sometimes wedges are still retained to adjust a large test-frame, as represented in Figure 265. This is placed on two 9-inch brick walls running along the fire-bridge wall and flue wall. It is then raised gradually to the height of 12 inches, and four pillars, each three bricks high, are placed beneath the bed-plate, which brings it up nearly to the compass-ring. By then driving flat wedges between the bed-plate and the brick posts the test is adjusted to its final position. A common method of supporting the test is by four screws (18 inches long, 1½ inches in diameter), working in two transverse bars placed 12 inches beneath the test-ring. This rests on a cast-iron plate into which the points of the four screws are set.

Four screws are also found in connection with a test-carriage, as shown in Figures 270–272.[1] Here the test is easily brought into

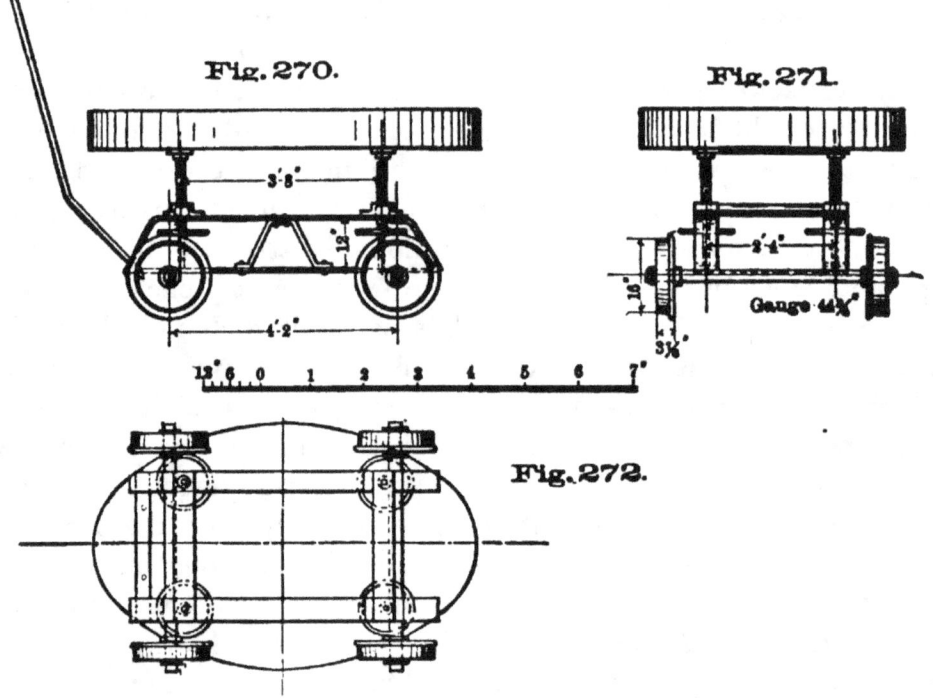

Fig. 270.

Fig. 271.

Fig. 272.

position, and then raised by means of the screws and brought up close against the compass-ring.

With the three arrangements described, the test, when once placed, is immovably fixed while the cupellation is proceeding.

[1] Taken from a drawing of Messrs. Fraser & Chalmers.

Further, when the furnace has become hot, the turning of the four screws often presents considerable difficulties. To obviate these disadvantages the two front screws have been removed and the test suspended or held by a support.

The Lynch [1] test-support is represented by Figures 273 and 274.

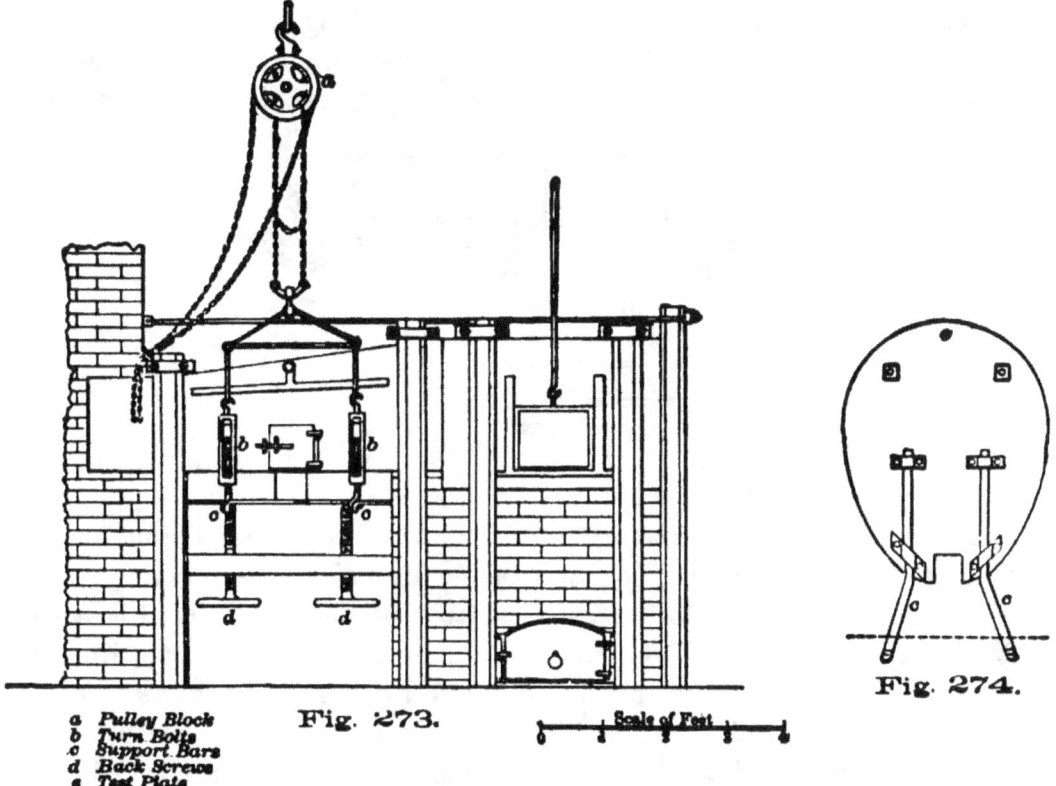

a *Pulley Block*
b *Turn Bolts*
c *Support Bars* Fig. 273.   Scale of Feet
d *Back Screws*
e *Test Plate*

Fig. 274.

The test plate *e*, shown bottom side up in Figure 274, has at the back the two sockets for the points of the two back screws *d* (Figure 273). Two support-bars *c*, diverging 27 inches when extending in front of the furnace, are clamped to the front half of the plate. When in place, they are hooked with the turn-bolts *b* to a triangle made of $\frac{3}{4}$-inch iron, which is attached to a differential pulley *a* hung from the roof. With the pulley the test can be raised or lowered quickly and evenly to regulate the flow of litharge without altering the depth of the litharge-gutter. The contents of the test (rich lead or silver) can also be poured. With the turn-bolts it can be moved sideways to counteract the action of the litharge, should this corrode one side of the hearth more than the other.

[1] Blake, *Trans. A. I. M. E.*, x., p. 220. Patent No. 275,282, April 3, 1883.

Another movable test-support is represented by Figures 261, 262, 275. The test is supported by a carriage *o*, with its movable upper frame *aa'*. This rests at the back on two screws *e* and *e'*, working in the blocks *d* and *d'*. At the front, it is supported by the screw *g*, working in the right arm *a* of the frame, which extends a short distance in front of the furnace, and is turned outward sufficiently for the wheel *h* to be to the right of the cupeller. By inserting a hook into one of its circular openings he turns the wheel to right or left, and thus raises or lowers the front of the

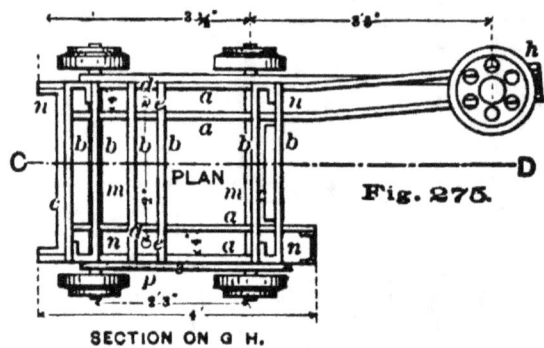

Fig. 275.

PLAN

SECTION ON G H.

frame *aa'* and with it that of the test. Into the upper frame cross-beams *b b' b''*, riveted to each other in pairs, are let in, which serve as support for the concave bottom of the test. If this were straight the surface of beam *a* would represent in the section an unbroken line, two more beams like *c* and *c'* being sufficient to give the frame the required stability. The wheels of the carriage are grooved. By having two rails running across the space *c* (Figure 256), all shifting of the carriage to right or left to get it into correct position is avoided.

The support at the back by the screws *e* and *e'* has the disadvantage that when the frame *aa'* has to be lowered there before the carriage is taken out, much difficulty is experienced in turning the screws. To avoid this the screws *e* and *e'* have in one instance been replaced by two pivots which, being fastened into the blocks *d* and *d'*, fit into circular openings of two blocks fastened to the beams *a* and *a'*. When the frame, resting at the back on these two pivots and supporting the test, has been wheeled into its correct position and is to be raised, this is done by lifting it a few inches with a crowbar ; then two U-shaped castings, of the correct height, and wide enough to enclose a pivot, are placed around these. On withdrawing the crowbar the frame sinks on these two castings, which then support it. When the test is to be exchanged, the frame is again slightly raised, the two U-shaped castings are thrown off, and the frame is slowly lowered, when it will again be supported by the pivots. The test-ring is then sufficiently low for the carriage to be withdrawn from under the compass-ring.

Another movable support with specially constructed test-ring has been invented by Roesing,[1] and is used at Tarnowitz, Silesia. Movable tests are in much favor when the test-ring is not water-cooled, *i.e.*, where the level of the lead is gradually lowered by cutting deeper the litharge-gutter ; with concentrating tests that are entirely jacketed they are not used, and with those that are water-cooled at the breast the regulating of the litharge-gutter has not caused such difficulties as require a movable test.

Whatever test-support may be in use, care is always taken to plaster over with clay the upper surface of the ring, so as to protect it against coming in direct contact with the flame, and at the same time to prevent the air from rushing in between compass-ring and test-ring.

§ 142. **The Blast.**—The blast was originally produced by a steam-jet. This has given way entirely to a blower. The machines in common use are the Baker and Root blowers, and the Sturtevant fan ; blower and engine are usually supported on the same bed-plate. The pressure of the blast is about four ounces per square inch. The blast-pipe is usually of sheet iron and is 3 inches in diameter. At some works it fits into a cast-iron nozzle, which delivers the air through an aperture 4 inches long and $\frac{1}{2}$ inch wide. This is better than simply flattening the end of the sheet-iron pipe.

§ 143. **The Tools.**—The tools required by the cupeller are few : 2 rods (7 feet long, of $\frac{1}{2}$-inch iron), bent to a hook ; 1 chisel-pointed bar (7 feet long, of $\frac{3}{4}$-inch iron); 1 saw (9 inches long, $\frac{1}{2}$ inch wide, 2 inches deep), attached to a handle (6 feet long, of $\frac{1}{2}$-inch iron), to cut down the breast ; 1 fire-hook (10 feet long, of 1-inch iron, with a 4 by 10-inch head); 1 scoop ; 1 ladle ; 10 bullion-moulds or silver-moulds, and 3 litharge-buggies having small kettles (13 inches in diameter and 8 inches deep) to hold the litharge.

§ 144. **Mode of Conducting the Process.**—The operations include filling the test and putting it in place, cupelling, and refining.

[1] *Berg- und Hüttenm. Ztg.*, 1883, p. 577 ; *Engineering and Mining Journal*, January 19, 1884.

1. *Filling the Test and Putting it in Place.*—The material originally used to fill the test was bone-ash ground fine enough to pass a 26-mesh sieve. This, being too expensive, has given way to a mixture of limestone and fire-clay ground through a 12-mesh sieve, the proportions being three parts by volume of limestone to two or sometimes only one part of clay, according to the plasticity. Portland cement is used at some works, and if of best quality is more durable than the limestone-clay filling. Sometimes a mixture of two-thirds cement and one-third coarsely ground fire-brick is used instead of pure cement. The test has also been brick-lined ; experiments, which are very promising, are going on with magnesia-brick.

In filling the test-ring with the limestone-clay mixture, this is moistened, as shown in § 134, and then tamped into the ring. Some works beat down the mass in three separate layers ; others add the necessary material all at once and begin then with the tamping. If the cast-iron test-ring (Figures 263–264*a*) is to be filled, a piece of wood having the form of the loop *d* is put in place, and then the filling beaten down. When finished, the wood is taken out, leaving open the slot for the discharge of the litharge. The wood is best withdrawn only when the test has somewhat dried, as then there is less danger of breaking off part of the filling. The tamping-irons are about 4 inches in diameter at the base. The test is filled entirely with the material, and the cavity then scooped out with a trowel. A very good way to insure a uniformly hard surface is to place a wooden frame on top of the test-ring, and then partly fill it with the hearth-material. When the frame has been removed, the excess of material is scraped off down to the test-ring, and the cavity then scooped out. A rim from 3 to 4 inches wide at back and sides, sometimes 6 and even 10 inches at the front, is left untouched. The cavity has its lowest point near the front to facilitate the dipping out of concentrated bullion or fine silver. The filling material should be at least 4 inches thick at the lowest point, and the depth of the cavity about 5 inches. Thus an oval test 4 feet 6 inches long, 2 feet 1 inch wide, and 5 inches deep, holds about 2,500 pounds of lead.

In filling the test with Portland cement or with a mixture of cement and ground brick, this is moistened and tamped down in the usual way, the cavity, however, being formed during the tamping. A quicker and better method is to place the test bottom upwards over a mould having the form of the cavity, and then to beat down the moistened cement. In using cement it is essential that the work be done quickly, as it must be finished before the cement shows any signs of setting.

The test frame *i* (Figures 261 and 262) is the only one that has a solid cast-iron bottom; it is slightly concave. The working bottom is made of fire-brick, which is set dry and then thoroughly grouted with fire-clay. To build up the sides, the cast-iron frame *k* is oiled and put in place, and moistened cement is beaten down in the intervening space *v*, to the top of the test-frame; the iron frame is then carefully removed. It is in this test that an entire lining of magnesia-brick promises good results. A test of the same description is used to concentrate copper matte, obtained from melting copper dross with galena (§ 119), into bottoms, and 60 per cent. matte, which is then converted into metallic copper.

When a test is filled, it has to stand for a fortnight and longer in a warm place (usually the cupelling-room) to dry. Before it is to be used in a cold furnace, a small charcoal fire is made on it. In a warm furnace the fire is kept low for three or four hours after the test is in place.

2. *Cupelling and Refining.*—When the test is in place and well warmed, the temperature of the furnace is gradually brought to a dark-red, and some lead introduced through the front and melted down. When this has become a cherry-red, the blast is let on and cupellation started. No distinction is made between dross, skimmings, and litharge, as in the German cupellation. The litharge is made to run off at the front, and fresh lead is supplied from the back, where one or two small bars protruding through openings into the furnace are melted down at such a rate as to keep the lead in the test always on the same level. The litharge is collected in a cast-iron pot (say 13 inches in diameter and 8 inches deep) running on wheels. The litharge-pot was at one time replaced by a water-box. This has two advantages. It reduces the temperature for the cupeller and presents the litharge in a granulated form, which is easily handled and sampled. It has, however, the disadvantage that a cupel-carriage cannot be used, and that in the blast furnace there is more loss in lead and silver when granular litharge is charged than if it is in lump form. Granulating has been abandoned.

With a stationary iron test-ring, the litharge is run off through a gutter cut into the filling. It is not often that one gutter serves for the passage of the litharge; generally there are three and often four, opened one after the other to prevent excessive corrosion. A movable test gives an additional way of regulating the flow of litharge by lowering and raising the front. With the Steitz water-jacket test the flow of the litharge is regulated only by the quantity of the lead that is melted off from the bars at the back of the furnace. The gutter can be closed for a short time by allowing litharge to accumulate there, or with a piece of clay.

The flow of the litharge is so regulated that about one-half the surface of the lead remains covered. The former practice of cupelling and fining in the same furnace has been abandoned at most works. At present it is common to concentrate the bullion to 60 or 70 per cent. of silver on one test and to fine it in a separate furnace. For this concentrating, the water-jacket test is excellent, as it can be run by an inexperienced man, while judgment and practice are necessary with a test where the litharge-gutter has to be regulated by the cupeller. By thus dividing the cupelling into the two operations of concentrating and finishing, a smaller number of experienced and reliable cupellers is necessary. When the bullion is concentrated to the desired degree, it is ladled out and goes to the finishing furnace, and the concentration furnace is again filled. Thus a concentrating furnace runs constantly. After a certain time the bottom becomes too thin and has to be exchanged. A test-ring filled with limestone-clay, if used four or five hours daily for finishing, lasts only thirty days ; a cement-test used for the same purpose lasts months. A water-jacket test filled with limestone-clay, used for concentrating, lasts only sixty days. A test-ring filled with cement and used for concentrating and refining lasts seven days.

The finishing is always done on a test having an iron test-ring. The operation is the same as in concentrating, but usually not continuous. Towards the end, when the silver-lead alloy becomes less readily fusible, the temperature has to be considerably raised. When the silver has become sufficiently concentrated, the addition of rich bullion is stopped. The last litharges are drawn off and the test remains almost filled with crude silver, which has now to be fined. It is not often that the brightening is seen. Samples taken from the metal bath show how far the cupellation has progressed. The fining consists usually in exposing the silver for some time to the action of the heat and the blast. Bone-ash is sometimes given in small quantities to absorb the impurities that float on the surface or have collected on the edge. At some works nitrate of soda is used in the furnace to make the silver at least 997 thousandths fine, as this has become the standard below which fine silver should not go. The nitre is spread on the silver, a shovelful (about twelve pounds) at a time. To prevent the soda from corroding the filling of the test-ring, some refiners spread finely ground brick over the nitre. The slag, floating on the silver, is removed only when the silver is fine and ready to be cast into moulds. About 15 shovelfuls of nitre are required for 50,000 ounces of silver.

The indications of fine silver are : a smooth, clean surface ; that stirring fails to bring impurities to the surface ; that a tool held over the silver is clearly reflected in it ; that a sample taken by dipping in a rod will show no spots whatever on the surface, and have a pure, silver-white color ; and that a sample taken with a spoon will spurt while cooling, although this is not a good test. Some refiners cast a small sample-bar, examine the surface, which should be smooth ; the fracture, which should be finely granular and show a silky lustre ; test for malleability by hammering, etc. The only way to know definitely how the fining is progressing is to make an assay. This is done in the dry way, by weighing out twice $\frac{1}{2}$ gramme of a granulated sample and $\frac{1}{2}$ gramme of c. p. silver as a check, and cupelling the three samples with the same amount of lead on three cupels placed in a row in the muffle. A second assay half an hour later will show whether any progress has been made. An assay in the wet way, with potassium-sulphocyanide, using ferric sulphate as an indicator, will give the same result quicker than cupelling.

When the silver is fine, that is, when it ranges between 997 and 999.5 thousandths, it is either ladled out into warmed moulds, or, if the Lynch test-support is used, it can be poured. Sometimes this is done into water, to be remelted at a lower temperature in a plumbago crucible or a new retort (heated in a tilting furnace), and then cast into moulds. This gives smoother bars than would be the case if ladled or poured from the test, but otherwise has no special advantage. Smooth bars are, however, demanded at present.

3. *Sampling and Assaying Fine Silver.*—The sample of the fine silver is best taken from the mould. When this has been filled, a long-handled iron spoon is inserted, the silver stirred with it, and the sample taken out and granulated in water. Taking chips from different parts of the bar is unsatisfactory, as the impurities are as a rule not evenly distributed.[1] In assaying the

[1] Blake, *Journal of Chemistry*, 1888, p. 71.

fine silver, either the Gay-Lussac method or the Vollhard method is used. The latter gives with the necessary care very satisfactory results, notwithstanding adverse criticism.[1] Sometimes the dry assay alone, as indicated above, is made to serve as a check for the mint returns. Details of these wet methods are given in every book on assaying. A steam bath used for dissolving the silver and an apparatus for shaking the assay bottles, have been fully described by Blake.[2] In assaying doré bullion for gold, it is advisable to weigh out and dissolve, without previous cupelling, a separate sample. The amount of gold present rarely exceeds 14 thousandths.

4. *By-Products.*—The by-products of the cupellation process are litharge, cupel-bottom, and flue-dust. Litharge coming from retort bullion runs, when it is pure, from 50 to 60 ounces to the ton ; when it is impure, *e.g.*, when drosses of the retort bullion are being scorified, often from 150 to 200 ounces. The cupel-bottom varies too much in lead and silver to give any average figure. A sample of flue-dust assayed 21.6 per cent. lead, 20 ounces silver, and 0.16 ounces gold. These by-products always go to the ore blast-furnace ; litharge is sometimes used, as already stated, to hasten the softening of base bullion that is especially hard.

5. *Results.*—In a test 4 feet 6 inches by 3 feet 6 inches, 7,000 pounds of retort bullion are cupelled by two men in twenty-four hours, using from 1½ to 2 tons of coal, according to the quality of the fuel. It is advisable to have 8-hour shifts for cupellers to prevent their becoming leaded.

The concentrating of 1,000 pounds of 70 per cent. bullion on a 33 by 28-inch test, 5 inches deep, and refining of the resulting silver (say 12,000 ounces silver), lasts about five hours, requires one man, and about 1,500 pounds of nut coal. When the finishing furnace is stopped for a day or two, the fire on the grate is kept going in order that the temperature of the test may not sink below a dull red heat ; charcoal is often kept aglow on the test, as it makes it again porous when it is much soaked with litharge.

The loss of lead in cupelling is generally given as 5 per cent.

§ 145. **Comparison of Methods.**—A comparison between the two methods of cupelling leads to the conclusion that the German method is, for purposes for which cupelling is generally used to-day, by far the more expensive. Although it forms litharge more

[1] Torrey, *Engineering and Mining Journal*, January 6, 1883.
[2] *Trans. A. I. M. E.*, x., p. 492.

rapidly, because the hearth is so much larger, it produces only a comparatively small amount of silver as the product of one operation. To remove the silver, the furnace has to be cooled, and the hearth torn out and replaced by a new one. This takes time, thus neutralizing the advantage of the quick formation of litharge, and costs much labor, fuel, and material, including a large amount of hearth-material, which has to be smelted in the blast-furnace for every cupellation. In the English cupelling furnace, especially with its American modifications, a cupel-bottom can be made to last for weeks ; the process is therefore less interrupted, and thus much expense for labor, fuel, and material saved, but it has the drawback that the litharge is always more apt to be rich and impure.

A German cupelling furnace may be in place when the resulting litharge is to be sold as such, and is therefore required to be pure, and poor in silver. The English furnace with American improvements is decidedly preferable if the bullion to be cupelled is so rich that the resulting litharge would in any case run too high in silver to be sold as such. In this case, and it is the common one to-day, it is not of much consequence whether the litharge be a little poorer or richer in silver, or if it be somewhat contaminated with impurities, as long as the advantages more than make up for such deficiencies.

INDEX.

(Figures refer to pages, not to sections.)

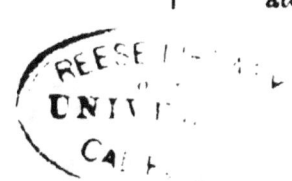

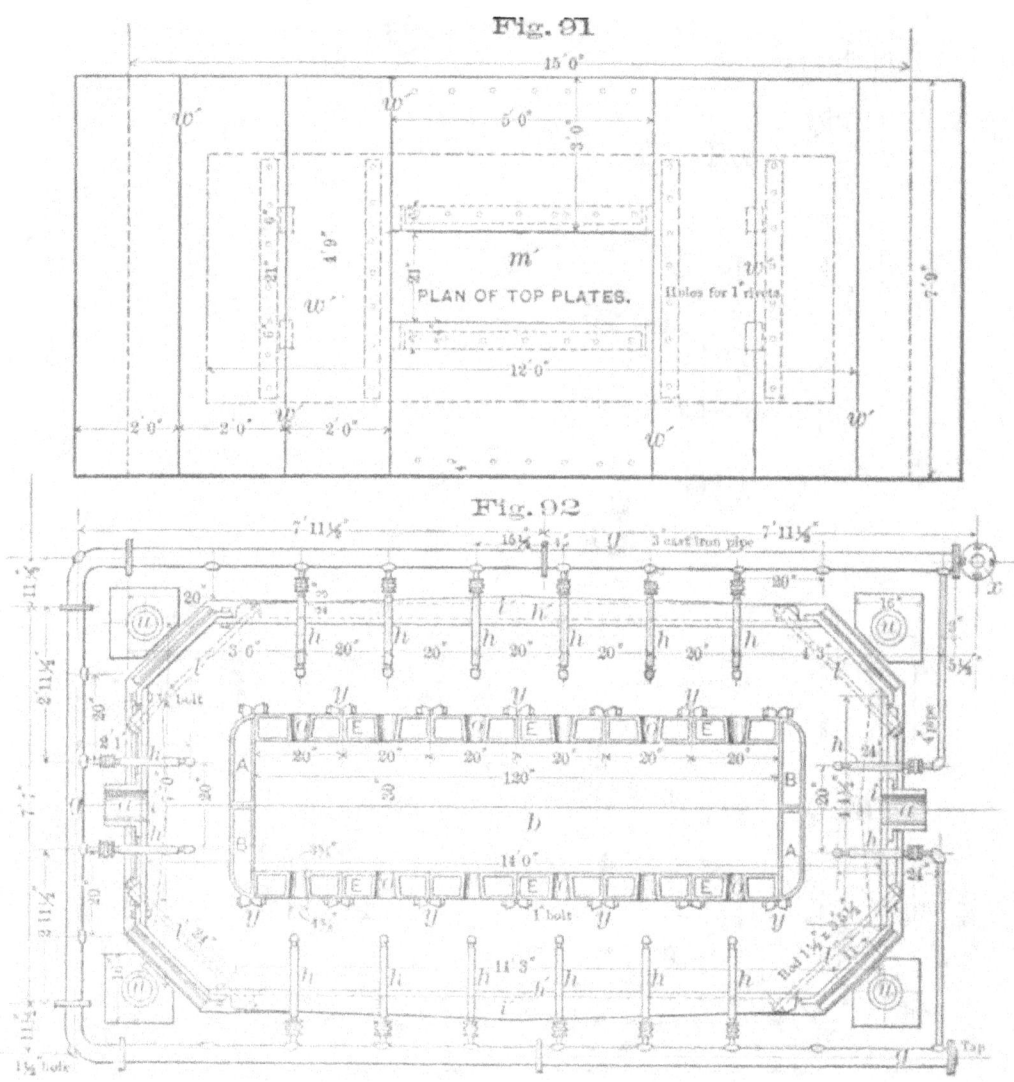

Fig. 91

PLAN OF TOP PLATES.

Fig. 92

PLAN THROUGH WATER JACKETS,

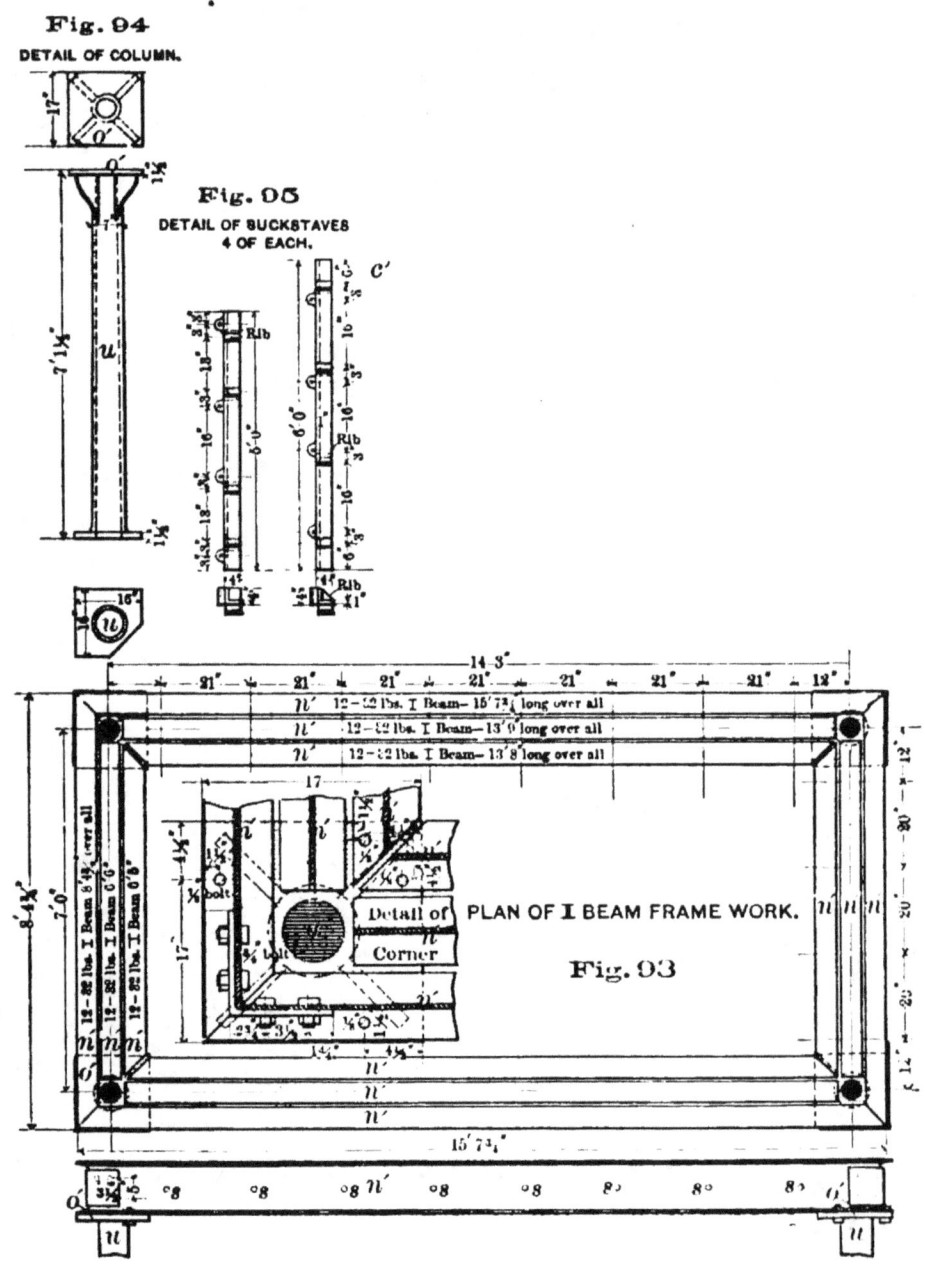

Fig. 94

DETAIL OF COLUMN.

Fig. 95

DETAIL OF BUCKSTAVES
4 OF EACH.

PLAN OF I BEAM FRAME WORK.

Fig. 93

Detail of Corner

Fig. 99

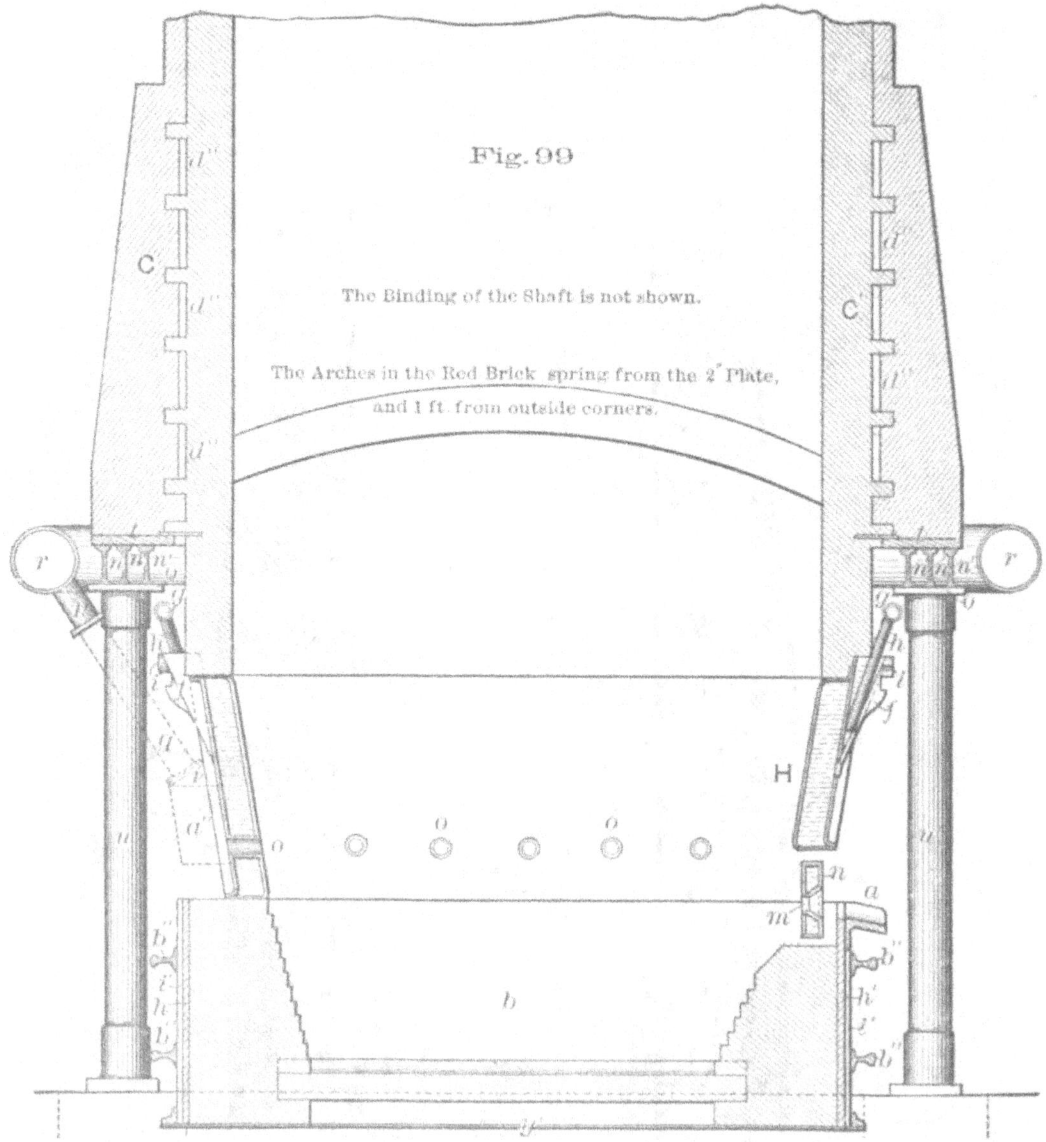

The Binding of the Shaft is not shown.

The Arches in the Red Brick spring from the 2 Plate,
and 1 ft. from outside corners.

H

Fig. 101

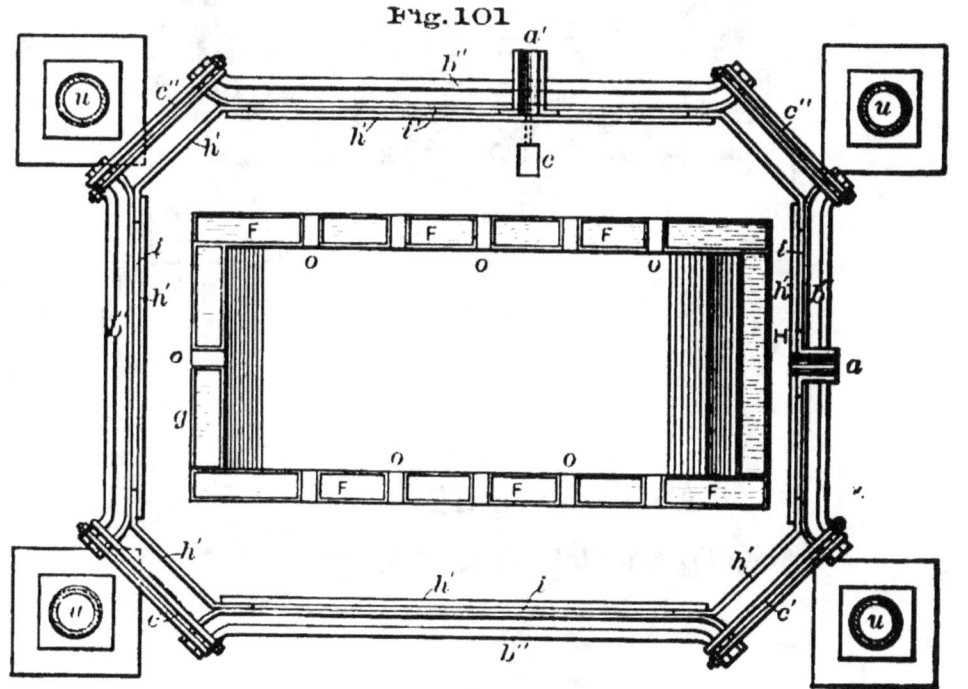

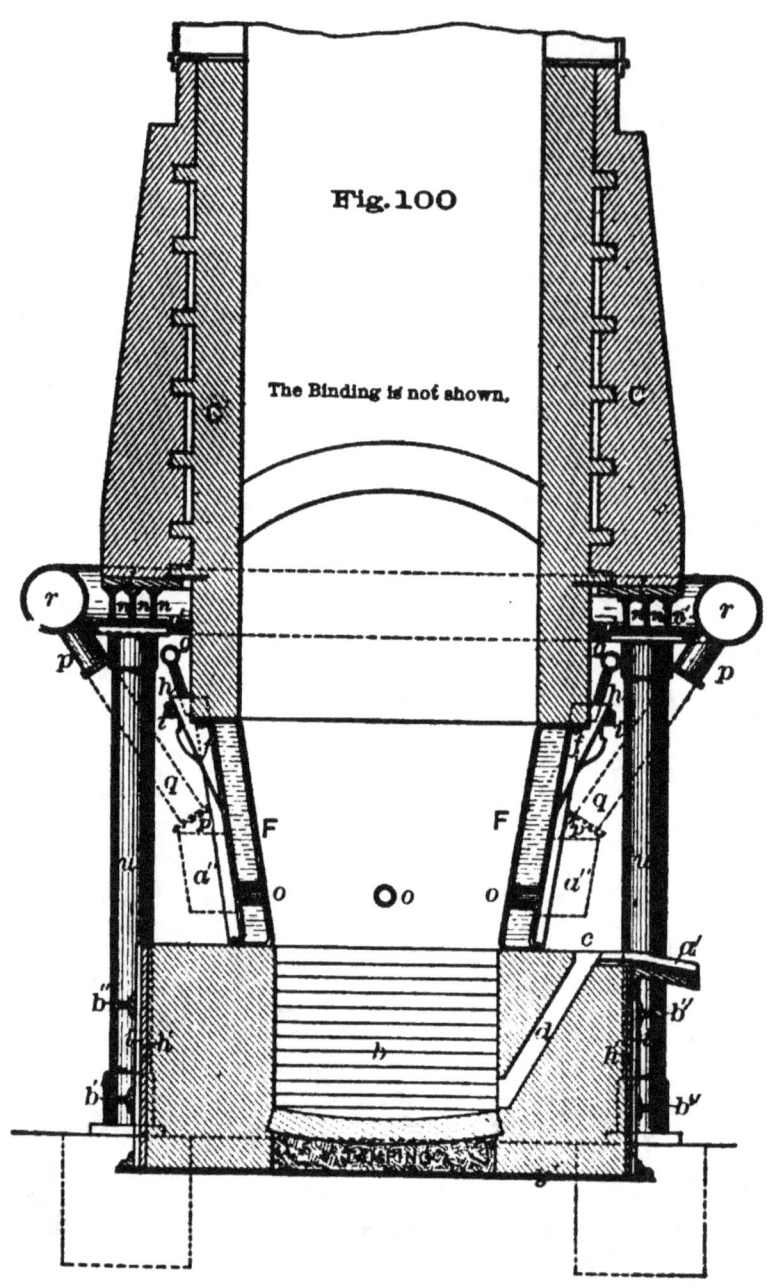

Fig. 100

The Binding is not shown.

BLAST FURNACE,

MONTANA SMELTING CO.

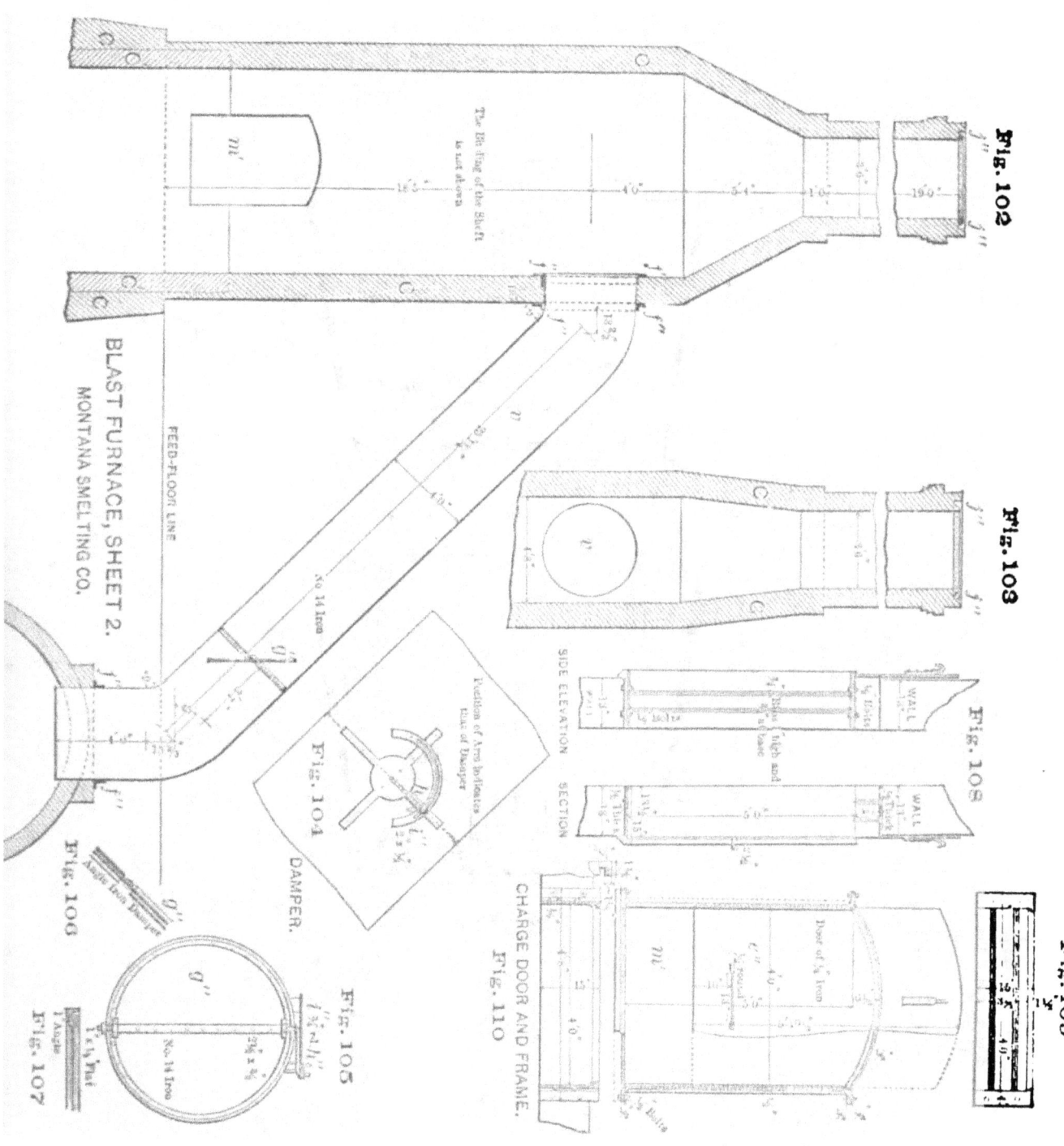

BLAST FURNACE, SHEET 2.

MONTANA SMELTING CO.

Fig. 102

Fig. 103

Fig. 108

SIDE ELEVATION   SECTION

CHARGE DOOR AND FRAME.

Fig. 110

Fig. 100

Fig. 104

DAMPER.

Position of Arms in relation to Face of Damper

Fig. 106

Fig. 105

Fig. 107

FEED-FLOOR LINE

The Bottom of the Shaft is laid down

Fig. 111

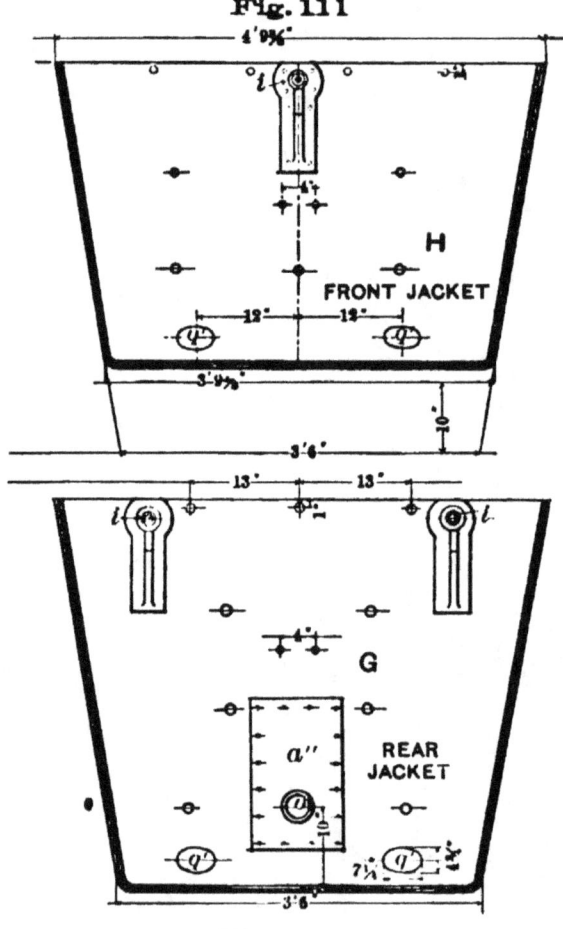

H

FRONT JACKET

Fig. 112

G

a″

REAR
JACKET

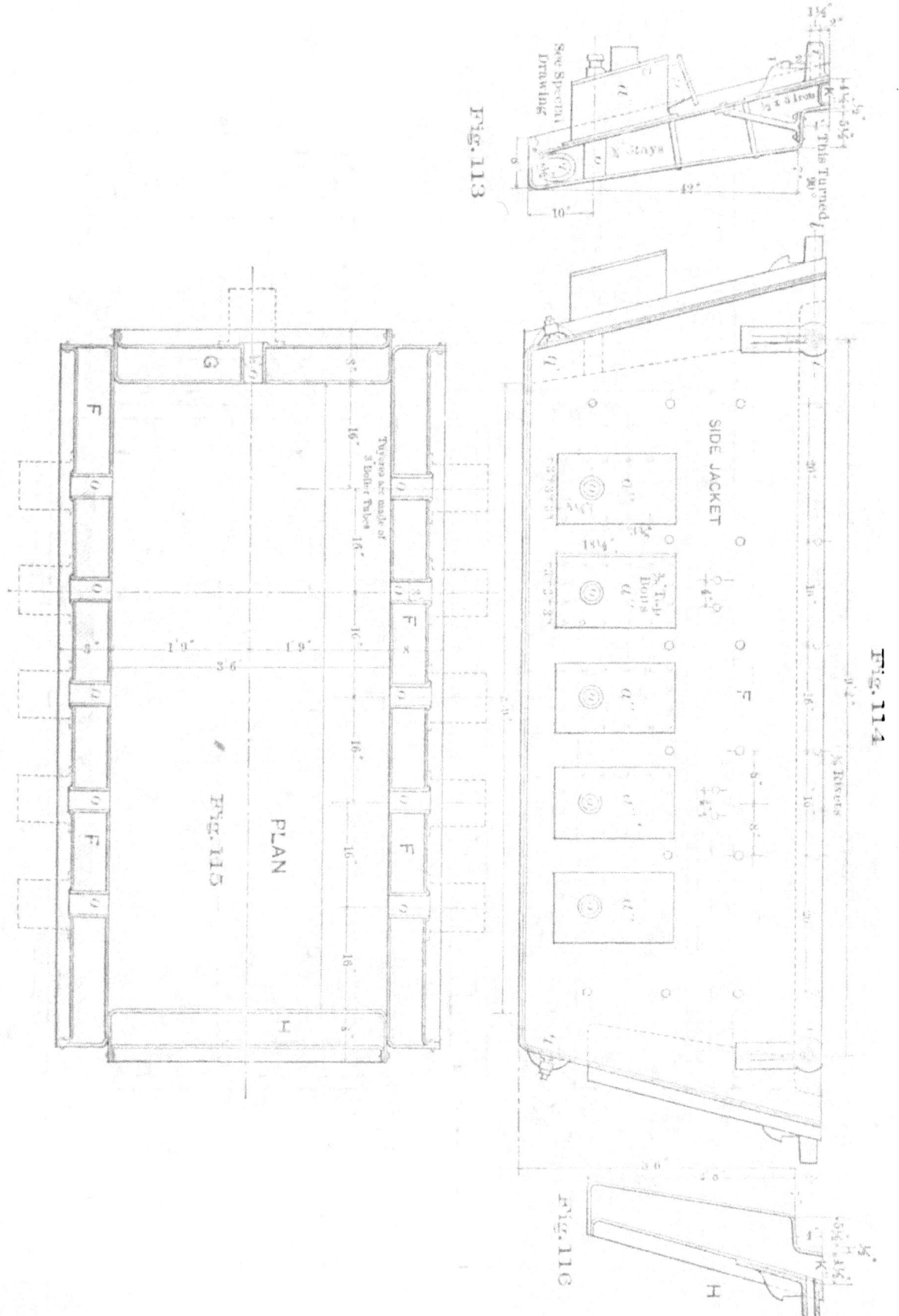

Fig. 113

Fig. 114

Fig. 115

PLAN

Fig. 116

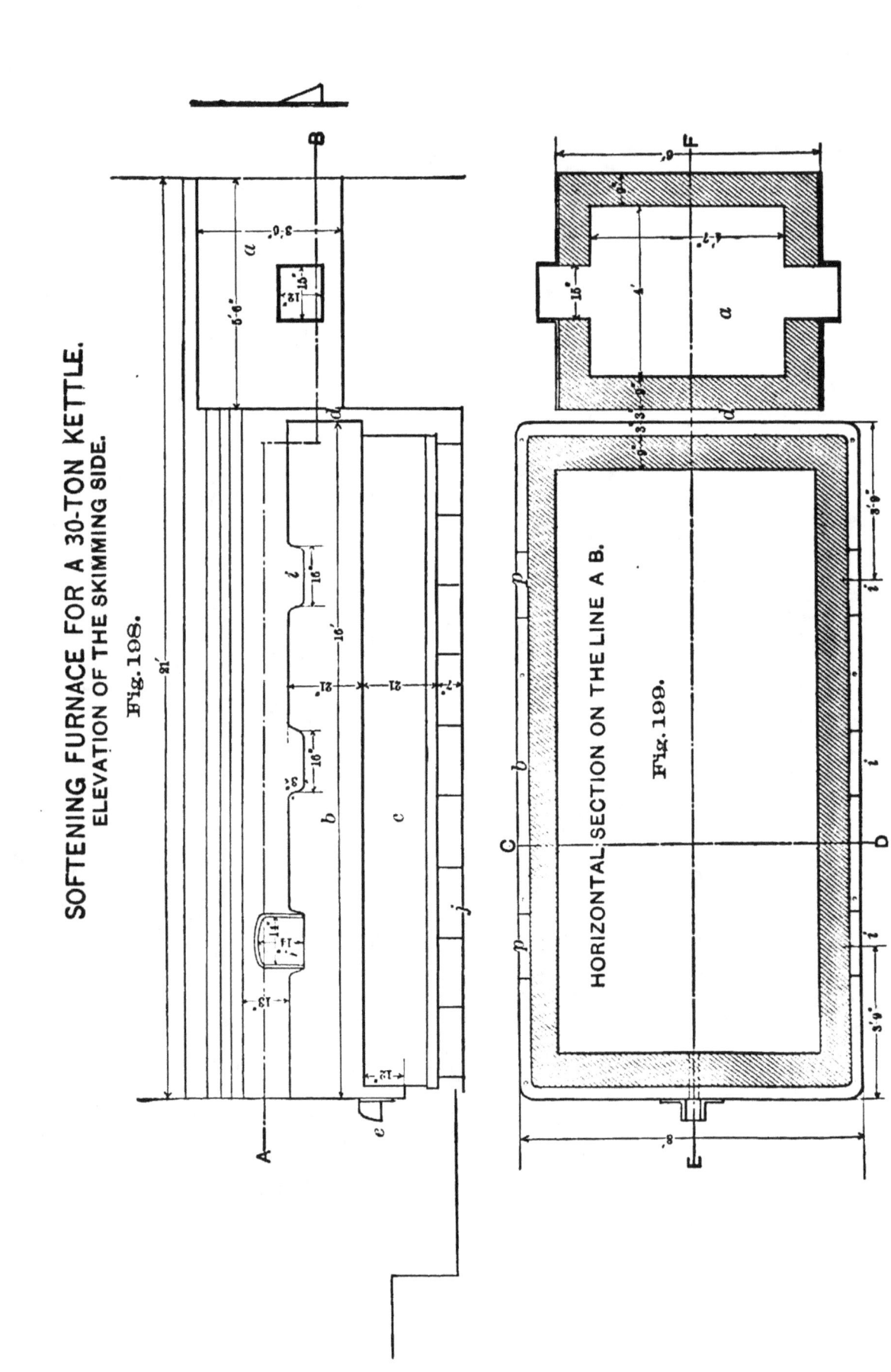

SOFTENING FURNACE FOR A 30-TON KETTLE.
ELEVATION OF THE SKIMMING SIDE.

Fig. 198.

HORIZONTAL SECTION ON THE LINE A B.

Fig. 199.

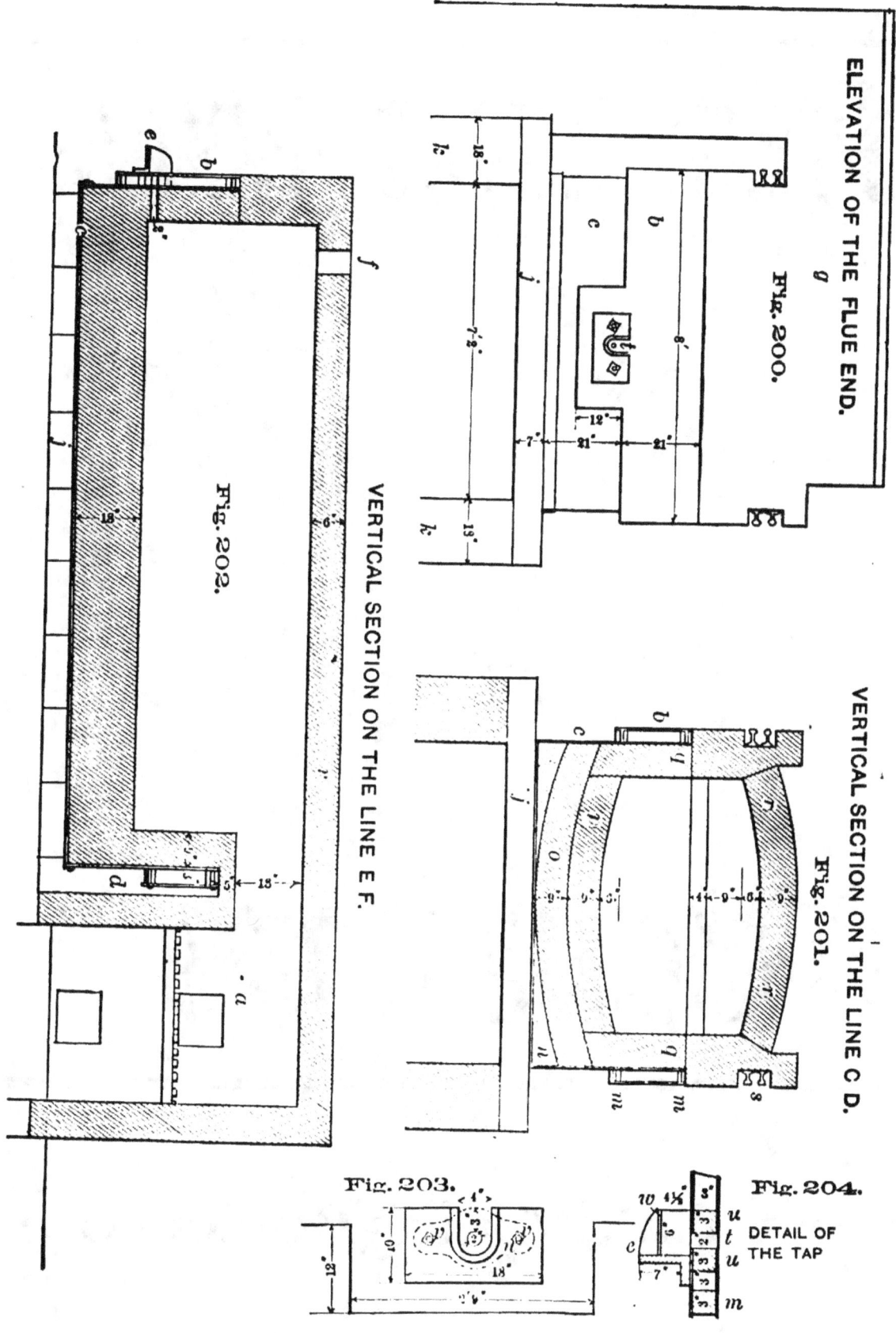

ELEVATION OF THE FLUE END.

Fig. 200.

VERTICAL SECTION ON THE LINE C D.

Fig. 201.

VERTICAL SECTION ON THE LINE E. F.

Fig. 202.

Fig. 203.

Fig. 204.

DETAIL OF THE TAP

GOLD RUSH BOOKS

OREGON, USA

www.GoldMiningBooks.com

Books On Mining

Visit: www.goldminingbooks.com to order your copies or ask your favorite book seller to offer them.

Mining Books by Kerby Jackson

Gold Dust: Stories From Oregon's Mining Years - Oregon mining historian and prospector, Kerby Jackson, brings you a treasure trove of seventeen stories on Southern Oregon's rich history of gold prospecting, the prospectors and their discoveries, and the breathtaking areas they settled in and made homes. 5" X 8", 98 ppgs. Retail Price: $11.99

The Golden Trail: More Stories From Oregon's Mining Years - In his follow-up to "Gold Dust: Stories of Oregon's Mining Years", this time around, Jackson brings us twelve tales from Oregon's Gold Rush, including the story about the first gold strike on Canyon Creek in Grant County, about the old timers who found gold by the pail full at the Victor Mine near Galice, how Iradel Bray discovered a rich ledge of gold on the Coquille River during the height of the Rogue River War, a tale of two elderly miners on the hunt for a lost mine in the Cascade Mountains, details about the discovery of the famous Armstrong Nugget and others. 5" X 8", 70 ppgs. Retail Price: $10.99

Oregon Mining Books

Geology and Mineral Resources of Josephine County, Oregon - Unavailable since the 1970's, this important publication was originally compiled by the Oregon Department of Geology and Mineral Industries and includes important details on the economic geology and mineral resources of this important mining area in South Western Oregon. Included are notes on the history, geology and development of important mines, as well as insights into the mining of gold, copper, nickel, limestone, chromium and other minerals found in large quantities in Josephine County, Oregon. 8.5" X 11", 54 ppgs. Retail Price: $9.99

Mines and Prospects of the Mount Reuben Mining District - Unavailable since 1947, this important publication was originally compiled by geologist Elton Youngberg of the Oregon Department of Geology and Mineral Industries and includes detailed descriptions, histories and the geology of the Mount Reuben Mining District in Josephine County, Oregon. Included are notes on the history, geology, development and assay statistics, as well as underground maps of all the major mines and prospects in the vicinity of this much neglected mining district. 8.5" X 11", 48 ppgs. Retail Price: $9.99

The Granite Mining District - Notes on the history, geology and development of important mines in the well known Granite Mining District which is located in Grant County, Oregon. Some of the mines discussed include the Ajax, Blue Ribbon, Buffalo, Continental, Cougar-Independence, Magnolia, New York, Standard and the Tillicum. Also included are many rare maps pertaining to the mines in the area. 8.5" X 11", 48 ppgs. Retail Price: $9.99

Ore Deposits of the Takilma and Waldo Mining Districts of Josephine County, Oregon - The Waldo and Takilma mining districts are most notable for the fact that the earliest large scale mining of placer gold and copper in Oregon took place in these two areas. Included are details about some of the earliest large gold mines in the state such as the Llano de Oro, High Gravel, Cameron, Platerica, Deep Gravel and others, as well as copper mines such as the famous Queen of Bronze mine, the Waldo, Lily and Cowboy mines. This volume also includes six maps and 20 original illustrations. 8.5" X 11", 74 ppgs. Retail Price: $9.99

Metal Mines of Douglas, Coos and Curry Counties, Oregon - Oregon mining historian Kerby Jackson introduces us to a classic work on Oregon's mining history in this important re-issue of Bulletin 14C Volume 1, otherwise known as the Douglas, Coos & Curry Counties, Oregon Metal Mines Handbook. Unavailable since 1940, this important publication was originally compiled by the Oregon Department of Geology and Mineral Industries includes detailed descriptions, histories and the geology of over 250 metallic mineral mines and prospects in this rugged area of South West Oregon. 8.5" X 11", 158 ppgs. Retail Price: $19.99

Metal Mines of Jackson County, Oregon - Unavailable since 1943, this important publication was originally compiled by the Oregon Department of Geology and Mineral Industries includes detailed descriptions, histories and the geology of over 450 metallic mineral mines and prospects in Jackson County, Oregon. Included are such famous gold mining areas as Gold Hill, Jacksonville, Sterling and the Upper Applegate. 8.5" X 11", 220 ppgs. **Retail Price: $24.99**

Metal Mines of Josephine County, Oregon - Oregon mining historian Kerby Jackson introduces us to a classic work on Oregon's mining history in this important re-issue of Bulletin 14C, otherwise known as the Josephine County, Oregon Metal Mines Handbook. Unavailable since 1952, this important publication was originally compiled by the Oregon Department of Geology and Mineral Industries includes detailed descriptions, histories and the geology of over 500 metallic mineral mines and prospects in Josephine County, Oregon. 8.5" X 11", 250 ppgs. **Retail Price: $24.99**

Metal Mines of North East Oregon - Oregon mining historian Kerby Jackson introduces us to a classic work on Oregon's mining history in this important re-issue of Bulletin 14A and 14B, otherwise known as the North East Oregon Metal Mines Handbook. Unavailable since 1941, this important publication was originally compiled by the Oregon Department of Geology and Mineral Industries and includes detailed descriptions, histories and the geology of over 750 metallic mineral mines and prospects in North Eastern Oregon. 8.5" X 11", 310 ppgs. **Retail Price: $29.99**

Metal Mines of North West Oregon - Oregon mining historian Kerby Jackson introduces us to a classic work on Oregon's mining history in this important re-issue of Bulletin 14D, otherwise known as the North West Oregon Metal Mines Handbook. Unavailable since 1951, this important publication was originally compiled by the Oregon Department of Geology and Mineral Industries and includes detailed descriptions, histories and the geology of over 250 metallic mineral mines and prospects in North Western Oregon. 8.5" X 11", 182 ppgs. **Retail Price: $19.99**

Mines and Prospects of Oregon - Mining historian Kerby Jackson introduces us to a classic mining work by the Oregon Bureau of Mines in this important re-issue of The Handbook of Mines and Prospects of Oregon. Unavailable since 1916, this publication includes important insights into hundreds of gold, silver, copper, coal, limestone and other mines that operated in the State of Oregon around the turn of the 19th Century. Included are not only geological details on early mines throughout Oregon, but also insights into their history, production, locations and in some cases, also included are rare maps of their underground workings. 8.5" X 11", 314 ppgs. **Retail Price: $24.99**

Lode Gold of the Klamath Mountains of Northern California and South West Oregon
(See California Mining Books)

Mineral Resources of South West Oregon - Unavailable since 1914, this publication includes important insights into dozens of mines that once operated in South West Oregon, including the famous gold fields of Josephine and Jackson Counties, as well as the Coal Mines of Coos County. Included are not only geological details on early mines throughout South West Oregon, but also insights into their history, production and locations. 8.5" X 11", 154 ppgs. **Retail Price: $11.99**

Chromite Mining in The Klamath Mountains of California and Oregon
(See California Mining Books)

Southern Oregon Mineral Wealth - Unavailable since 1904, this rare publication provides a unique snapshot into the mines that were operating in the area at the time. Included are not only geological details on early mines throughout South West Oregon, but also insights into their history, production and locations. Some of the mining areas include Grave Creek, Greenback, Wolf Creek, Jump Off Joe Creek, Granite Hill, Galice, Mount Reuben, Gold Hill, Galls Creek, Kane Creek, Sardine Creek, Birdseye Creek, Evans Creek, Foots Creek, Jacksonville, Ashland, the Applegate River, Waldo, Kerby and the Illinois River, Althouse and Sucker Creek, as well as insights into local copper mining and other topics. 8.5" X 11", 64 ppgs. **Retail Price: $8.99**

Geology and Ore Deposits of the Takilma and Waldo Mining Districts - Unavailable since the 1933, this publication was originally compiled by the United States Geological Survey and includes details on gold and copper mining in the Takilma and Waldo Districts of Josephine County, Oregon. The Waldo and Takilma mining districts are most notable for the fact that the earliest large scale mining of placer gold and copper in Oregon took place in these two areas. Included in this report are details about some of the earliest large gold mines in the state such as the Llano de Oro, High Gravel, Cameron, Platerica, Deep Gravel and others, as well as copper mines such as the famous Queen of Bronze mine, the Waldo, Lily and Cowboy mines. In addition to geological examinations, insights are also provided into the production, day to day operations and early histories of these mines, as well as calculations of known mineral reserves in the area. This volume also includes six maps and 20 original illustrations. 8.5" X 11", 74 ppgs. **Retail Price: $9.99**

Gold Mines of Oregon - Oregon mining historian Kerby Jackson introduces us to a classic work on Oregon's mining history in this important re-issue of Bulletin 61, otherwise known as "Gold and Silver In Oregon". Unavailable since 1968, this important publication was originally compiled by geologists Howard C. Brooks and Len Ramp of the Oregon Department of Geology and Mineral Industries and includes detailed descriptions, histories and the geology of over 450 gold mines Oregon. Included are notes on the history, geology and gold production statistics of all the major mining areas in Oregon including the Klamath Mountains, the Blue Mountains and the North Cascades. While gold is where you find it, as every miner knows, the path to success is to prospect for gold where it was previously found. **8.5" X 11", 344 ppgs. Retail Price: $24.99**

Mines and Mineral Resources of Curry County Oregon - Originally published in 1916, this important publication on Oregon Mining has not been available for nearly a century. Included are rare insights into the history, production and locations of dozens of gold mines in Curry County, Oregon, as well as detailed information on important Oregon mining districts in that area such as those at Agness, Bald Face Creek, Mule Creek, Boulder Creek, China Diggings, Collier Creek, Elk River, Gold Beach, Rock Creek, Sixes River and elsewhere. Particular attention is especially paid to the famous beach gold deposits of this portion of the Oregon Coast. **8.5" X 11", 140 ppgs. Retail Price: $11.99**

Chromite Mining in South West Oregon - Originally published in 1961, this important publication on Oregon Mining has not been available for nearly a century. Included are rare insights into the history, production and locations of nearly 300 chromite mines in South Western Oregon. **8.5" X 11", 184 ppgs. Retail Price: $14.99**

Mineral Resources of Douglas County Oregon - Originally published in 1972, this important publication on Oregon Mining has not been available for nearly forty years. Included are rare insights into the geology, history, production and locations of numerous gold mines and other mining properties in Douglas County, Oregon. **8.5" X 11", 124 ppgs. Retail Price: $11.99**

Mineral Resources of Coos County Oregon - Originally published in 1972, this important publication on Oregon Mining has not been available for nearly forty years. Included are rare insights into the geology, history, production and locations of numerous gold mines and other mining properties in Coos County, Oregon. **8.5" X 11", 100 ppgs. Retail Price: $11.99**

Mineral Resources of Lane County Oregon - Originally published in 1938, this important publication on Oregon Mining has not been available for nearly seventy five years. Included are extremely rare insights into the geology and mines of Lane County, Oregon, in particular in the Bohemia, Blue River, Oakridge, Black Butte and Winberry Mining Districts. **8.5" X 11", 82 ppgs. Retail Price: $9.99**

Mineral Resources of the Upper Chetco River of Oregon: Including the Kalmiopsis Wilderness - Originally published in 1975, this important publication on Oregon Mining has not been available for nearly forty years. Withdrawn under the 1872 Mining Act since 1984, real insight into the minerals resources and mines of the Upper Chetco River has long been unavailable due to the remoteness of the area. Despite this, the decades of battle between property owners and environmental extremists over the last private mining inholding in the area has continued to pique the interest of those interested in mining and other forms of natural resource use. Gold mining began in the area in the 1850's and has a rich history in this geographic area, even if the facts surrounding it are little known. Included are twenty two rare photographs, as well as insights into the Becca and Morning Mine, the Emmly Mine (also known as Emily Camp), the Frazier Mine, the Golden Dream or Higgins Mine, Hustis Mine, Peck Mine and others. **8.5" X 11", 64 ppgs. Retail Price: $8.99**

Gold Dredging in Oregon - Originally published in 1939, this important publication on Oregon Mining has not been available for nearly seventy five years. Included are extremely rare insights into the history and day to day operations of the dragline and bucketline gold dredges that once worked the placer gold fields of South West and North East Oregon in decades gone by. Also included are details into the areas that were worked by gold dredges in Josephine, Jackson, Baker and Grant counties, as well as the economic factors that impacted this mining method. This volume also offers a unique look into the values of river bottom land in relation to both farming and mining, in how farm lands were mined, re-soiled and reclamated after the dredges worked them. Featured are hard to find maps of the gold dredge fields, as well as rare photographs from a bygone era. **8.5" X 11", 86 ppgs. Retail Price: $8.99**

Quick Silver Mining in Oregon - Originally published in 1963, this important publication on Oregon Mining has not been available for over fifty years. This publication includes details into the history and production of Elemental Mercury or Quicksilver in the State of Oregon. **8.5" X 11", 238 ppgs. Retail Price: $15.99**

Mines of the Greenhorn Mining District of Grant County Oregon - Originally published in 1948, this important publication on Oregon Mining has not been available for over sixty five years. In this publication are rare insights into the mines of the famous Greenhorn Mining District of Grant County, Oregon, especially the famous Morning Mine. Also included are details on the Tempest, Tiger, Bi-Metallic, Windsor, Psyche, Big Johnny, Snow Creek, Banzette and Paramount Mines, as well as prospects in the vicinities in the famous mining areas of Mormon Basin, Vinegar Basin and Desolation Creek. Included are hard to find mine maps and dozens of rare photographs from the bygone era of Grant County's rich mining history. **8.5" X 11", 72 ppgs. Retail Price: $9.99**

Geology of the Wallowa Mountains of Oregon: Part I (Volume 1) - Originally published in 1938, this important publication on Oregon Mining has not been available for nearly seventy five years. Included are details on the geology of this unique portion of North Eastern Oregon. This is the first part of a two book series on the area. Accompanying the text are rare photographs and historic maps.8.5" X 11", 92 ppgs. **Retail Price: $9.99**

Geology of the Wallowa Mountains of Oregon: Part II (Volume 2) - Originally published in 1938, this important publication on Oregon Mining has not been available for nearly seventy five years. Included are details on the geology of this unique portion of North Eastern Oregon. This is the first part of a two book series on the area. Accompanying the text are rare photographs and historic maps.8.5" X 11", 94 ppgs. **Retail Price: $9.99**

Field Identification of Minerals For Oregon Prospectors - Originally published in 1940, this important publication on Oregon Mining has not been available for nearly seventy five years. Included in this volume is an easy system for testing and identifying a wide range of minerals that might be found by prospectors, geologists and rockhounds in the State of Oregon, as well as in other locales. Topics include how to put together your own field testing kit and how to conduct rudimentary tests in the field. This volume is written in a clear and concise way to make it useful even for beginners. 8.5" X 11", 158 ppgs. **Retail Price: $14.99**

The Bohemia Mining District of Oregon - Originally published in 1900, this important publication on Oregon Mining has not been available for over a century. Included in this volume are important insights into the famous Bohemia Mining District of Oregon, including the histories and locations of important gold mines in the area such as the Ophir Mine, Clarence, Acturas, Peek-a-boo, White Swan, Combination Mine, the Musick Mine, The California, White Ghost, The Mystery, Wall Street, Vesuvius, Story, Lizzie Bullock, Delta, Elsie Dora, Golden Slipper, Broadway, Champion Mine, Knott, Noonday, Helena, White Wings, Riverside and others. Also included are notes on the nearby Blue River Mining District. 8.5" X 11", 58 ppgs. **Retail Price: $9.99**

The Gold Fields of Eastern Oregon - Unavailable since 1900, this publication was originally compiled by the Baker City Chamber of Commerce Offering important insights into the gold mining history of Eastern Oregon, "The Gold Fields of Eastern Oregon" sheds a rare light on many of the gold mines that were operating at the turn of the 19th Century in Baker County and Grant County in North Eastern Oregon. Some of the areas featured include the Cable Cove District, Baisely-Elhorn, Granite, Red Boy, Bonanza, Susanville, Sparta, Virtue, Vaughn, Sumpter, Burnt River, Rye Valley and other mining districts. Included is basic information on not only many gold mines that are well known to those interested in Eastern Oregon mining history, but also many mines and prospects which have been mostly lost to the passage of time. Accompanying are numerous rare photos **8.5" X 11", 78 ppgs. Retail Price: $10.99**

Gold Mining in Eastern Oregon - Originally published in 1938, this important publication on Oregon Mining has not been available for over a century. Included in this volume are important insights into the famous mining districts of Eastern Oregon during the late 1930's. Particular attention is given to those gold mines with milling and concentrating facilities in the Greenhorn, Red Boy, Alamo, Bonanza, Granite, Cable Cove, Cracker Creek, Virtue, Keating, Medical Springs, Sanger, Sparta, Chicken Creek, Mormon Basin, Connor Creek, Cornucopia and the Bull Run Mining Districts. Some of the mines featured include the Ben Harrison, North Pole-Columbia, Highland Maxwell, Baisley-Elkhorn, White Swan, Balm Creek, Twin Baby, Gem of Sparta, New Deal, Gleason, Gifford-Johnson, Cornucopia, Record, Bull Run, Orion and others. Of particular interest are the mill flow sheets and descriptions of milling operations of these mines. 8.5" X 11", 68 ppgs. **Retail Price: $8.99**

The Gold Belt of the Blue Mountains of Oregon - Originally published in 1901, this important publication on Oregon Mining has not been available for over a century. Included in this volume are rare insights into the gold deposits of the Blue Mountains of North East Oregon, including the history of their early discovery and early production. Extensive details are offered on this important mining area's mineralogy and economic geology, as well as insights into nearby gold placers, silver deposits and copper deposits. Featured are the Elkhorn and Rock Creek mining districts, the Pocahontas district, Auburn and Minersville districts, Sumpter and Cracker Creek, Cable Cove, the Camp Carson district, Granite, Alamo, Greenhorn, Robinsonville, the Upper Burnt River Valley and Bonanza districts, Susanville, Quartzburg, Canyon Creek, Virtue, the Copper Butte district, the North Powder River, Sparta, Eagle Creek, Cornucopia, Pine Creek, Lower Powder River, the Upper Snake River Canyon, Rye Valley, Lower Burnt River Valley, Mormon Basin, the Malheur and Clarks Creek districts, Sutton Creek and others. Of particular interest are important details on numerous gold mines and prospects in these mining districts, including their locations, histories, geology and other important information, as well as information on silver, copper and fire opal deposits. 8.5" X 11", 250 ppgs. **Retail Price: $24.99**

<u>Mining in the Cascades Range of Oregon</u> - Originally published in 1938, this important publication on Oregon Mining has not been available for over seventy five years. Included in this volume are rare insights into the gold mines and other types of metal mines in the Cascades Mountain Range of Oregon. Some of the important mining areas covered include the famous Bohemia Mining District, the North Santiam Mining District, Quartzville Mining District, Blue River Mining District, Fall Creek Mining District, Oakridge District, Zinc District, Buzzard-Al Sarena District, Grand Cove, Climax District and Barron Mining District. Of particular interest are important details on over 100 mines and prospects in these mining districts, including their locations, histories, geology and other important information. **8.5" X 11", 170 ppgs. Retail Price: $14.99**

<u>Beach Gold Placers of the Oregon Coast</u> - Originally published in 1934, this important publication on Oregon Mining has not been available for over 80 years. Included in this volume are rare insights into the beach gold deposits of the State of Oregon, including their locations, occurance, composition and geology. Of particular interest is information on placer platinum in Oregon's rich beach deposits. Also included are the locations and other information on some famous Oregon beach mines, including the Pioneer, Eagle, Chickamin, Iowa and beach placer mines north of the mouth of the Rogue River. **8.5" X 11", 60 ppgs. Retail Price: $8.99**

Idaho Mining Books

<u>**Gold in Idaho**</u> - Unavailable since the 1940's, this publication was originally compiled by the Idaho Bureau of Mines and includes details on gold mining in Idaho. Included is not only raw data on gold production in Idaho, but also valuable insight into where gold may be found in Idaho, as well as practical information on the gold bearing rocks and other geological features that will assist those looking for placer and lode gold in the State of Idaho. This volume also includes thirteen gold maps that greatly enhance the practical usability of the information contained in this small book detailing where to find gold in Idaho. **8.5" X 11", 72 ppgs. Retail Price: $9.99**

<u>**Geology of the Couer D'Alene Mining District of Idaho**</u> - Unavailable since 1961, this publication was originally compiled by the Idaho Bureau of Mines and Geology and includes details on the mining of gold, silver and other minerals in the famous Coeur D'Alene Mining District in Northern Idaho. Included are details on the early history of the Coeur D'Alene Mining District, local tectonic settings, ore deposit features, information on the mineral belts of the Osburn Fault, as well as detailed information on the famous Bunker Hill Mine, the Dayrock Mine, Galena Mine, Lucky Friday Mine and the infamous Sunshine Mine. This volume also includes sixteen hard to find maps. **8.5" X 11", 70 ppgs. Retail Price: $9.99**

<u>**The Gold Camps and Silver Cities of Idaho**</u> - Originally published in 1963, this important publication on Idaho Mining has not been available for nearly fifty years. Included are rare insights into the history of Idaho's Gold Rush, as well as the mad craze for silver in the Idaho Panhandle. Documented in fine detail are the early mining excitements at Boise Basin, at South Boise, in the Owyhees, at Deadwood, Long Valley, Stanley Basin and Robinson Bar, at Atlanta, on the famous Boise River, Volcano, Little Smokey, Banner, Boise Ridge, Hailey, Leesburg, Lemhi, Pearl, at South Mountain, Shoup and Ulysses, Yellow Jacket and Loon Creek. The story follows with the appearance of Chinese miners at the new mining camps on the Snake River, Black Pine, Yankee Fork, Bay Horse, Clayton, Heath, Seven Devils, Gibbonsville, Vienna and Sawtooth City. Also included are special sections on the Idaho Lead and Silver mines of the late 1800's, as well as the mining discoveries of the early 1900's that paved the way for Idaho's modern mining and mineral industry. Lavishly illustrated with rare historic photos, this volume provides a one of a kind documentary into Idaho's mining history that is sure to be enjoyed by not only modern miners and prospectors who still scour the hills in search of nature's treasures, but also those enjoy history and tromping through overgrown ghost towns and long abandoned mining camps. **8.5" X 11", 186 ppgs. Retail Price: $14.99**

<u>**Ore Deposits and Mining in North Western Custer County Idaho**</u> - Unavailable since 1913, this important publication was originally published by the Us Department of the Interior and has been unavailable for a century. Included are fine details on the geology, geography, gold placers and gold and silver bearing quartz veins of the mining region of North West Custer County, Idaho. Of particular interest is a rare look at the mines and prospects of the region, including those such as the Ramshorn Mine, SkyLark, Riverview, Excelsior, Beardsley, Pacific, Hoosier, Silver Brick, Forest Rose and dozens of others in the Bay Horse Mining District. Also covered are the mines of the Yankee Fork District such as the Lucky Boy, Badger, Black, Enterprise, Charles Dickens, Morrison, Golden Sunbeam, Montana, Golden Gate and others, as well as those in the Loon Mining District. **8.5" X 11", 126 ppgs. Retail Price: $12.99**

Gold Rush To Idaho - Unavailable since 1963, this important publication was originally published by the Idaho Bureau of Mines and has been unavailable for 50 years. "Gold Rush To Idaho" revisits the earliest years of the discovery of gold in Idaho Territory and introduces us to the conditions that the pioneer gold seekers met when they blazed a trail through the wilderness of Idaho's mountains and discovered the precious yellow metal at Oro Fino and Pierce. Subsequent rushes followed at places like Elk City, Newsome, Clearwater Station, Florence, Warrens and elsewhere. Of particular interest is a rare look at the hardships that the first miners in Idaho met with during their day to day existences and their attempts to bring law and order to their mining camps. 8.5" X 11", 88 ppgs. **Retail Price: $9.99**

The Geology and Mines of Northern Idaho and North Western Montana - Unavailable since 1909, this important publication was originally published by the Us Department of the Interior and has been unavailable for a century. Included are fine details on the geology and geography of the mining regions of Northern Idaho and North Western Montana. Of particular interest is a rare look at the mines and prospects of the region, including those in the Pine Creek Mining District, Lake Pend Oreille district, Troy Mining District, Sylvanite District, Cabinet Mining District, Prospect Mining District and the Missoula Valley. Some of the mines featured include the Iron Mountain, Silver Butte, Snowshoe, Grouse Mountain Mine and others. 8.5" X 11", 142 ppgs. **Retail Price: $12.99**

Mining in the Alturas Quadrangle of Blaine County Idaho - Unavailable since 1922, this important publication was originally published by the Idaho Bureau of Mines and has been unavailable for ninety years. Topics include the geology, rock formations and the formation of ore deposits in this important mining area of Idaho. Of particular focus is information on the local geology, quartz veins and ore deposits of this portion of Idaho. Included are hard to find details, including the descriptions and locations of numerous gold and silver mines in the area including the Silver King, Pilgrim, Columbia, Lone Jack, Sunbeam, Pride of the West, Lucky Boy, Scotia, Atlanta, Beaver-Bidwell and others mines and prospects. 8.5" X 11", 56 ppgs. **Retail Price: $8.99**

Mining in Lemhi County Idaho - Originally published in 1913, this important book on Idaho Mining has not been available to miners for over a century. Included are rare insights into hundreds of gold, silver, copper and other mines in this famous Idaho mining area. Details include the locations, geology, history, production and other facts of the mines of this region, not only gold and silver hardrock mines, but also gold placer mines, lead-silver deposits, copper mines, cobalt-nickel deposits, tungsten and tin mines . It is lavishly illustrated with hard to find photos of the period and rare mining maps. Some of the vicinities featured include the Nicholia Mining District, Spring Mountain District, Texas District, Blue Wing District, Junction District, McDevitt District, Pratt Creek, Eldorado District, Kirtley Creek, Carmen Creek, Gibbonsville, Indian Creek, Mineral Hill District, Mackinaw, Eureka District, Blackbird District, YellowJacket District, Gravel Range District, Junction District, Parker Mountain and other mining districts. 8.5" X 11", 226 ppgs. **Retail Price: $19.99**

Utah Mining Books

Fluorite in Utah - Unavailable since 1954, this publication was originally compiled by the USGS, State of Utah and U.S. Atomic Energy Commission and details the mining of fluorspar, also known as fluorite in the State of Utah. Included are details on the geology and history of fluorspar (fluorite) mining in Utah, including details on where this unique gem mineral may be found in the State of Utah. 8.5" X 11", 60 ppgs. **Retail Price: $8.99**

California Mining Books

The Tertiary Gravels of the Sierra Nevada of California - Mining historian Kerby Jackson introduces us to a classic mining work by Waldemar Lindgren in this important re-issue of The Tertiary Gravels of the Sierra Nevada of California. Unavailable since 1911, this publication includes details on the gold bearing ancient river channels of the famous Sierra Nevada region of California. 8.5" X 11", 282 ppgs. **Retail Price: $19.99**

The Mother Lode Mining Region of California - Unavailable since 1900, this publication includes details on the gold mines of California's famous Mother Lode gold mining area. Included are details on the geology, history and important gold mines of the region, as well as insights into historic mining methods, mine timbering, mining machinery, mining bell signals and other details on how these mines operated. Also included are insights into the gold mines of the California Mother Lode that were in operation during the first sixty years of California's mining history. 8.5" X 11", 176 ppgs. **Retail Price: $14.99**

Lode Gold of the Klamath Mountains of Northern California and South West Oregon - Unavailable since 1971, this publication was originally compiled by Preston E. Hotz and includes details on the lode mining districts of Oregon and California's Klamath Mountains. Included are details on the geology, history and important lode mines of the French Gulch, Deadwood, Whiskeytown, Shasta, Redding, Muletown, South Fork, Old Diggings, Dog Creek (Delta), Bully Choop (Indian Creek), Harrison Gulch, Hayfork, Minersville, Trinity Center, Canyon Creek, East Fork, New River, Denny, Liberty (Black Bear), Cecilville, Callahan, Yreka, Fort Jones and Happy Camp mining districts in California, as well as the Ashland, Rogue River, Applegate, Illinois River, Takilma, Greenback, Galice, Silver Peak, Myrtle Creek and Mule Creek districts of South Western Oregon. Also included are insights into the mineralization and other characteristics of this important mining region. 8.5" X 11", 100 ppgs. **Retail Price: $10.99**

Mines and Mineral Resources of Shasta County, Siskiyou County, Trinity County: California - Unavailable since 1915, this publication was originally compiled by the California State Mining Bureau and includes details on the gold mines of this area of Northern California. Also included are insights into the mineralization and other characteristics of this important mining region, as well as the location of historic gold mines. 8.5" X 11", 204 ppgs. Retail Price: $19.99

Geology of the Yreka Quadrangle, Siskiyou County, California - Unavailable since 1977, this publication was originally compiled by Preston E. Hotz and includes details on the geology of the Yreka Quadrangle of Siskiyou County, California. Also included are insights into the mineralization and other characteristics of this important mining region. 8.5" X 11", 78 ppgs. Retail Price: $7.99

Mines of San Diego and Imperial Counties, California - Originally published in 1914, this important publication on California Mining has not been available for a century. This publication includes important information on the early gold mines of San Diego and Imperial County, which were some of the first gold fields mined in California by early Spanish and Mexican miners before the 49ers came on the scene. Included are not only details on early mining methods in the area, production statistics and geological information, but also the location of the early gold mines that helped make California "The Golden State". Also included are details on the mining of other minerals such as silver, lead, zinc, manganese, tungsten, vanadium, asbestos, barite, borax, cement, clay, dolomite, fluospar, gem stones, graphite, marble, salines, petroleum, stronium, talc and others. 8.5" X 11", 116 ppgs. Retail Price: $12.99

Mines of Sierra County, California - Unavailable since 1920, this publication was originally compiled by the California State Mining Bureau and includes details on the gold mines of Sierra County, California. Also included are insights into the mineralization and other characteristics of this important mining region, as well as the location of historic gold mines. 8.5" X 11", 156 ppgs. Retail Price: $19.99

Mines of Plumas County, California - Unavailable since 1918, this publication was originally compiled by the California State Mining Bureau and includes details on the gold mines of Plumas County, California. Also included are insights into the mineralization and other characteristics of this important mining region, as well as the location of historic gold mines. 8.5" X 11", 200 ppgs. Retail Price: $19.99

Mines of El Dorado, Placer, Sacramento and Yuba Counties, California - Originally published in 1917, this important publication on California Mining has not been available for nearly a century. This publication includes important information on the early gold mines of El Dorado County, Placer County, Sacramento County and Yuba County, which were some of the first gold fields mined by the Forty-Niners during the California Gold Rush. Included are not only details on early mining methods in the area, production statistics and geological information, but also the location of the early gold mines that helped make California "The Golden State". Also included are insights into the early mining of chrome, copper and other minerals in this important mining area. 8.5" X 11", 204 ppgs. Retail Price: $19.99

Mines of Los Angeles, Orange and Riverside Counties, California - Originally published in 1917, this important publication on California Mining has not been available for nearly a century. This publication includes important information on the early gold mines of Los Angeles County, Orange County and Riverside County, which were some of the first gold fields mined in California by early Spanish and Mexican miners before the 49ers came on the scene. Included are not only details on early mining methods in the area, production statistics and geological information, but also the location of the early gold mines that helped make California "The Golden State". 8.5" X 11", 146 ppgs. Retail Price: $12.99

Mines of San Bernadino and Tulare Counties, California - Originally published in 1917, this important publication on California Mining has not been available for nearly a century. This publication includes important information on the early gold mines of San Bernadino and Tulare County, which were some of the first gold fields mined in California by early Spanish and Mexican miners before the 49ers came on the scene. Included are not only details on early mining methods in the area, production statistics and geological information, but also the location of the early gold mines that helped make California "The Golden State". Also included are details on the mining of other minerals such as copper, iron, lead, zinc, manganese, tungsten, vanadium, asbestos, barite, borax, cement, clay, dolomite, fluospar, gem stones, graphite, marble, salines, petroleum, stronium, talc and others. 8.5" X 11", 200 ppgs. Retail Price: $19.99

Chromite Mining in The Klamath Mountains of California and Oregon - Unavailable since 1919, this publication was originally compiled by J.S. Diller of the United States Department of Geological Survey and includes details on the chromite mines of this area of Northern California and Southern Oregon. Also included are insights into the mineralization and other characteristics of this important mining region, as well as the location of historic mines. Also included are insights into chromite mining in Eastern Oregon and Montana. 8.5" X 11", 98 ppgs. Retail Price: $9.99

Mines and Mining in Amador, Calaveras and Tuolumne Counties, California - Unavailable since 1915, this publication was originally compiled by William Tucker and includes details on the mines and mineral resources of this important California mining area. Included are details on the geology, history and important gold mines of the region, as well as insights into other local mineral resources such as asbestos, clay, copper, talc, limestone and others. Also included are insights into the mineralization and other characteristics of this important portion of California's Mother Lode mining region. **8.5" X 11", 198 ppgs. Retail Price: $14.99**

The Cerro Gordo Mining District of Inyo County California - Unavailable since 1963, this publication was originally compiled by the United States Department of Interior. Included are insights into the mineralization and other characteristics of this important mining region of Southern California. Topics include the mining of gold and silver in this important mining district in Inyo County, California, including details on the history, production and locations of the Cerro Gordo Mine, the Morning Star Mine, Estelle Tunnel, Charles Lease Tunnel, Ignacio, Hart, Crosscut Tunnel, Sunset, Upper Newtown, Newtown, Ella, Perseverance, Newsboy, Belmont and other silver and gold mines in the Cerro Gordo Mining District. This volume also includes important insights into the fossil record, geologic formations, faults and other aspects of economic geology in this California mining district. **8.5" X 11", 104 ppgs. Retail Price: $10.99**

Mining in Butte, Lassen, Modoc, Sutter and Tehama Counties of California - Unavailable since 1917, this publication was originally compiled by the United States Department of Interior. Included are insights into the mineralization and other characteristics of this important mining region of California. Topics include the mining of asbestos, chromite, gold, diamonds and manganese in Butte County, the mining of gold and copper in the Hayden Hill and Diamond Mountain mining districts of Lassen County, the mining of coal, salt, copper and gold in the High Grade and Winters mining districts of Modoc County, gold mining in Sutter County and the mining of gold, chromite, manganese and copper in Tehama County. This volume also includes the production records and locations of numerous mines in this important mining region. **8.5" X 11", 114 ppgs. Retail Price: $11.99**

Mines of Trinity County California - Originally published in 1965, this important publication on California Mining has not been available for nearly fifty years. This publication includes important information on mines and mining in Trinity County, California, as well insights into the mineralization and geology of this important mining area in Northern California. Included are extensive details on hardrock and placer gold mines and prospects, including charts showing the locations of these historic mines.. **8.5" X 11", 144 ppgs. Retail Price: $12.99**

Mines of Kern County California - Originally published in 1962, this important publication on California Mining has not been available for nearly fifty years. This publication includes important information on mines and mining in Kern County, California, as well insights into the mineralization and geology of this important mining area in California. Included are extensive details on hardrock and placer gold mines and prospects, including charts showing the locations of these historic mines. **8.5" X 11", 398 ppgs. Retail Price: $24.99**

Mines of Calaveras County California - Originally published in 1962, this important publication on California Mining has not been available for nearly fifty years. This publication includes important information on mines and mining in Calaveras County, California, as well insights into the mineralization and geology of this important mining area in Northern California. Included are extensive details on hardrock and placer gold mines and prospects, including charts showing the locations of these historic mines. **8.5" X 11", 236 ppgs. Retail Price: $19.99**

Lode Gold Mining in Grass Valley California - Unavailable since 1940, this publication was originally compiled by the United States Department of Interior. Included are insights into the gold mineralization and other characteristics of this important mining region of Nevada County, California. This volume also includes important insights into the geologic formations, faults and other aspects of economic geology in this California mining district. Of particular interest are the fine details on many hardrock gold mines in the area, including their locations, histories, development and mineralization. Some of the mines featured include the Gold Hill Mine, Massachusetts Hill, Boundary, Peabody, Golden Center, North Star, Omaha, Lone Jack, Homeward Bound, Hartery, Wisconsin, Allison Ranch, Phoenix, Kate Hayes, W.Y.O.D., Empire, Rich Hill, Daisy Hill, Orleans, Sultana, Centennial, Conlin, Ben Franklin, Crown Point and many others. **8.5" X 11", 148 ppgs. Retail Price: $12.99**

Lode Mining in the Alleghany District of Sierra County California - Unavailable since 1913, this publication was originally compiled by the United States Department of Interior. Included are insights into the mineralization and other characteristics of this important mining region of Sierra County. Included are details on the history, production and locations of numerous hardrock gold mines in this famous California area, including the Tightner Mine, Minnie D., Osceola, Eldorado, Twenty One, Sherman, Kenton, Oriental, Rainbow, Plumbago, Irelan, Gold Canyon, North Fork, Federal, Kate Hardy and others. This volume also includes important insights into the fossil record, geologic formations, faults and other aspects of economic geology in this California mining district. **8.5" X 11", 48 ppgs. Retail Price: $7.99**

Six Months In The Gold Mines During The California Gold Rush - Unavailable since 1850, this important work is a first hand account of one "49'ers" personal experience during the great California Gold Rush, shedding important light on one of the most exciting periods in the history of not only California, but also the world. Compiled from journals written between 1847 and 1849 by E. Gould Buffum, a native of New York, "Six Months In The Gold Mines During The California Gold Rush" offers a rare look into the day to day lives of the people who came to California to work in her gold mines when the state was still a great frontier. 8.5" X 11", 290 ppgs. Retail Price: $19.99

Quartz Mines of the Grass Valley Mining District of California - Unavailable since 1867, this important publication has not been available since those days. This rare publication offers a short dissertation on the early hardrock mines in this important mining district in the California Mother Lode region between the 1850's and 1860's. Also included are hard to find details on the mineralization and locations of these mines, as well as how they were operated in those day. 8.5" X 11", 44 ppgs. Retail Price: $8.99

Alaska Mining Books

Ore Deposits of the Willow Creek Mining District, Alaska - Unavailable since 1954, this hard to find publication includes valuable insights into the Willow Creek Mining District near Hatcher Pass in Alaska. The publication includes insights into the history, geology and locations of the well known mines in the area, including the Gold Cord, Independence, Fern, Mabel, Lonesome, Snowbird, Schroff-O'Neil, High Grade, Marion Twin, Thorpe, Webfoot, Kelly-Willow, Lane, Holland and others. 8.5" X 11", 96 ppgs. Retail Price: $9.99

The Juneau Gold Belt of Alaska - Unavailable since 1906, this hard to find publication includes valuable insights into the gold mines around Juneau, Alaska. The publication includes important details into the history, geology and locations of the well known gold mines and prospects in the area, including those around Windham Bay, Holkham Bay, Port Snettisham, on Grindstone and Rhine Creeks, Gold Creek, Douglas Island, Salmon Creek, Lemon Creek, Nugget Creek, from the Mendenhall River to Berners Bay, McGinnis Creek, Montana Creek, Peterson Creek, Windfall Creek, the Eagle River, Yankee Basin, Yankee Curve, Kowee Creek and elsewhere. Not only are gold placer mines included, but also hardrock gold mines. 8.5" X 11", 224 ppgs. Retail Price: $19.99

Arizona Mining Books

Mines and Mining in Northern Yuma County Arizona - Originally published in 1911, this important publication on Arizona Mining has not been available for over a hundred years. Included are rare insights into the gold, silver, copper and quicksilver mines of Yuma County, Arizona together with hard to find maps and photographs. Some of the mines and mining districts featured include the Planet Copper Mine, Mineral Hill, the Clara Consolidated Mine, Viati Mine, Copper Basin prospect, Bowman Mine, Quartz King, Billy Mack, Carnation, the Wardwell and Osbourne, Valensuella Copper, the Mariquita, Colonial Mine, the French American, the New York-Plomosa, Guadalupe, Lead Camp, Mudersbach Copper Camp, Yellow Bird, the Arizona Northern (Salome Strike), Bonanza (Harqua Hala), Golden Eagle, Hercules, Socorro and others. 8.5" X 11", 144 ppgs. Retail Price: $11.99

The Aravaipa and Stanley Mining Districts of Graham County Arizona - Originally published in 1925, this important publication on Arizona Mining has not been available for nearly ninety years. Included are rare insights into the gold and silver mines of these two important mining districts, together with hard to find maps. 8.5" X 11", 140 ppgs. Retail Price: $11.99

Gold in the Gold Basin and Lost Basin Mining Districts of Mohave County, Arizona - This volume contains rare insights into the geology and gold mineralization of the Gold Basin and Lost Basin Mining Districts of Mohave County, Arizona that will be of benefit to miners and prospectors. Also included is a significant body of information on the gold mines and prospects of this portion of Arizona. This volume is lavishly illustrated with rare photos and mining maps. 8.5" X 11", 188 ppgs. Retail Price: $19.99

Mines of the Jerome and Bradshaw Mountains of Arizona - This important publication on Arizona Mining has not been available for ninety years. This volume contains rare insights into the geology and ore deposits of the Jerome and Bradshaw Mountains of Arizona that will be of benefit to miners and prospectors who work those areas. Included is a significant body of information on the mines and prospects of the Verde, Black Hills, Cherry Creek, Prescott, Walker, Groom Creek, Hassayampa, Bigbug, Turkey Creek, Agua Fria, Black Canyon, Peck, Tiger, Pine Grove, Bradshaw, Tintop, Humbug and Castle Creek Mining Districts. This volume is lavishly illustrated with rare photos and mining maps. 8.5" X 11", 218 ppgs. Retail Price: $19.99

The Ajo Mining District of Pima County Arizona - This important publication on Arizona Mining has not been available for nearly seventy years. This volume contains rare insights into the geology and mineralization of the Ajo Mining District in Pima County, Arizona and in particular the famous New Cornelia Mine. 8.5" X 11", 126 ppgs. Retail Price: $11.99

Mining in the Santa Rita and Patagonia Mountains of Arizona - Originally published in 1915, this important publication on Arizona Mining has not been available for nearly a century. Included are rare insights into hundreds of gold, silver, copper and other mines in this famous Arizona mining area. Details include the locations, geology, history, production and other facts of the mines of this region. **8.5" X 11", 394 ppgs. Retail Price: $24.99**

Mining in the Bisbee Quadrangle of Arizona - Originally published in 1906, this important publication on Arizona Mining has not been available for nearly a century. Included are rare insights into hundreds of gold, silver, copper and other mines in this famous Arizona mining area. Details include the locations, geology, history, production and other facts of the mines of this important mining region. **8.5" X 11", 188 ppgs. Retail Price: $14.99**

Montana Mining Books

A History of Butte Montana: The World's Greatest Mining Camp - First published in 1900 by H.C. Freeman, this important publication sheds a bright light on one of the most important mining areas in the history of The West. Together with his insights, as well as rare photographs of the periods, Harry Freeman describes Butte and its vicinity from its early beginnings, right up to its flush years when copper flowed from its mines like a river. At the time of publication, Butte, Montana was known worldwide as "The Richest Mining Spot On Earth" and produced not only vast amounts of copper, but also silver, gold and other metals from its mines. Freeman illustrates, with great detail, the most important mines in the vicinity of Butte, providing rare details on their owners, their history and most importantly, how the mines operated and how their treasures were extracted. Of particular interest are the dozens of rare photographs that depict mines such as the famous Anaconda, the Silver Bow, the Smoke House, Moose, Paulin, Buffalo, Little Minah, the Mountain Consolidated, West Greyrock, Cora, the Green Mountain, Diamond, Bell, Parnell, the Neversweat, Nipper, Original and many others. **8.5" X 11", 142 ppgs. Retail Price: $12.99**

The Butte Mining District of Montana - This important publication on Montana Mining has not been available for over a century. Included are rare insights into the gold, copper and silver mines of Butte, Montana together with hard to find maps and photographs. Some of the topics include the early history of gold, silver and copper mining in the Butte area, insight into the geology of its mining areas, the local distribution of gold, silver and copper ores, as well their composition and how to identify them. Also included are detailed facts about the mines in the Butte Mining District, including the famous Anaconda Mine, Gagnon, Parrot, Blue Vein, Moscow, Poulin, Stella, Buffalo, Green Mountain, Wake Up Jim, the Diamond-Bell Group, Mountain Consolidated, East Greyrock, West Greyrock, Snowball, Corra, Speculator, Adirondack, Miners Union, the Jessie-Edith May Group, Otisco, Iduna, Colorado, Lizzie, Cambers, Anderson, Hesperus, Preferencia and dozens of others. **8.5" X 11", 298 ppgs. Retail Price: $24.99**

Mines of the Helena Mining Region of Montana - This important publication on Montana Mining has not been available for over a century. Included are rare insights into the gold, copper and silver mines of the vicinity of Helena, Montana, including the Marysville Mining District, Elliston Mining District, Rimini Mining District, Helena Mining District, Clancy Mining District, Wickes Mining District, Boulder and Basin Mining Districts and the Elkhorn Mining District. Some of the topics include the early history of gold, silver and copper mining in the Helena area, insight into the geology of its mining areas, the local distribution of gold, silver and copper ores, as well their composition and how to identify them. Also included are detailed facts, history, geology and locations of over one hundred gold, silver and copper mines in the area . **8.5" X 11", 162 ppgs, Retail Price: $14.99**

Mines and Geology of the Garnet Range of Montana - This important publication on Montana Mining has not been available for over a century. Included are rare insights into the gold, copper and silver mines of the vicinity of this important mining area of Montana. Some of the topics include the early history of gold, silver and copper mining in the Garnet Mountains, insight into the geology of its mining areas, the local distribution of gold, silver and copper ores, as well their composition and how to identify them. Also included are detailed facts, history, geology and locations of numerous gold, silver and copper mines in the area . **8.5" X 11", 100 ppgs, Retail Price: $11.99**

Mines and Geology of the Philipsburg Quadrangle of Montana - This important publication on Montana Mining has not been available for over a century. Included are rare insights into the gold, copper and silver mines of the vicinity of this important mining area of Montana. Some of the topics include the early history of gold, silver and copper mining in the Philipsburg Quadrangle, insight into the geology of its mining areas, the local distribution of gold, silver and copper ores, as well their composition and how to identify them. Also included are detailed facts, history, geology and locations of over one hundred gold, silver and copper mines in the area **8.5" X 11", 290 ppgs, Retail Price: $24.99**

Geology of the Marysville Mining District of Montana - Included are rare insights into the mining geology of the Marysville Mining District. Some of the topics include the early history of gold, silver and copper mining in the area, insight into the geology of its mining areas, the local distribution of gold, silver and copper ores, as well their composition and how to identify them. Also included are detailed facts, history, geology and locations of gold, silver and copper mines in the area **8.5" X 11", 198 ppgs, Retail Price: $19.99**

<u>The Geology and Mines of Northern Idaho and North Western Montana</u>

See listing under Idaho.

Nevada Mining Books

The Bull Frog Mining District of Nevada - Unavailable since 1910, this publication was originally compiled by the United States Department of Interior. This volume also includes important insights into the geologic formations, faults and other aspects of economic geology in this Nevada mining district. Of particular interest are the fine details on many mines in the area, including their locations, histories, development and mineralization. Some of the mines featured include the National Bank Mine, Providence, Gibraltor, Tramps, Denver, Original Bullfrog, Gold Bar, Mayflower, Homestake-King and other mines and prospects. **8.5" X 11", 152 ppgs, Retail Price: $14.99**

History of the Comstock Lode - Unavailable since 1876, this publication was originally released by John Wiley & Sons. This volume also includes important insights into the famous Comstock Lode of Nevada that represented the first major silver discovery in the United States. During its spectacular run, the Comstock produced over 192 million ounces of silver and 8.2 million ounces of gold. Not only did the Comstock result in one of the largest mining rushes in history and yield immense fortunes for its owners, but it made important contributions to the development of the State of Nevada, as well as neighboring California. Included here are important details on not only the early development and history of the Comstock, but also rare early insight into its mines, ore and its geology.**8.5" X 11", 244 ppgs, Retail Price: $19.99**

Colorado Mining Books

Ores of The Leadville Mining District - Unavailable since 1926, this publication was originally compiled by the United States Department of Interior. This volume also includes important insights into the ores and mineralization of the Leadville Mining District in Colorado. Topics include historic ore prospecting methods, local geology, insights into ore veins and stockworks, the local trend and distribution of ore channels, reverse faults, shattered rock above replacement ore bodies, mineral enrichment in oxidized and sulphide zones and more. **8.5" X 11", 66 ppgs, Retail Price: $8.99**

Mining in Colorado - Unavailable since 1926, this publication was originally compiled by the United States Department of Interior. This volume also includes important insights into the mining history of Colorado from its early beginnings in the 1850's right up to the mid 1920's. Not only is Colorado's gold mining heritage included, but also its silver, copper, lead and zinc mining industry. Each mining area is treated separately, detailing the development of Colorado's mines on a county by county basis. **8.5" X 11", 284 ppgs, Retail Price: $19.99**

Gold Mining in Gilpin County Colorado - Unavailable since 1876, this publication was originally compiled by the Register Steam Printing House of Central City, Colorado. A rare glimpse at the gold mining history and early mines of Gilpin County, Colorado from their first discovery in the 1850's up to the "flush years" of the mid 1870's. Of particular interest is the history of the discovery of gold in Gilpin County and details about the men who made those first strikes. Special focus is given to the early gold mines and first mining districts of the area, many of which are not detailed in other books on Colorado's gold mining history. **8.5" X 11", 156 ppgs, Retail Price: $12.99**

Mining in the Gold Brick Mining District of Colorado - Important insights into the history of the Gold Brick Mining District, as well as its local geography and economic geology. Also included are the histories and locations of historic mines in this important Colorado Mining District, including the Cortland, Carter, Raymond, Gold Links, Sacramento, Bassick, Sandy Hook, Chronicle, Grand Prize, Chloride, Granite Mountain, Lucille, Gray Mountain, Hilltop, Maggie Mitchell, Silver Islet, Revenue, Roosevelt, Carbonate King and others. In addition to hardrock mining, are also included are details on gold placer mining in this portion of Colorado. **8.5" X 11", 140 ppgs, Retail Price: $12.99**

Washington Mining Books

The Republic Mining District of Washington - Unavailable since 1910, this important publication was originally published by the Washington Geologic Survey and has been unavailable for a century. Topics include the geology, rock formations and the formation of ore deposits in this important mining area of Washington State. Also included are hard to find details on the geology, history and locations of dozens of mines in the area. Some of the mines featured include the New Republic Mine, Ben Hur, Morning Glory, the South Republic Mine, Quilp, Surprise, Black Tail, Lone Pine, San Poil, Mountain Lion, Tom Thumb, Elcaliph and many others. **8.5" X 11", 94 ppgs, Retail Price: $10.99**

The Myers Creek and Nighthawk Mining Districts of Washington - Unavailable since 1911, this important publication was originally published by the Washington Geologic Survey and has been unavailable for a century. Topics include the geology, rock formations and the formation of ore deposits in these important mining areas of Washington State. Also included are hard to find details on the geology, history and locations of dozens of mines in the area. Some of the mines featured include the Grant Mine, Monterey, Nip and Tuck, Myers Creek, Number Nine, Neutral, Rainbow, Aztec, Crystal Butte, Apex, Butcher Boy, Molson, Mad River, Olentangy, Delate, Kelsey, Golden Chariot, Okanogan, Ohio, Forty-Ninth Parallel, Nighthawk, Favorite, Little Chopaka, Summit, Number One, California, Peerless, Caaba, Prize Group, Ruby, Mountain Sheep, Golden Zone, Rich Bar, Similkameen, Kimberly, Triune, Hiawatha, Trinity, Hornsilver, Maquae, Bellevue, Bullfrog, Palmer Lake, Ivanhoe, Copper World and many others.
8.5" X 11", 136 ppgs, Retail Price: $12.99

The Blewett Mining District of Washington - Unavailable since 1911, this important publication was originally published by the Washington Geologic Survey and has been unavailable for a century. Topics include the geology, rock formations and the formation of ore deposits in this important mining area of Washington State. Also included are hard to find details on the geology, history and locations of dozens of mines in the area. Some of the mines featured include the Washington Meteor, Alta Vista, Pole Pick, Blinn, North Star, Golden Eagle, Tip Top, Wilder, Golden Guinea, Lucky Queen, Blue Bell, Prospect, Homestake, Lone Rock, Johnson, and others. 8.5" X 11", 134 ppgs, Retail Price: $12.99

Silver Mining In Washington - Unavailable since 1955, this important publication was originally published by the Washington Geologic Survey. Featured are the hard to find locations and details pertaining to Washington's silver mines. 8.5" X 11", 180 ppgs, Retail Price: $15.99

The Mines of Snohomish County Washington - Unavailable since 1942, this important publication was originally published by the Washington Geologic Survey and has been unavailable for seventy years. Featured are details on a large number of gold, silver, copper, lead and other metallic mineral mines. Included are the locations of each historic mine, along with information on the commodity produced. 8.5" X 11", 98 ppgs, Retail Price: $10.99

The Mines of Chelan County Washington - Unavailable since 1943, this important publication was originally published by the Washington Geologic Survey and has been unavailable for seventy years. Featured are details on a large number of gold, silver, copper, lead and other metallic mineral mines. Included are the locations of each historic mine, along with information on the commodity. 8.5" X 11", 88 ppgs, Retail Price: $9.99

Metal Mines of Washington - Unavailable since 1921, this important publication was originally published by the Washington Geologic Survey and has been unavailable for nearly ninety years. Widely considered a masterpiece on the Washington Mining Industry, "Metal Mines of Washington" sheds light on the important details of Washington's early mining years. Featured are details on hundreds of gold, silver, copper, lead and other metallic mineral mines. Included are hard to find details on the mineral resources of this state, as well as the locations of historic mines. Lavishly illustrated with maps and historic photos and complete with a glossary to explain any technical terms found in the text, this is one of the most important works on mining in the State of Washington. No prospector or miner should be without it if they are interested in mining in Washington. 8.5" X 11", 396 ppgs, Retail Price: $24.99

Gem Stones In Washington - Unavailable since 1949, this important publication was originally published by the Washington Geologic Survey and has been unavailable since first published. Included are details on where to find naturally occurring gem stones in the State of Washington, including quartz crystal, amethyst, smoky quartz, milky quartz, agates, bloodstone, carnelian, chert, flint, jasper, onyx, petrified wood, opal, fire opal, hyalite and others. 8.5" X 11", 54 ppgs, Retail Price: $8.99

The Covada Mining District of Washington - Unavailable since 1913, this important publication was originally published by the Washington Geologic Survey and has been unavailable for a century. Topics include the geology, rock formations and the formation of ore deposits in this important mining area of Washington State. Also included are hard to find details on the geology, history and locations of dozens of mines in the area. Some of the mines featured include the Admiral, Advance, Algonkian, Big Bug, Big Chief, Big Joker, Black Hawk, Black Tail, Black Thorn, Captain, Cherokee Strip, Colorado, Dan Patch, Dead Shot, Etta, Good Ore, Greasy Run, Great Scott, Idora, IXL, Jay Bird, Kentucky Bell, King Solomon, Laurel, Laura S, Little Jay, Meteor, Neglected, Northern Light, Old Nell, Plymouth Rock, Polaris, Quandary, Reserve, Shoo Fly, Silver Plume, Three Pines, Vernie, White Rose and dozens of others. 8.5" X 11", 114 ppgs, Retail Price: $10.99

The Index Mining District of Washington - Unavailable since 1912, this important publication was originally published by the Washington Geologic Survey and has been unavailable for a century. Topics include the geology, rock formations and the formation of ore deposits in this important mining area of Washington State. Also included are hard to find details on the geology, history and locations of dozens of mines in the area. Some of the mines featured include the Sunset, Non-Pareil, Ethel Consolidated, Kittaning, Merchant, Homestead, Co-operative, Lost Creek, Uncle Sam, Calumet, Florence-Rae, Bitter Creek, Index Peacock, Gunn Peak, Helena, North Star, Buckeye, Copper Bell, Red Cross and others. 8.5" X 11", 114 ppgs, Retail Price: $11.99

Mining & Mineral Resources of Stevens County Washington - Unavailable since 1920, this important publication was originally published by the Washington Geologic Survey and has been unavailable for a century. Topics include the geology, rock formations and the formation of ore deposits in these important mining areas of Washington State. Also included are hard to find details on the geology, history and locations of hundreds of mines in the area. **8.5" X 11", 372 ppgs, Retail Price: $24.99**

The Mines and Geology of the Loomis Quadrangle Okanogan County, Washington - Unavailable since 1972, this important publication was originally published by the Washington Geologic Survey and has been unavailable for a century. Topics include the geology, rock formations and the formation of ore deposits in this important mining area of Washington State. Also included are hard to find details on the geology, history and locations of dozens of gold, copper, silver and other mines in the area. **8.5" X 11", 150 ppgs, Retail Price: $12.99**

The Conconully Mining District of Okanogan County Washington - Unavailable since 1973, this important publication was originally published by the Washington Geologic Survey and has been unavailable for a century. Topics include the geology, rock formations and the formation of ore deposits in this important mining area of Washington State, which also includes Salmon Creek, Blue Lake and Galena. Also included are hard to find details on the geology, mining history and locations of dozens of mines in the area. Some of the mines include Arlington, Fourth of July, Sonny Boy, First Thought, Last Chance, War Eagle-Peacock, Wheeler, Mohawk, Lone Star, Woo Loo Moo Loo, Keystone, Hughes, Plant-Callahan, Johnny Boy, Leuena, Gubser, John Arthur, Tough Nut, Homestake, Key and many others **8.5" X 11", 68 ppgs, Retail Price: $8.99**

Wyoming Mining Books

Mining in the Laramie Basin of Wyoming - Unavailable since 1909, this publication was originally compiled by the United States Department of Interior. Also included are insights into the mineralization and other characteristics of this important mining region, especially in regards to coal, limestone, gypsum, bentonite clay, cement, sand, clay and copper. **8.5" X 11", 104 ppgs, Retail Price: $11.99**

New Mexico Mining Books

The Mogollon Mining District of New Mexico - Unavailable since 1927, this important publication was originally published by the US Department of Interior and has been unavailable for 80 years. Topics include the geology, rock formations and the formation of ore deposits in this important mining area in New Mexico. Of particular focus is information on the history and production of the ore deposits in this area, their form and structure, vein filling, their paragenesis, origins and ore shoots, as well as oxidation and supergene enrichment. Also included are hard to find details, including the descriptions and locations of numerous gold, silver and other types of mines, including the Eureka, Pacific, South Alpine, Great Western, Enterprise, Buffalo, Mountain View, Floride, Gold Dust, Last Chance, Deadwood, Confidence, Maud S., Deep Down, Little Fanney, Trilby, Johnson, Alberta, Comet, Golden Eagle, Cooney, Queen, the Iron Crown, Eberle, Clifton, Andrew Jackson mine, Mascot and others. **8.5" X 11", 144 ppgs, Retail Price: $12.99**

The Percha Mining District of Kingston New Mexico - Unavailable since 1883, this important publication was originally published by the Kingston Tribune and has been unavailable for over one hundred and thirty five years. Having been written during the earliest years of gold and silver mining in the Percha Mining District, unlike other books on the subject, this work offers the unique perspective of having actually been written while the early mining history of this area was still being made. In fact, the work was written so early in the development of this area that many of the notable mines in the Percha District were less than a few years old and were still being operated by their original discoverers with the same enthusiasm as when they were first located. Included are hard to find details on the very earliest gold and silver mines of this important mining district near Kingston in Sierra County, New Mexico. **8.5" X 11", 68 ppgs, Retail Price: $9.99**

East Coast Mining Books

The Gold Fields of the Southern Appalachians - Unavailable since 1895, this important publication was originally published by the US Department of Interior and has been unavailable for nearly 120 years. Topics include the geology, rock formations and the formation of ore deposits in this important mining area of the American South. Of particular focus is information on the history and statistics of the ore deposits in this area, their form and structure and veins. Also included are details on the placer gold deposits of the region. The gold fields of the Georgian Belt, Carolinian Belt and the South Mountain Mining District of North Carolina are all treated in descriptive detail. Included are hard to find details, including the descriptions and locations of numerous gold mines in Georgia, North Carolina and elsewhere in the American South. Also included are details on the gold belts of the British Maritime Provinces and the Green Mountains. **8.5" X 11", 104 ppgs, Retail Price: $9.99**

Gold Rush Tales Series

Millions in Siskiyou County Gold - In this first volume of the "Gold Rush Tales" series, leading mining historian and editor Kerby Jackson, introduces us to the story of how millions of dollars worth of gold was discovered in Siskiyou County during the California Gold Rush. Lavishly illustrated with photos from the 19th Century, this hard to find information was first published in 1897 and sheds important light onto the gold rush era in Siskiyou County, California and the experiences of the men who dug for the gold and actually found it. 8.5" X 11", 82 ppgs, Retail Price: $9.99

The California Rand in the Days of '49 - In this second volume of the "Gold Rush Tales" series, leading mining historian and editor Kerby Jackson, introduces us to four tales from the California Gold Rush. Lavishly illustrated with photos from the 19th Century, this hard to find information was first published in 1890's and includes the stories of "California's Rand", details about Chinese miners, how one early miner named Baker struck it rich and also the story of Alphonzo Bowers, who invented the first hydraulic gold dredge. 8.5" X 11", 54 ppgs, Retail Price: $9.99

More Mining Books

Prospecting and Developing A Small Mine - Topics covered include the classification of varying ores, how to take a proper ore sample, the proper reduction of ore samples, alluvial sampling, how to understand geology as it is applied to prospecting and mining, prospecting procedures, methods of ore treatment, the application of drilling and blasting in a small mine and other topics that the small scale miner will find of benefit. 8.5" X 11", 112 ppgs, Retail Price: $11.99

Timbering For Small Underground Mines - Topics covered include the selection of caps and posts, the treatment of mine timbers, how to install mine timbers, repairing damaged timbers, use of drift supports, headboards, squeeze sets, ore chute construction, mine cribbing, square set timbering methods, the use of steel and concrete sets and other topics that the small underground miner will find of benefit. This volume also includes twenty eight illustrations depicting the proper construction of mine timbering and support systems that greatly enhance the practical usability of the information contained in this small book. 8.5" X 11", 88 ppgs. Retail Price: $10.99

Timbering and Mining - A classic mining publication on Hard Rock Mining by W.H. Storms. Unavailable since 1909, this rare publication provides an in depth look at American methods of underground mine timbering and mining methods. Topics include the selection and preservation of mine timbers, drifting and drift sets, driving in running ground, structural steel in mine workings, timbering drifts in gravel mines, timbering methods for driving shafts, positioning drill holes in shafts, timbering stations at shafts, drainage, mining large ore bodies by means of open cuts or by the "Glory Hole" system, stoping out ore in flat or low lying veins, use of the "Caving System", stoping in swelling ground, how to stope out large ore bodies, Square Set timbering on the Comstock and its modifications by California miners, the construction of ore chutes, stoping ore bodies by use of the "Block System", how to work dangerous ground, information on the "Delprat System" of stoping without mine timbers, construction and use of headframes and much more. This volume provides a reference into not only practical methods of mining and timbering that may be employed in narrow vein mining by small miners today, but also rare insights into how mines were being worked at the turn of the 19th Century. 8.5" X 11", 288 ppgs. Retail Price: $24.99

A Study of Ore Deposits For The Practical Miner - Mining historian Kerby Jackson introduces us to a classic mining publication on ore deposits by J.P. Wallace. First published in 1908, it has been unavailable for over a century. Included are important insights into the properties of minerals and their identification, on the occurrence and origin of gold, on gold alloys, insights into gold bearing sulfides such as pyrites and arsenopyrites, on gold bearing vanadium, gold and silver tellurides, lead and mercury tellurides, on silver ores, platinum and iridium, mercury ores, copper ores, lead ores, zinc ores, iron ores, chromium ores, manganese ores, nickel ores, tin ores, tungsten ores and others. Also included are facts regarding rock forming minerals, their composition and occurrences, on igneous, sedimentary, metamorphic and intrusive rocks, as well as how they are geologically disturbed by dikes, flows and faults, as well as the effects of these geologic actions and why they are important to the miner. Written specifically with the common miner and prospector in mind, the book will help to unlock the earth's hidden wealth for you and is written in a simple and concise language that anyone can understand. 8.5" X 11", 366 ppgs. Retail Price: $24.99

Mine Drainage - Unavailable since 1896, this rare publication provides an in depth look at American methods of underground mine drainage and mining pump systems. This volume provides a reference into not only practical methods of mining drainage that may be employed in narrow vein mining by small miners today, but also rare insights into how mines were being worked at the turn of the 19th Century. 8.5" X 11", 218 ppgs. Retail Price: $24.99

Fire Assaying Gold, Silver and Lead Ores - Unavailable since 1907, this important publication was originally published by the Mining and Scientific Press and was designed to introduce miners and prospectors of gold, silver and lead to the art of fire assaying. Topics include the fire assaying of ores and products containing gold, silver and lead; the sampling and preparation of ore for an assay; care of the assay office, assay furnaces; crucibles and scorifiers; assay balances; metallic ores; scorification assays; cupelling; parting' crucible assays, the roasting of ores and more. This classic provides a time honored method of assaying put forward in a clear, concise and easy to understand language that will make it a benefit to even beginners. **8.5" X 11", 96 ppgs. Retail Price: $11.99**

Methods of Mine Timbering - Originally published in 1896, this important publication on mining engineering has not been available for nearly a century. Included are rare insights into historical methods of timbering structural support that were used in underground metal mines during the California that still have a practical application for the small scale hardrock miner of today. **8.5" X 11", 94 ppgs. Retail Price: $10.99**

The Enrichment of Copper Sulfide Ores - First published in 1913, it has been unavailable for over a century. Topics include the definition and types of ore enrichment, the oxidation of copper ores, the precipitation of metallic sulfides. Also included are the results of dozens of lab experiments pertaining to the enrichment of sulfide ores that will be of interest to the practical hard rock mine operator in his efforts to release the metallic bounty from his mine's ore. **8.5" X 11", 92 ppgs. Retail Price: $9.99**

A Study of Magmatic Sulfide Ores - Unavailable since 1914, this rare publication provides an in depth look at magmatic sulfide ores. Some of the topics included are the definition and classification of magmatic ores, descriptions of some magmatic sulfide ore deposits known at the time of publication including copper and nickel bearing pyrrohitic ore bodies, chalcopyrite-bornite deposits, pyritic deposits, magnetite-ileminite deposits, chromite deposits and magmatic iron ore deposits. Also included are details on how to recognize these types of ore deposits while prospecting for valuable hardrock minerals. **8.5" X 11", 138 ppgs. Retail Price: $11.99**

The Cyanide Process of Gold Recovery - Unavailable since 1894 and released under the name "The Cyanide Process: Its Practical Application and Economical Results", this rare publication provides an in depth look at the early use of cyanide leaching for gold recovery from hardrock mine ores. This volume provides a reference into the early development and use of cyanide leaching to recover gold. **8.5" X 11", 162 ppgs. Retail Price: $14.99**

California Gold Milling Practices - Unavailable since 1895 and released under the name "California Gold Practices", this rare publication provides an in depth look at early methods of milling used to reduce gold ores in California during the late 19th century. This volume provides a reference into the early development and use of milling equipment during the earliest years of the California Gold Rush up to the age of the Industrial Revolution. Much of the information still applies today and will be of use to small scale miners engaging in hardrock mining. **8.5" X 11", 104 ppgs. Retail Price: $10.99**

www.ingramcontent.com/pod-product-compliance
Lightning Source LLC
Chambersburg PA
CBHW080756180526
45168CB00006B/2231